Michael Chapman · Gedis Grudzinskas
Tim Chard (Eds)

IMPLANTATION

Biological and Clinical Aspects

With 74 Figures

Springer-Verlag Berlin Heidelberg GmbH

M.G. Chapman, MRCOG
Department of Obstetrics and Gynaecology,
Guy's Hospital, London, SE1 9RT, UK

J.G. Grudzinskas, MRCOG, FRACOG
Academic Unit of Obstetrics and Gynaecology,
The London Hospital, Whitechapel,
London, E1 1BB, UK

T. Chard, MD, FRCOG
Academic Unit of Reproductive Physiology,
St Bartholomew's Hospital Medical College,
London, EC1A 7BE, UK

Front cover: Immunohistological localisation of α_2-PEG in the endometrium during the menstrual cycle employing monoclonal antibodies.

ISBN 978-1-4471-3531-9

British Library Cataloguing in Publication Data
Chapman, M.G.
Implantation biological and clinical aspects.
1. Women. Ova. Implantation I. Title II. Grudzinskas, J.G. (Jurgis Gediminas)
III. Chard, T. (Tim)
612′.62

Library of Congress Cataloging-in-Publication Data
Implantation: biological and clinical aspects/Michael Chapman, Gedis Grudzinskas, and Tim Chard (eds).
p. cm.
Includes bibliographies and index.
ISBN 978-1-4471-3531-9 ISBN 978-1-4471-3529-6 (eBook)
DOI 10.1007/978-1-4471-3529-6
1.Ovum implantation. I. Chapman, Michael, 1949–
II. Grudzinskas, J. G. (Jurgis Gediminas) III. Chard, T.
[DNLM: 1. Ovulation. 2. Ovum Implantation. 3. Pregnancy—physiology.
WQ 205 I333] QP275.I45 1988 599.8′0433—dc19 DNLM/DLC

Originally published by Springer-Verlag Berlin Heidelberg New York in 1988
Softcover reprint of the hardcover 1st edition 1988

Filmset by Goodfellow and Egan, Cambridge, UK

2128/3916–543210—Printed on acid-free paper.

Preface

The advent of assisted conception procedures such as in-vitro fertilisation (IVF) has provided the impetus for exploration of the factors that lead to the establishment of pregnancy. This collection of papers from leading research workers brings together current concepts of the processes which may be of importance in implantation.

The complex signals from the embryo to the ovary, endometrium and myometrium are now being revealed through studies in both primates and other mammalian species. This book addresses the interrelationship of pituitary and ovarian hormones in controlling ovulation and the preparation of the intrauterine environment for implantation. Once fertilisation has occurred and trophoblast has formed, the next vital step is the production of materials which signal the presence of the pregnancy to the rest of the body. Trophoblastic proteins and other early-pregnancy factors are prime candidates for this role.

Recent studies have emphasised the importance of the intrauterine environment in implantation. Specific secretory products of the endometrium have great potential in this process. The prostaglandins also play an essential part.

Immunological adjustments are now considered a condition for the successful establishment of pregnancy. The possible use of immunotherapy in the treatment of recurrent abortion has highlighted interest in this area. The use of immunological techniques for contraception are in their infancy but offer much hope for the future.

Clinical information on implantation failure and early pregnancy loss has grown rapidly with the intensive observation of pregnancies resulting from IVF, gamete intrafallopian transfer (GIFT) and other assisted fertility procedures. However, clinical intervention to improve the chances of success remains controversial.

London
1988

Michael Chapman
Gedis Grudzinskas
Tim Chard

Contents

Contributors

S.C. Bell, PhD
Department of Obstetrics and Gynaecology and Biochemistry, University of Leicester, UK

A.E. Bolton, PhD
Department of Biological Sciences, Sheffield City Polytechnic, Sheffield, UK

S. Campbell, FRCOG
Academic Unit of Obstetrics and Gynaecology, King's College Hospital, London, UK

M.G. Chapman, MRCOG
Department of Obstetrics and Gynaecology, Guy's Hospital, London, UK

T. Chard, FRCOG
Academic Unit of Reproductive Physiology, St Bartholomew's Hospital Medical College, London, UK

D. Clark, PhD
Department of Medicine, Obstetrics and Gynaecology, Molecular Virology and Reproductive Biology Programme, McMaster University, Hamilton, Ontario, Canada

R.G. Forman, MRCOG
Nuffield Department of Obstetrics and Gynaecology, The Radcliffe Hospital, Headington, Oxford, UK

R. Frydman, MD
Hôpital Antoine Béclère, Clamart, France

A.A. Gidley-Baird, PhD
MRC/AFRC Comparative Physiology Research Group, Institute of Zoology, London, UK

J.G. Grudzinskas, FRACOG
Academic Unit of Obstetrics and Gynaecology, The London Hospital, London, UK

J. Hau
Institute of Medical Microbiology, University of Odense, Odense, Denmark

J.P. Hearn, PhD
MRC/AFRC Comparative Physiology Research Group, Institute of Zoology, London, UK

M.-L. Huhtala, PhD
Labsystems Research Laboratories, Helsinki, Finland

W.R. Jones, FRACOG
Department of Obstetrics and Gynaecology, Flinders Medical Centre, Bedford Park, South Australia

M. Julkunen
Labsystems Research Laboratories, Helsinki, Finland

R.W. Kelly, PhD
MRC Reproductive Biology Unit, Centre for Reproductive Biology, Edinburgh, UK

R. Koistinen
Labsystems Research Laboratories, Helsinki, Finland

E.A. Lenton, PhD
Harris Birthright Research Centre for Reproductive Medicine, Jessop Hospital, Sheffield, UK

B.A. Mason, MB
Hallam Street Clinic, London, UK

E.A. Mowles
Department of Biology and Chemistry, North East London Polytechnic, London, UK

C. O'Neill, PhD
Human Reproduction Unit, Royal North Shore Hospital, St Leonards, New South Wales, Australia

A.G. Pockley
Department of Biological Sciences, Sheffield City Polytechnic, Sheffield, UK

H.K. Poulsen
Institute of Medical Microbiology, University of Odense, Odense, Denmark

L. Regan, MRCOG
Department of Obstetrics and Gynaecology, Addenbrooke's Hospital, Cambridge, UK

A.F. Riddle, MRCOG
Hallam Street Clinic, London, UK

S. Seppälä
Labsystems Research Laboratories, Helsinki, Finland

V. Sharma, MRCOG
Hallam Street Clinic, London, UK

M.J. Sinosich, MSc
RIA Laboratory, Department of Obstetrics and Gynaecology, Royal North Shore Hospital, St Leonards, New South Wales, Australia

S.K. Smith, MRCOG
MRC Reproductive Biology Unit, Centre for Reproductive Biology, Edinburgh, UK

N. Spinks
Human Reproduction Unit, Royal North Shore Hospital, St Leonards, New South Wales, Australia

I. Stabile, PhD
Academic Units of Obstetrics and Gynaecology, King's College Hospital and The London Hospital, London, UK

R.J. Stoker
Department of Biology and Chemistry, North East London Polytechnic, London, UK

B. Teisner, PhD
Institute of Medical Microbiology, University of Odense, Odense, Denmark

J. Testart, PhD
Hôpital Antoine Béclère, Clamart, France

D. Tornehave
Institute of Medical Microbiology, University of Odense, Odense, Denmark

G.E. Webley, PhD
MRC/AFRC Comparative Physiology Research Group, Institute of Zoology, London, UK

J.G. Westergaard, MD
Department of Obstetrics and Gynaecology, Odense University Hospital, Odense, Denmark

O.M.R. Westwood
Department of Obstetrics and Gynaecology, Guy's Hospital, London, UK

P.G. Whittaker, PhD
University Department of Obstetrics and Gynaecology, Princess Mary Maternity Hospital, Newcastle, UK

J.L. Yovich, FRACOG
PIVET Medical Centre, Perth, Western Australia

Section I
GENERAL

1. Embryo Implantation in Primates

J. P. Hearn, G. E. Webley and A. A. Gidley-Baird

The Regulation of Implantation

The attachment of the blastocyst to the maternal endometrium, with subsequent invasion of trophoblast and the establishment of nutrient channels for the embryo, is a critical period in early pregnancy. Over the space of a few days an embryo-maternal dialogue must be established to sustain the life of the corpus luteum, which would otherwise decline at the end of the cycle. The endocrine products of the corpus luteum, principally progesterone and facilitatory oestrogen, are required to transform the endometrium, which in turn provides a variety of proteins and other substances. The precise inventory of substances and their physiological effects are as yet unknown, since it has proved difficult in primates to obtain data on the local interactions at the site of implantation and in the corpus luteum.

The recent advancement of knowledge in early human embryology and the physiology of the preimplantation embryo, associated with improvements in in-vitro fertilisation (IVF) procedures, means that we now know more about the early development of man than other primates (Edwards 1985). However, this knowledge is largely restricted to preimplantation stages of pregnancy. For ethical reasons (with regard to human beings) and because suitable material from non-human primates is sparse, the mechanisms controlling implantation in primates are still virtually unknown.

IVF treatment for infertile couples is the culmination of many years of basic research in rodents and other species. Yet the control of corpus luteum function and the endocrinology of early pregnancy differ considerably in primates and non-primates. There is also considerable variation in the morphology of early implantation in primate species (Fig. 1.1) and in the timing of early embryonic events around the time of attachment and the first detectable appearance of chorionic gonadotrophin (CG) in the peripheral circulation (Table 1.1) (Hearn 1986).

We have studied the control of implantation by examining the production of CG by the embryo both in vivo and in vitro; the secretion of possible preimplan-

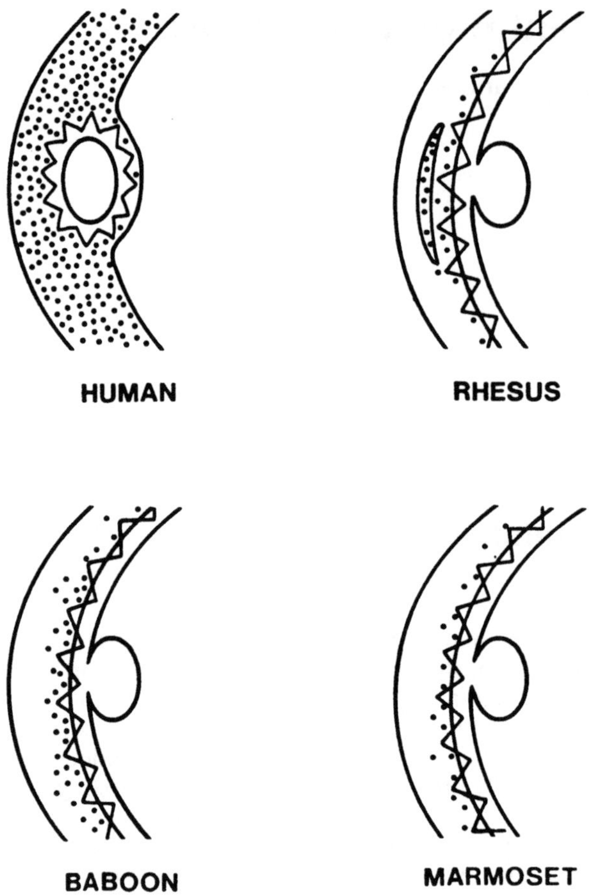

Fig. 1.1. Implantation in primates. The human trophoblast sinks under the endometrial epithelium and there is a massive endometrial reaction. In the monkey species studied implantation is superficial, although rapid contact is made with the maternal vasculature. The degree of endometrial response (*dotted area*) varies in intensity and timing according to the species. The trophoblast–maternal interface is shown by a *zig-zag line*.

Table 1.1. First appearance of chorionic gonadotrophin in peripheral plasma in relation to the time of embryo attachment for four primate species. Values refer to days after ovulation (modified from Hearn 1986)

Species	Embryo attachment	CG first detected	Reference
Human	7–8	9–10	Lenton (1988)
Baboon	8–10	12–14	Shaikh (1978)
Rhesus	8–10	12–14	Atkinson et al (1975)
Marmoset	11–12	14–16	Hearn (1983); Hearn et al. (1987); Moore et al. (1985)

tation signals by the embryo; and endocrine interactions in the corpus luteum around the time of its "rescue" by luteotrophins thought to be secreted by the embryo. We have concentrated on implantation and corpus luteum function in the marmoset monkey (*Callithrix jacchus*), comparing, where possible, data from Old World primate species including man.

Early Embryonic Signals

After Implantation

The first well-defined secretion of the early embryo known to be essential for its survival is CG. This gonadotrophin, composed of α and β subunits and an aminoacid sequence very similar to luteinising hormone (LH) (Canfield et al. 1971; Puett 1986), takes over the luteotrophic support of the corpus luteum and is now thought to cause a reduction in the pituitary secretion of LH (Lenton and Woodward, in press). Implantation in the marmoset monkey commences on day 11–12 after ovulation (Moore et al. 1985; Smith et al. 1987), and the first clear rise in measurable CG in the peripheral circulation is on day 16 (Fig. 1.2). Studies of embryos maintained in culture over the peri-implantation period and allowed to attach to a monolayer of marmoset fibroblasts indicate that the embryos start secreting CG at or immediately after attachment on the equivalent of day 11–12 in vivo (Hearn et al. 1987). The possibility that CG is secreted before attachment is now being tested, but our studies to date suggest that this is unlikely.

The physiological function of CG is thought to be primarily to support the corpus luteum. Whether it has in addition a local function at the site of implantation, perhaps playing a part in the invasion of trophoblast and in early embryonic differentiation, is as yet unknown. Studies reported elsewhere showed that either active or passive immunisation of marmoset monkeys against human CG (hCG) β-subunit during the first 6 weeks of pregnancy disrupted implantation. The animals remained infertile for as long as antibody titres were high (Hearn 1978). More recent work indicated that when marmoset blastocysts were cultured in vitro in the presence or absence of antisera raised in marmosets against hCG β-subunit the embryos were unable to progress to attachment and outgrowth (Hearn et al., in press). The results from the above studies suggest that CG is an essential requirement for normal implantation and corpus luteum function in primates.

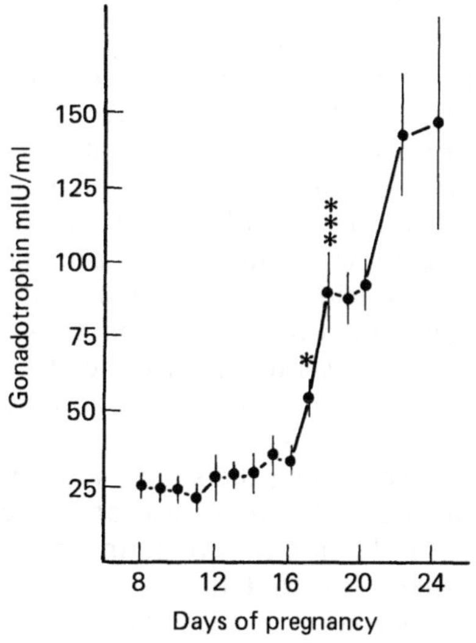

Fig. 1.2. The levels of chorionic gonadotrophin in the peripheral plasma of marmoset monkeys ($n=10$) during the first 24 days of pregnancy (days after ovulation). Significance of increase in gonadotrophin secretion after day 16 of pregnancy: $^{*}P < 0.05$. $^{***}P < 0.001$.

Before Implantation

There is increasing evidence that the preimplantation embryos of several species secrete substances that assist in implantation. These include blastocyst-secreted oestrogen in the pig (Flint et al. 1979); early pregnancy factor in a number of species (Morton et al. 1983); and trophoblastin in the sheep (Martel et al. 1979; Godkin et al. 1982). Studies by O'Neill and colleagues suggest that the embryo secretes a platelet activating factor that causes a thrombocytopenia in early pregnancy (O'Neill 1985). The difficulties encountered to date in the monitoring and measurement of early pregnancy factor and of platelet activating factor show that far more precise and robust assays are required to confirm and extend these possibilities. While early pregnancy factor has been an intractable problem for several years, studies of platelet activating factor are making considerable progress towards more precise confirmation (see Chapter 6).

A study of early pregnancy associated thrombocytopenia was carried out on ten marmoset monkeys during the conception cycle, and six monkeys were monitored through non pregnant cycles as controls. Blood samples of 0.1 ml taken from the femoral vein for 3 days before and 20 days after ovulation were analysed for the numbers of circulating platelets using the methods of O'Neill (1985). The presence of embryos was confirmed on days 8–10 after ovulation by midventral laparotomy and recovery of the embryos by flushing the uterine cavity.

Table 1.2. Reduction of circulating platelets associated with early pregnancy in the marmoset monkey. Percentage reduction is calculated from the mean of 3 daily preovulatory samples compared with 8 postovulatory samples

Condition	Animals (*n*)	Platelet reduction		
		15%–30%	<15%	None
Conception cycles	10	7	2*	1
Non-pregnant cycles	6	0	2	4

*Abnormal embryos were recovered from 2 animals

The results obtained from these studies showed that the numbers of circulating platelets were reduced in the period immediately after fertilisation (Table 1.2), but there was considerable intra-individual and between-individual variation. Non-pregnant control animals showed less variation. If results from the experimental group were calculated as a mean value for each day of pregnancy and compared with non-pregnant controls, a significant ($P<0.05$) depression of platelets was found in the pregnant animals. When the results for each animal were considered on their own in comparison with a matched control, platelet reduction was observed in most of the pregnant animals, but the individual variation was too high either to allow the system to be used as a diagnosis of pregnancy or to inspire confidence in the feasibility and precision of what is a very laborious methodology. In addition, the variation reported from three observers using the same method on the same samples was not acceptable. We concluded that there was an association between thrombocytopenia manifested as a reduction in circulating platelets during early pregnancy in the marmoset monkey but that the methodology was as yet far too imprecise to allow any interpretation of cause and effect in this study.

Function of the Corpus Luteum

There is now substantial evidence that the primate corpus luteum is dependent on luteotrophic support provided by pituitary LH (Fraser et al. 1985; Healy et al. 1984). There is less certainty about the mechanism of luteal regression, which is thought to occur as luteolytic factors gain dominance towards the end of the cycle. This process is reversed if the implanting embryo secretes adequate amounts of CG (Ross 1979; Baird 1985; Hearn 1986). A likely luteolytic agent is prostaglandin $F_{2\alpha}$ ($PGF_{2\alpha}$) of intraovarian origin; intraluteal administration of $PGF_{2\alpha}$ to the rhesus monkey results in luteolysis (Knobil 1973; Auletta et al. 1984). There is, however, a difference between the peripheral action of $PGF_{2\alpha}$ in New World and Old World primates. A single intramuscular injection of a $PGF_{2\alpha}$ analogue, cloprostenol, is luteolytic when administered to the marmoset, a New World monkey, after day 8 of the luteal phase or during pregnancy (Summers et al. 1985) but is not luteolytic in the baboon, an Old World monkey, at any stage of the cycle or pregnancy (Eley et al. 1987).

Control of the life-span of the primate corpus luteum appears to depend on the balance between luteotrophic and luteolytic agents. Progress in understanding their identities and relative contributions may have been restricted by interactions occurring at a local level which cannot be monitored at the systemic level. To investigate interactions within the corpus luteum we have developed a perfusion cannula system to monitor the direct action of agents on the corpus luteum of the marmoset monkey in vivo (Webley and Hearn 1987; Hearn and Webley 1987). This system offers the additional advantage of tissue integrity and preserved luteal innervation, which are considered to be important for demonstrating the luteolytic action of $PGF_{2\alpha}$ on the human corpus luteum (Hamberger et al. 1980; Bennegard et al. 1984).

The perfusion system employed a double-sleeved silastic cannula, which was passed through the exposed corpus luteum of the anaesthetised animal. The secretion of progesterone was monitored by its measurement in 10-min fractions of buffer perfused through the corpus luteum with a peristaltic minipump. The progesterone responses were determined after addition of CG or cloprostenol in the perfusion buffer for 30 min, as examples of known luteotrophic and luteolytic agents, respectively. The system was used to test the effect of melatonin, which has been shown to stimulate progesterone production by human granulosa cells (Webley and Luck 1986), and the action of a luteolytic agent potentially useful for fertility control, deglycosylated hCG (DGhCG). Interactions between the hormones were investigated after their inclusion either together in the same perfusion or in consecutive perfusions. The progesterone responses to treatment perfusions were compared with the response to buffer alone.

Perfusion with hCG significantly ($P<0.01$) stimulated progesterone secretion, in contrast to cloprostenol, which caused an immediate and significant fall in progesterone production (Figs. 1.3, 1.4). Melatonin, perfused at a physiological concentration of 860 pmol/l, significantly ($P<0.01$) stimulated progesterone secretion, as did DGhCG, which gave a similar response to that of hCG. Perfusion of hCG together with cloprostenol prevented the inhibition of progesterone observed with cloprostenol alone. In contrast, perfusion of hCG through corpora lutea previously exposed to cloprostenol did not significantly stimulate progesterone production. Melatonin, when perfused either together

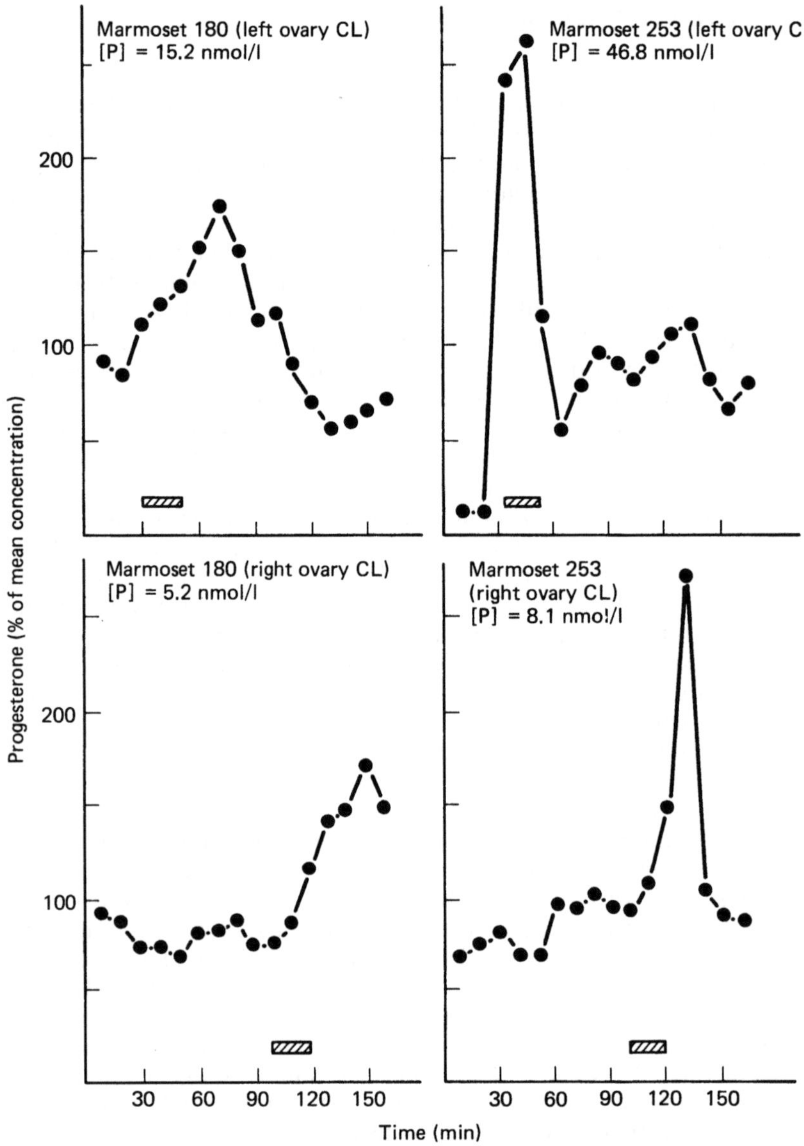

Fig. 1.3. Progesterone concentrations in 10-min fractions of Krebs–Ringer bicarbonate buffer before and after perfusion of 25 IU/ml of hCG for 30 min (*horizontal bars*). Progesterone is expressed as a percentage of the mean [P]. CL, corpus luteum. (From Webley and Hearn 1987.)

with or following cloprostenol, prevented inhibition of progesterone by cloprostenol and instead stimulated production (Fig. 1.5).

This in-vivo system provides a method for investigating the interactions which occur either at the time of luteolysis or during "rescue" of the corpus luteum, the proximate target organ for embryo-derived luteotrophins. The results so far provide further evidence that $PGF_{2\alpha}$ may be the natural intraluteal luteolysin,

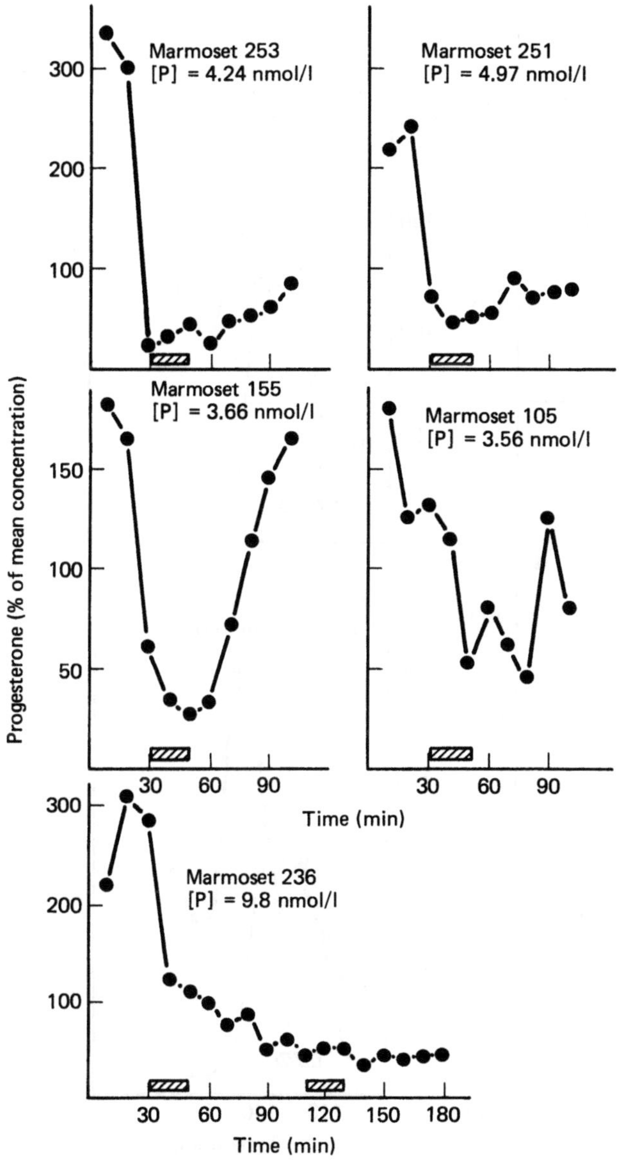

Fig. 1.4. Progesterone concentrations, expressed as a percentage of mean concentration [P], in 10-min fractions of buffer perfused through corpora lutea exposed to cloprostenol (0.5 μg/ml) for 30 min (*horizontal bars*). (From Hearn and Webley 1987.)

acting both directly on progesterone secretion and indirectly by preventing the luteotrophic action of CG. Our findings of a luteotrophic action for melatonin both in man and in the marmoset are somewhat unexpected and indicate a peripheral site of action in the primate in addition to the central site proposed for other species (Bittman et al. 1985). The ability of melatonin to prevent the

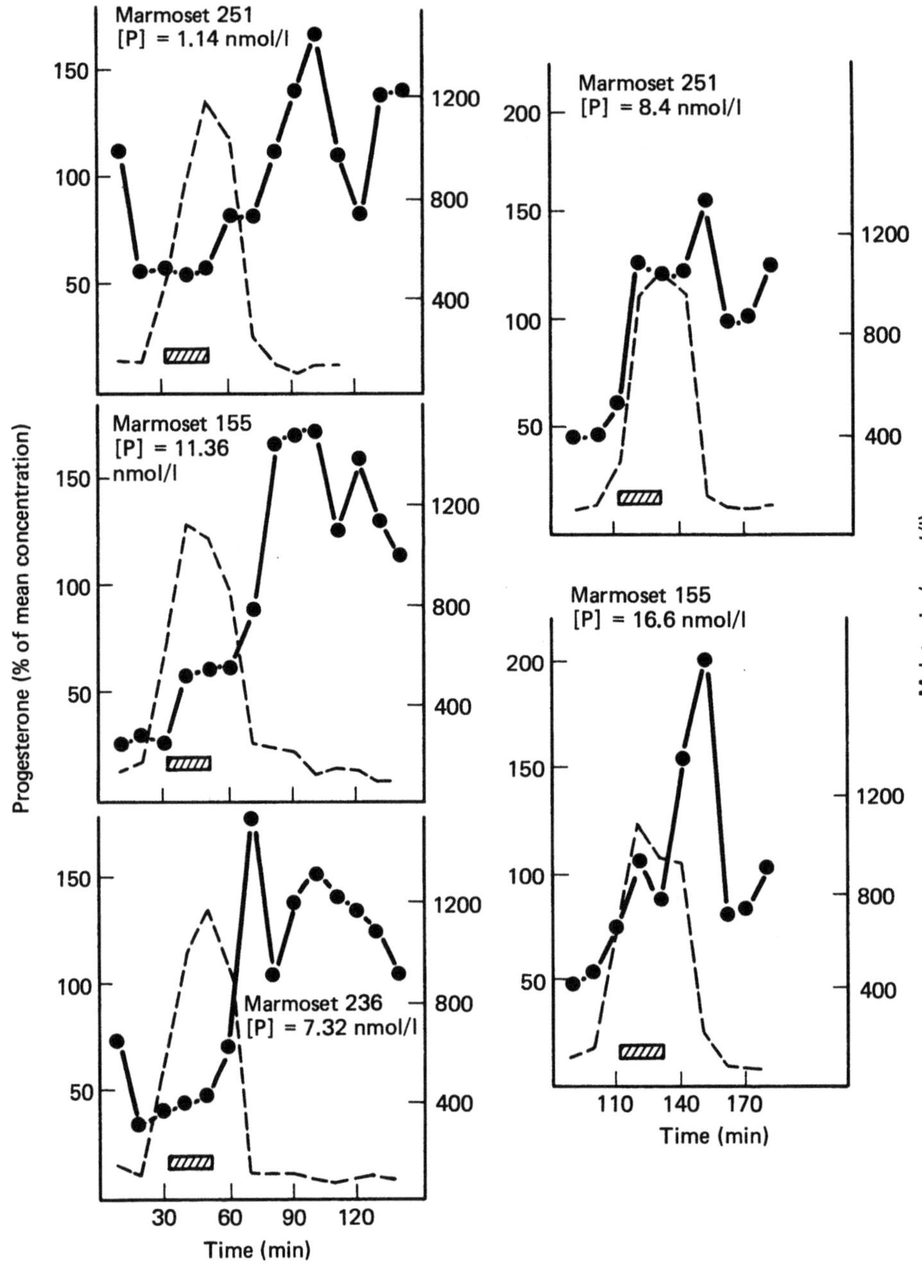

Fig. 1.5. Progesterone concentrations, expressed as a percentage of the mean [P], in 10-min fractions of buffer perfused through corpora lutea exposed to (*left*) cloprostenol (0.5 µg/ml) and melatonin (860 pmol/l) for 30 min (*horizontal bars*), 20–50 min from the start of perfusion, and (*right*) melatonin (860 pmol/1) for 30 min (*horizontal bar*), 100–130 min after the start of perfusion through corpora lutea previously exposed to cloprostenol. The concentration of melatonin in the fractions is shown by the *dotted lines*.

luteolytic action of $PGF_{2\alpha}$ might indicate a common site of action. One possible site of action is the β-adrenergic system, since it has been suggested that the luteolytic action of $PGF_{2\alpha}$ depends on β-adrenergic activity (Bennegard et al. 1984) and melatonin has been shown to increase β-adrenergic binding at the pineal gland (Sweat 1986).

The application of the perfusion system to the study of potential agents for control of fertility and infertility was demonstrated with DGhCG. We were unable to distinguish an antagonistic effect of DGhCG on hCG action, indicating that the preparation was probably of insufficient purity but also suggesting that deglycosylation of hCG does not render the molecule inactive in terms of its direct steroidogenic activity at the marmoset corpus luteum in vivo. Melatonin might have greater potential for clinical application as a facilitatory agent in supporting corpus luteum function. It is relatively non-toxic and apparently stimulates progesterone secretion without enhancing oestradiol production (Webley and Luck 1986). However, demonstration of this potential through peripheral administration has yet to be achieved.

Conclusions

CG is the first measurable signal from the embryo to the mother and it is secreted by the embryo from the time of its attachment to the endometrial epithelium. CG appears in the peripheral circulation, in significant quantities, 2 to 3 days after attachment. Antisera to the hCG β-subunit prevent implantation and disrupt early pregnancy in primates, presumably by blocking the embryonic message that supports the corpus luteum. The hypothesis that CG has also a local function at the site of implantation, and perhaps in early trophoblastic differentiation, requires considerable further testing.

It seems likely that the embryo secretes other substances during the preimplantation stages of pregnancy, but the presence and the physiological role of such signals have yet to be confirmed. Platelet activating factor is one candidate as an early embryonic signal. Far more sensitive and robust assays are required to examine this possibility, which, if confirmed, has considerable clinical application as a diagnostic test and a monitor of embryo viability.

Perfusion of the corpus luteum in vivo provides a method for the study of local endocrine interactions. The corpus luteum is a target organ for early embryonic messages and its "rescue" is essential for the survival of the embryo. The finding that melatonin has a capacity for luteal support, under the experimental conditions described above, suggests that this hormone might prove to have some potential as a facilitatory system in maintenance of the corpus luteum.

There is still a great deal to be done in developing the methodology to allow the study of local production and interaction of the embryo–maternal signals responsible for the initiation of implantation and the support of the corpus luteum of early pregnancy. The difficulties to be overcome are in sampling from local sites at the embryo–endometrial junction in vivo and in vitro without disrupting normal morphological and physiological integrity.

The ability to monitor the corpus luteum during its transformation, presumably by embryonic signals, into the corpus luteum of pregnancy open up some interesting possibilities. Measurement of local interactions in the corpus luteum at this time should allow a sensitive and rapid screening of potential luteotrophic and luteolytic agents relevant both to the treatment of infertility and to the development of new methods of controlling fertility. The results of the studies reported here provide some encouragement that the approach is feasible.

References

Atkinson LE, Hotchkiss J, Fritz GR, Surve AH, Neill JD, Knobil E (1975) Circulating levels of steroids and chorionic gonadotrophin during pregnancy in the rhesus monkey, with special attention to the rescue of the corpus luteum in early pregnancy. Biol Reprod 12: 335–345

Auletta FJ, Kamps DL, Pories S, Bisset J, Gibson M (1984) An intra-corpus luteum site for the luteolytic action of prostaglandin $F_{2\alpha}$ in the rhesus monkey. Prostaglandins 27: 285–298

Baird DT (1985) Control of luteolysis. In: Jeffcoate SL (ed) The luteal phase. John Wiley & Sons, Chichester, pp 25–43

Bennegard N, Dennefors B, Hamberger L (1984) Interaction between catecholamines and prostaglandin $F_{2\alpha}$ in human luteolysis. Acta Endocr 106: 532–537

Bittman EL, Kaynard AK, Olster DH, Robinson JE, Yellon SM, Karsch FJ (1985) Pineal melatonin mediates photoperiodic control of pulsatile luteinizing hormone secretion in the ewe. Neuroendocrinology 40: 409–418

Canfield RE, Morgan FJ, Kammerman S, Bell JJ, Agosto GM (1971) Studies of human gonadotrophin. Rec Prog Hormone Res 27: 121–156

Edwards RG (1985) Current status of human conceptions in vitro Proc R Soc Lond B 223: 417–448

Eley RM, Summers PM, Hearn JP (1987) Failure of prostaglandin $F_{2\alpha}$ analogue, cloprostenol, to induce functional luteolysis in the olive baboon (*Papio cynocephalus anubis*). J Med Prim 16: 1–12

Flint APF, Burton RD, Gadsby JE, Saunders PTK, Heap RB (1979) Blastocyst oestrogen synthesis and the maternal recognition of pregnancy. In: Whelan J (ed) Maternal recognition of pregnancy (Ciba Foundation Symposium no. 64). Excerpta Medica, Amsterdam, pp 209–228

Fraser HM, Baird DT, McRae GI, Nestor JJ, Vickery BH (1985) Suppression of luteal progesterone secretion in the stumptailed macaque by an antagonist analogue of luteinising hormone-releasing hormone. J Endocr 104: R1–R4

Godkin JD, Bazer FW, Moffatt J, Lessions F, Roberts RM (1982) Purification and properties of a major low molecular weight protein released by the trophoblast of sheep blastocysts at day 13–21. J Reprod Fert 65: 141–150

Hamberger L, Dennefors B, Hamberger B, et al. (1980) Is vascular innervation a prerequisite for PG-induced luteolysis in the human corpus luteum? In: Samuelsson B, Ramwell PW, Paoletti R (eds) Advances in prostaglandin and thromboxane research, vol 8. Raven Press, New York, pp 1365–1368

Healy DL, Schenken RS, Lynch A, Williams RF, Hodgen GD (1984) Pulsatile progesterone secretion: its relevance to clinical evaluation of corpus luteum function. Fert Steril 41: 114–121

Hearn JP (1978) Immunological interference with the maternal recognition of pregnancy in primates. In: Whelan J (ed) Maternal recognition of pregnancy (Ciba Foundation Symposium no. 64). Excerpta Medica, Amsterdam, pp 353–376

Hearn JP (1983) The common marmoset (*Callithrix jacchus*). In: Hearn JP (ed) Reproduction in New World primates. MTP Press, Lancaster, pp 181–216

Hearn JP (1986) The embryo–maternal dialogue during early pregnancy in primates. J Reprod Fert 76: 809–819

Hearn JP, Webley GE (1987) Regulation of the corpus luteum of early pregnancy in the marmoset monkey: local interactions of luteotrophic and luteolytic hormones in vivo and their effects on the secretion of progesterone. J Endocrinol 231: 231–239

Hearn JP, Summers PM, Webley GE (1987) Intraembryonic and luteal effects of chorionic gonadotrophin during the peri-implantation period in a primate, *Callithrix jacchus*. In: Christiansen C, Riis BJ (eds) Highlights on endocrinology, Proc 1st Eur Congr Endocrinol, Norhaven Bogtrykkeri, Copenhagen, pp 281–286

Hearn JP, Gidley-Baird AA, Hodges JK, Summers PM, Webley GE (in press) Embryonic signals during the peri-implantation period in primates (Valedictory symposium for Professor A Klopper). J Reprod Fert Suppl (in press)

Knobil E (1973) On the regulation of the primate corpus luteum. Biol Reprod 8: 246–258

Lenton EA, Woodward AJ (in press) The endocrinology of conceptual cycles and implantation (Valedictory symposium for Professor A Klopper). J Reprod Fert Suppl (in press)

Martel J, Lacroix MC, Lauder C, Saunier M, Wintenberger-Torres S (1979) Trophoblastin, an antiluteolytic protein present in early pregnant sheep. J Reprod Fert 56: 63–73

Moore HDM, Gems S, Hearn JP (1985) Early implantation stages in the marmoset monkey (*Callithrix jacchus*). Am J Anat 172: 265–278

Morton H, Morton DJ, Ellendorf F (1983) The appearances and characteristics of early pregnancy factor in the pig. J Reprod Fert 69: 437–466

O'Neill C (1985) Examination of the causes of early pregnancy associated thrombocytopenia in mice. J Reprod Fert 73: 567–577

Puett D (1986) Human choriogonadotrophin. BioEssays 4: 70–75

Ross GT (1979) Human chorionic gonadotrophin and maternal recognition of pregnancy. In: Whelan J (ed) Maternal Recognition of Pregnancy (Ciba Foundation Symposium no. 64). Excerpta Medica, Amsterdam, pp 191–208

Shaikh AA (1978) Animals models for research in human reproduction. National Institutes of Health, Bethesda, invited report

Smith CA, Moore HDM, Hearn JP (1987) The ultrastructure of early implantation in the marmoset monkey (*Callithrix jacchus*). Anat Embryol 175: 399–410

Summers PM, Wennink CJ, Hodges JK (1985) Cloprostenol-induced luteolysis in the marmoset monkey (*Callithrix jacchus*). J Reprod Fert 73: 133–138

Sweat FW (1986) Beta-adrenergic binding is increased by melatonin and alpha-adrenergic compounds. Biochem Biophys Res Comm 138: 1196–1202

Webley GE, Hearn JP (1987) Local production of progesterone by the corpus luteum of the marmoset monkey in response to perfusion with chorionic gonadotrophin and melatonin in vivo. J Endocr 112: 449–457

Webley GE, Luck MR (1986) Melatonin directly stimulates the secretion of progesterone by human and bovine granulosa cells luteinized in vitro. J Reprod Fert 78: 711–717

Section II
PITUITARY AND OVARIAN HORMONES

2. Pituitary and Ovarian Hormones in Implantation and Early Pregnancy

E.A. Lenton

Preparation for Implantation

There are many aspects to the preparation for implantation, such as fertilisation of the ovum, cleavage of the embryo, transportation to the site of implantation and optimum endometrial secretory activity, all of which are critical to the establishment of a successful pregnancy. Many of these factors are considered in some detail elsewhere in this book, and only those aspects involving pituitary and ovarian hormones will be discussed here. It is important, though, to realise that the endocrine changes do not occur in isolation but are merely one aspect of the total integration of all the critical events regulating normal human implantation.

Conception and Non-conception Cycles

The cyclical changes in the pituitary hormones (luteinising hormone [LH] and follicle stimulating hormone [FSH]) and the ovarian steroids (oestradiol and progesterone) throughout the cycles in which conception occurred, or where conception was either not desired or did not occur, are shown in Fig. 2.1 (and see Lenton et al. 1982c). There were no significant differences in any of the hormones measured up to the time of implantation in the mid-luteal phase except for plasma progesterone. Following implantation, as expected, ovarian steroid concentrations rose steadily, whereas FSH levels remained at mid-luteal phase levels and did not show the normal non-pregnant late luteal rise. LH concentrations paradoxically appeared to increase after implantation but, as will be discussed later, this was almost certainly due to cross-reaction with human chorionic gonadotrophin (hCG) and not hypersecretion of pituitary LH.

Preimplantation Progesterone Concentrations

The only significant preimplantation hormone concentration differences between conception and non-conception cycles observed was in the plasma pro-

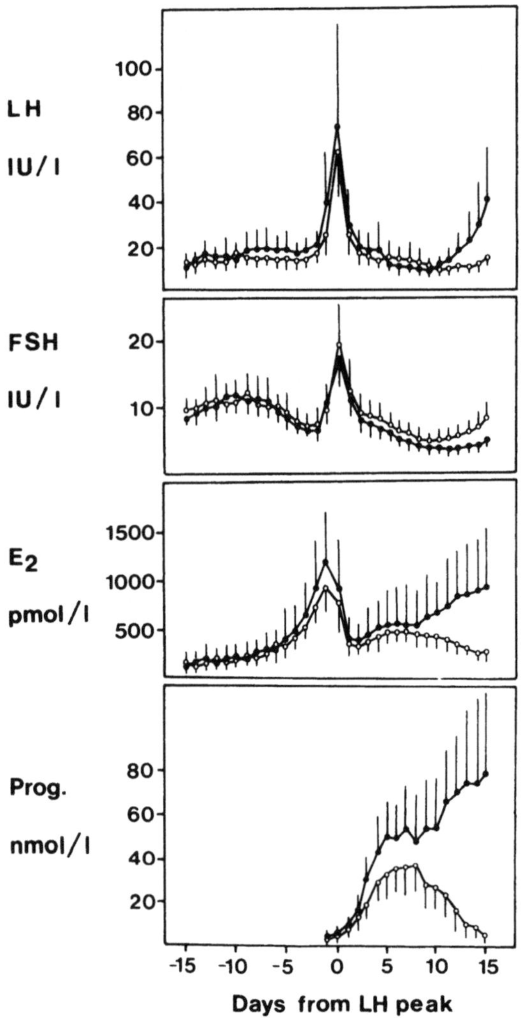

Fig. 2.1. Daily geometric means (and 68% confidence limits) of LH, FSH, oestradiol (E_2) and progesterone (Prog.) during spontaneous conception (●–●) and non-conception cycles (○–○). (Redrawn from Lenton et al. 1982c.)

gesterone profile. This is an intriguing observation and has been interpreted as supporting the existence of preimplantation embryo-associated signals which are able to influence luteal function. Whilst this is an attractive hypothesis, it must be acknowledged that there are other plausible explanations. Firstly it is important to appreciate that the data shown in Fig. 2.1 are the results of comparing two populations – namely, cycles from women who conceived, with cycles where no fertilised embryo was present. Whilst the data presented in this manner are valid to show that overall in cycles where conception occurs progesterone levels are higher than in non-conception cycles, they do not permit the conclusion that the higher levels are due to preimplantation signals. This is because the two populations were not selected with the same rigour and so are not directly equivalent. To clarify this statement one must consider that a conception cycle represents the optimum reproductive cycle, the cycle where all of the endocrine signals and the associated factors necessary for implantation have synchronised perfectly. Thus a group of conception cycles must by definition represent a self-selected group of the most favourable cycles. Such a group would be biased in their selection in comparison with a group of much less highly selected cycles where conception was not desired (so no possibility of optimum self-selection) or did not occur (owing to inappropriate coital timing, an infertile cycle etc.).

Unfortunately, unless non-conception cycles can be as carefully selected as conception cycles are self-selected, then they will remain a heterogeneous group and thus not be directly comparable. This point is clearly illustrated by the alternative method of describing population data shown in Table 2.1. In this situation the medians and confidence intervals have been obtained after horizontal pooling of the data per individual (using the "progesterone index", defined as the mean progesterone concentration per subject over the period 5 to 8 days, inclusive, after the LH surge). The data in Fig. 2.1 were obtained by vertical pooling and so only describe the behaviour of the group and can give little information about individuals within the group. From Table 2.1 it is obvious that while the control non-conception cycles (obtained from normal women) show progesterone indices that cover the entire physiological ovulatory range (i.e., 16–62 nmol/l), the highly selected conception cycles occupy only the top half of the physiological range (31–73 nmol/l). Also included in Table 2.1 are data from a large series of women with longstanding unexplained infertility to demonstrate that this group, which are self-selected by failure to conceive, tend to have progesterone indices which fall in the lower part of the physiological range (16–49 nmol/l). This does not mean that all women with unexplained infertility will exhibit low luteal progesterone concentrations; rather that cycles with low progesterone will be more common in this group than in the normal non-pregnant controls. Conversely, low progesterone cycles will be most uncommon amongst spontaneous conception cycles. Thus, although there may well be luteotrophic preimplantation embryonic signals, data such as those presented in Fig. 2.1 should be interpreted with caution as giving scientific support to this hypothesis.

Table 2.1. Distribution of progesterone indices (over the interval LH+5 to LH+8) in conception and control cycles and in cycles from women with unexplained infertility

Cycle	−48%	−34%	Median	+34%	+48%
Conception cycles (n= 27)	31	38	47	59	73
Control cycles (n = 62)	16	28	40	53	62
Infertile cycles (n = 127)	16	25	33	41	49

Implantation

The best available evidence on the probable time of implantation in human beings comes from the many detailed studies by Hertig and his colleagues, summarised by Hertig (1975). The likely sequence of events, taken directly from their observations is as follows. The fertilised ovum remains within the fallopian tube for the first 3 postovulatory days before entering the uterine cavity at the 8 to 12 cell stage. After very rapid cell division over the next 48 h, the blastocyst is ready to begin implanting. The phase of "early implantation" covers the period 6 to 8 days after ovulation, and during this phase the embryo merges with and gradually sinks into the maternal endometrium. During "midimplantation" (days 9 to 10) the embryo continues to sink further into the endometrium, which by now has started to develop decidua. Lacunae within the trophoblast communicate with each other and contain maternal blood. Gradually they coalesce with the lumina of maternal blood vessels to form the beginning of the utero-placental circulation. By "late implantation" (days 11 and 12) there are moderate to marked gestational hyperplasia and multiple connections between the lacunae and the surrounding vessels.

These times refer to the postovulation age of the embryo, which cannot be easily determined in spontaneous conceptions. A more convenient reference point is the day of the LH surge (Day 0), which precedes ovulation by approximately 24 h. Thus the timings of implantation given by Hertig (1975), when expressed relative to the LH surge, become: 7 days for the earliest attachment, 10 to 11 days for the time when maternal blood is in contact with the trophoblast and contains its secretory products and 12 to 13 days for establishment of full uteroplacental communication.

Monitoring Implantation

Any protein produced by the trophoblast which is directed into the maternal circulation can conveniently be used to monitor implantation. Two proteins which have been used for this purpose are human chorionic gonadotrophin (hCG) (Lenton et al. 1981b) and pregnancy-specific β_1-glycoprotein (SP_1) (Lenton et al. 1981a). An essential assumption is that these proteins do not appear in the maternal circulation until after implantation has begun (Catt et al. 1975). This assumption is likely to be valid, because when embryos are cultured in vitro, no hCG can be detected in the culture medium until the blastocyst begins to hatch from the zona pellucida (Fishel et al. 1984). Hatching is closely associated in vivo with the actual process of implantation.

Assays used to monitor the early stages of implantation and pregnancy must be sensitive, and currently the best hCG assays have greater sensitivity than SP_1 assays and so hCG is the hormone of choice for this purpose. In practice, any assay that can detect hCG (for example, an LH radioimmunoassay [RIA]) can be used to monitor early pregnancy, but the precision with which the first increase in hCG is detected will depend on the specificity of the systems. This point is illustrated in Fig. 2.2, which shows the daily profile of LH/hCG through a spontaneous conception cycle as detected by three different assay systems. These were, first, a conventional radioimmunoassay for LH (Lenton et al. 1978), calibrated with respect to the 2nd International Reference Preparation for human menopausal gonadotrophin (2 IRP-hMG) as an LH standard. This assay detects LH (100%) and hCG (30%). The second assay was an hCG radio-receptor assay based on bovine corpus-luteum cell membranes (Biocept-G, Wampole Laboratories, USA), calibrated with respect to the 1st International Reference Preparation for hCG (75/537) (Boyko and Russell, 1979). In our hands this assay was able to detect hCG (100%) and LH (50%), although

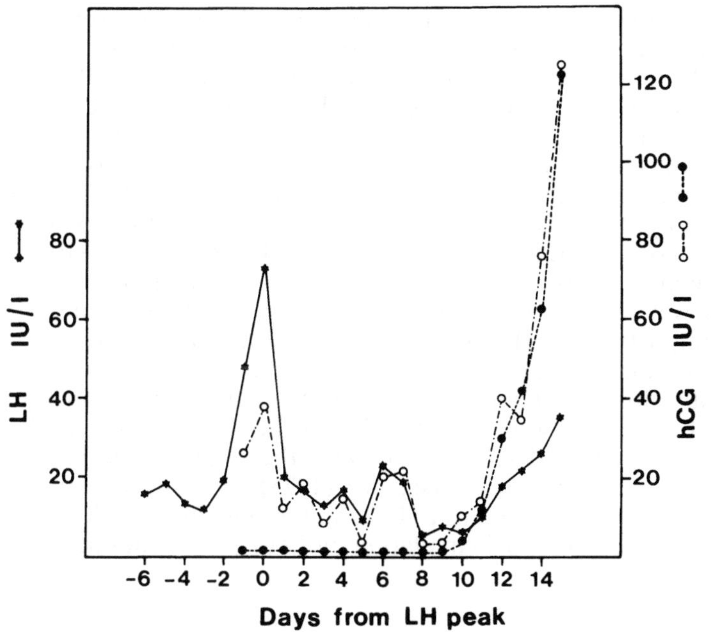

Fig. 2.2. Daily concentrations of LH (LH-RIA,*–*) LH/hCG (hCG–RRA; ○–·–○) and hCG (β-hCG-RIA; ●---●) during a spontaneous normal conception cycle. The increase in hCG following implantatin was observed directly (hCG-RIA) or indirectly (LH-RIA, hCG-RRA) to occur from about LH +10 in this woman. (See text for details of the assays.)

slightly different cross-reactivities have been reported by others (Boyko 1979). The third assay was virtually specific for hCG and has been reported elsewhere (Lenton et al. 1982b). This assay was an hCG radioimmunoassay with an antibody raised against the β-subunit of hCG. Calibrated against the same hCG standard (1 IRP-hCG), it detected hCG (100%) but virtually no LH (0.1%). In the normal conception cycle shown in Fig. 2.2, the first detectable increase due to circulating hCG was seen on LH+10 or LH+11 in all three assays. However, only in the specific β subunit system was this a clear increase over non-detectable preimplantation concentrations. In both the other systems, cross-reaction with LH confused the early luteal phase profile. The sensitivity of this hCG assay was 2 IU/l (1 IRP-hCG), which would have been 1 IU/l with respect to the 2nd International Standard for hCG (2 IS-hCG). Since maternal hCG concentrations are already rising rapidly from the time of detection, it is possible that hCG had in fact reached the maternal circulation somewhat earlier but that the current assay was insufficiently sensitive to detect it.

Time of Implantation

Using the β-hCG assay described above, the day on which hCG was first significantly greater than the assay sensitivity (2 IU/l) was LH+8 (5.3%), LH+9 (10.5%), LH+10 (47.4%) and LH+11 (36.8%) in 19 normal conception cycles

(Lenton et al. 1982b). However, as it seemed likely that these times were not directly related to the time of implantation because of the lack of sensitivity of the hCG assay, an ultrasensitive version was developed (Lenton et al. 1982b). Sensitivity of this assay varied between 0.1 and 0.3 IU/l. With the ultrasensitive assay, hCG concentrations were monitored daily throughout the luteal phase in 28 successful spontaneous pregnancies. Despite the increase in sensitivity, hCG was not detected before LH+8 in any conception cycle (Table 2.2). In fact the only difference shown by the 10-fold increase in sensitivity was that the time of implantation became more precisely located to 3 days (LH+8, +9 and +10) in the luteal phase. A similar, but possibly more consistent, method of establishing the time when hCG was present is to obtain the time at which hCG concentrations were uniformly 1.0 IU/l. (This is relatively easy to do graphically, since hCG concentrations are increasing exponentially.) However, as shown in Table 2.2, this does not really alter the timing or duration of the implantation window. Thus, it seems unlikely, even if it were possible, that further improvement in the sensitivity of an hCG assay would substantially alter these timings. From the morphological data of Hertig and his colleagues (1975) the time of earliest attachment of the embryo was localised to LH+7, and in our studies hCG itself was already detectable in the maternal circulation by LH+8 in about one-third of the conception cycles.

Table 2.2. Cumulative frequency of detection of hCG in maternal circulation in spontaneous successful pregnancies

Day following LH surge	hCG first detected (%)		hCG concentration equivalent to 1.0 IU/l[c]
	Concentration ≥ 5 IU/l[a]	Concentration ≥ 0.5 IU/l[b]	
LH +7			
LH +8	5.3	32.1	26.9
LH +9	15.8	89.2	88.4
LH +10	63.2	100.0	100.0
LH +11	100.0		

hCG assays of two sensitivities (2 IU/l and 0.3 IU/l) were used and the day of first detection was defined as the day following the LH surge (day 0) on which hCG concentrations first exceeded the sensitivity threshold of the assay. An alternative method based on assessing the day on which all hCG concentrations were equivalent (at 1.0 IU/l) is also presented.

[a] $n = 19$.
[b] $n = 28$.
[c] $n = 26$.

The Consequences of Implantation

Although it is possible that there are preimplantation embryo-associated signals, the first irrefutable evidence of an embryonic endocrine message which has a clearly demonstrable effect is the appearance of hCG and the resultant rescue of the corpus luteum. These events are closely associated in time (Fig. 2.3). Both oestradiol and progesterone concentrations are beginning to respond by LH+9,

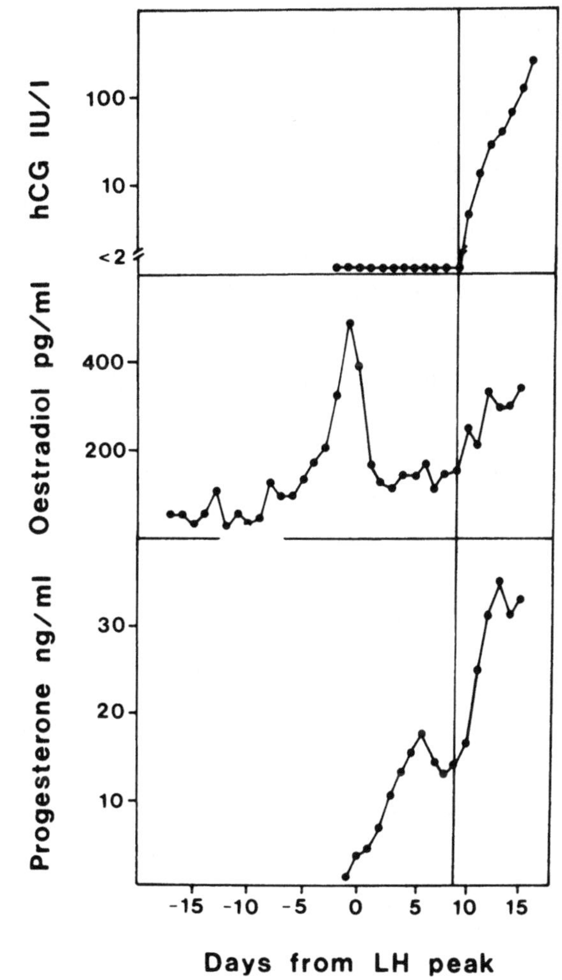

Fig. 2.3. Daily concentrations of oestradiol and progesterone throughout the spontaneous conception cycle from the same subject as in Fig. 2.2. The changes in steroid concentrations relative to the first detected change in hCG (assay sensitivity 2 IU/l) are indicated. However, when an assay of greater sensitivity was used, hCG was first detected 24 h earlier than indicated.

even though hCG is not detected with the routine assay until LH+10. However, with the ultrasensitive assay, hCG is first detected (at a concentration of only 0.5 IU/l) on LH+9. The mid-luteal phase corpus luteum seems to be extremely sensitive to low concentrations of hCG, and the start of the rescue response is seen within a few hours of hCG reaching the maternal circulation. Over the first 48 h of the peri-implantation period steroid levels do not rise markedly, but rather the expected decline from mid-luteal peak levels is halted. However, by the time hCG concentrations have risen to about 5 IU/l, both oestradiol and progesterone concentrations have increased dramatically. In the example illustrated (Fig. 2.3) this rise occurred between LH+10 and LH+11. It is tempting to speculate that the steady exponential rise in hCG (Lenton et al. 1982b) and the striking luteal response occurring some 2–3 days after hCG was first detected mark the end of the implantation process itself (Hertig 1975) and the beginning of an efficient fetal-maternal communication.

Early Pregnancy

The collection of sufficient prospective data on early human pregnancy is difficult, but developments in in-vitro fertilisation have stimulated awareness of this important area of reproductive biology.

We have accumulated a data bank of well-documented spontaneous conception cycles in normal women, and in many of these cycles daily blood sampling was continued well into the early stages of pregnancy. In order to look serially and synchronously at a number of endocrine events without the difficulties inherent in repeated sample analysis, the samples were pooled on a daily basis, using the documented day of the LH surge as a reference point. In this way we obtained 50 early-pregnancy plasma pools, each of sufficient volume for multiparameter analysis, covering the first 50 days following ovulation in 17 spontaneous successful pregnancies. (Hormone concentrations in each of these plasma pools will be equivalent to the arithmetic mean of the concentrations in the 17 constituent cycles, not the geometric mean which might possibly be more correct, particularly as many of the parameters measured increase exponentially.)

Pituitary Function in Early Pregnancy

Plasma LH and the β subunit of LH were measured in the 50 pregnancy pools with standard radioimmunoassays and reagents obtained from the Medical Research Council Unit for Biological Standards and Control, London. As

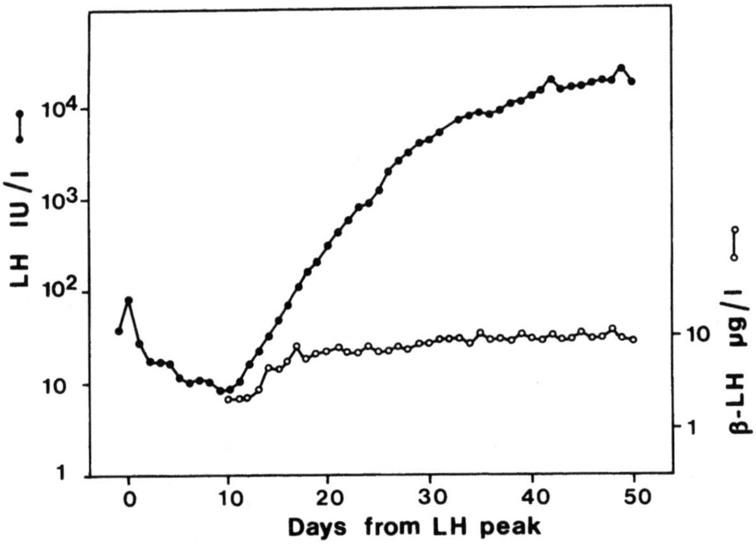

Fig. 2.4. Changing concentrations of LH and the β-subunit of LH during the first 50 days after ovulation in daily plasma pools from 17 successful spontaneous pregnancies. LH levels rise because of a 30% cross-reaction with hCG in this LH radioimmunoassay. The LH standard was the 2nd International Reference Preparation for human menopausal gonadotrophin.

expected, LH concentrations rose sharply after implantation because of the cross-reaction with hCG (Fig. 2.4). The β subunit of LH was measured to see if there was any additional pituitary LH secretion in early pregnancy that had been masked by the high concentrations of hCG. The impression gained was that although there might have been a small amount of residual pituitary LH activity, the relative change in β-LH concentration was small in comparison with the change in hCG. Since this work was done, new LH and FSH immunoradiometric assays have become available which do not cross-react with hCG (LH and FSH Maiaclone, Serono Diagnostics). With these sensitive assays it has been possible to show that secretion of both LH and FSH is almost totally suppressed from the time of implantation (Woodward and Lenton, unpublished observations).

HCG in Early Pregnancy

Although there was clear cross-reaction with hCG in the LH assay, the actual concentrations measured cannot accurately reflect the hCG levels, because the cross-reaction is only partial (approximately 30%) Reanalysis of the same plasma pools in the same LH radioimmunoassay with an hCG standard (1 IRP-hCG) yielded concentrations identical to those obtained using either the hCG radioreceptor assay (RRA) or the β-hCG-RIA, both of which measured hCG efficiently (Fig. 2.5). It is reassuring that all the assay systems, whether totally specific (β-hCG), non-specific (LH-RIA) or working on the principle of competitive binding to receptors (hCG-RRA), gave indistinguishable results with

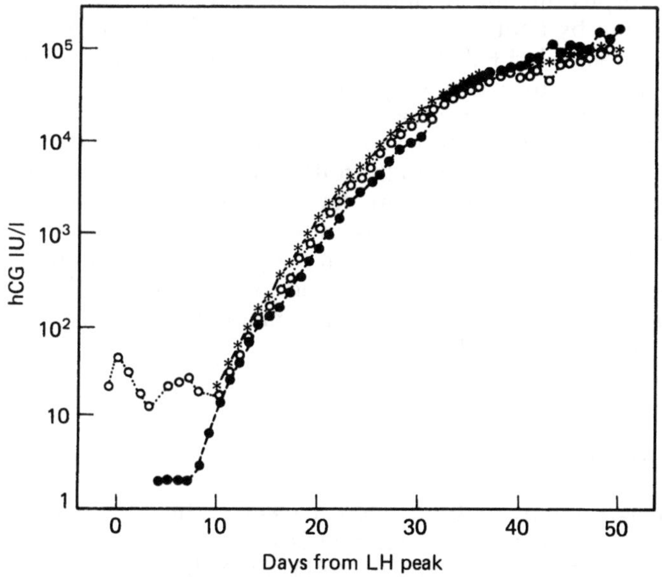

Fig. 2.5. Concentrations of hCG in the same early pregnancy plasma pools as shown in Fig. 2.4. hCG was measured with specific (β-hCG-RIA, ●—●) and non-specific (LH-RIA, *–*; hCG-RRA, ○–○) assay systems, which were all calibrated with respect to a common hCG standard (1st International Reference Preparation for hCG; 75/537).

respect to the amount of hCG measured and its profile during early pregnancy (Batzer 1980).

Another important feature of Fig. 2.5 is that although hCG concentrations rise rapidly during the first 30 days of pregnancy, the rise is not truly exponential over the whole of this time. In fact doubling times for the increase in hCG ranged from 16 h (hCG concentrations between 2 and 20 IU/l), slowing to 1.3 days (between 20 and 400 IU/l) and decreasing still further to 1.8 days (between 400 and 4000 IU/l) (Marshall et al. 1968; Chartier et al. 1979). Eventually hCG concentrations plateau at about 50 days and then gradually decline (Batzer et al. 1981) throughout the remainder of the pregnancy.

Steroid Changes in Early Pregnancy

The pattern of steroid changes during very early pregnancy has been well described in the rhesus monkey (Neill et al. 1969) but only in individual instances in man before about 6 weeks' gestation (Manganiello et al. 1981). Consequently, we looked at steroid changes in the early pregnancy plasma pools to examine the role of the corpus luteum at this time.

Oestradiol concentrations increase serially from implantation (Fig. 2.6) and follow a curve which is approximately exponential. The doubling time for the rise in oestradiol concentrations was found to be 14 days. This contrasts with the rise in prolactin during early pregnancy which, although again exponential, is somewhat slower, with a doubling time of 24 days (Lenton et al. 1982a).

Progesterone concentrations show a completely different pattern (Fig. 2.6). Following rescue of the corpus luteum on about LH+10, progesterone levels increase to reach a maximum by about LH+14. Levels are maintained for the next few days, but from about LH+20 they begin declining, to reach new minimum levels (which are still in excess of preimplantation concentrations) around LH+24. Progesterone levels then remain stable for the remainder of the first 50 days, except for perhaps a small transient rise around LH+32 to +34. 17-Hydroxyprogesterone concentrations initially follow progesterone concentrations quite closely again reaching maximum levels between LH+14 and LH+20 and then declining. This decline is transiently interrupted by a small rise between LH+32 and +34 (similar to that seen in the progesterone profile), but from about LH+36 17-hydroxyprogesterone and progesterone concentrations diverge. Whilst progesterone levels are maintained, those of 17-hydroxyprogesterone continue to decline.

Role of the Corpus Luteum

17-Hydroxyprogesterone is thought to be a good marker of luteal function, since, unlike progesterone, 17-hydroxyprogesterone is not synthesised by the placenta (Yoshimi et al. 1969; Manganiello et al. 1981). Thus the divergence in the concentration of these two steroids after LH+36 suggests that by this time secretory function of the corpus luteum is rapidly waning, and the fact that progesterone levels overall do not fall suggests that the placenta is already making a significant contribution by this time (Yoshimi et al. 1969).

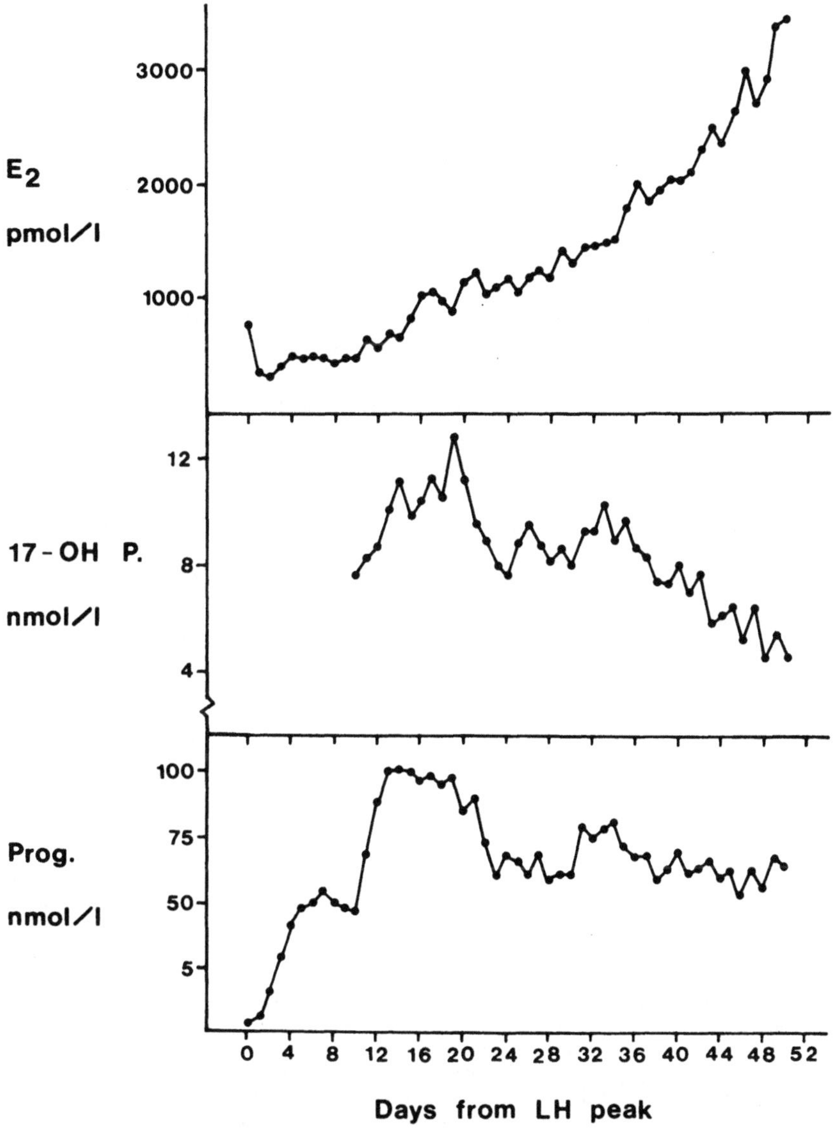

Fig. 2.6. Daily concentrations of oestradiol, 17-hydroxyprogesterone (17-OH P.) and progesterone over the first 50 days of pregnancy in the same early pregnancy pools as in Figs. 2.4 and 2.5.

The significance of the small secondary increase in both 17-hydroxyprogesterone and progesterone around LH+34 is not known, although it is clearly present at the same time in early pregnancy in the monkey (Neill et al. 1969). There are no obvious concurrent changes in LH or hCG associated with this secondary rise. Nor is it clear why oestradiol concentrations follow such a different pattern from progesterone. Both steroids appear to "rescue" following implantation (see Fig. 2.3), but whereas oestradiol levels continue to rise (in response to continuing increases in hCG or due to placental oestradiol production?) the progesterone response to hCG is clearly transient and the duration of the corpus-luteum-dominated phase of early pregnancy relatively short (Csapo et al. 1972; Goodman and Hodgen 1979).

Summary

There are significant differences in progesterone concentrations before implantation between conception and non-conception cycles. Although these differences may reflect putative preimplantation luteotrophic signals, this is by no means clearly established and other explanations are possible. Implantation as monitored by the first detection of hCG in maternal circulation (using an ultrasensitive hCG assay) occurs over a narrow window of only 3 days in the mid-luteal phase in successful spontaneous (i.e., unstimulated) conception cycles. Rescue of the corpus luteum occurs very soon after the first appearance of hCG in the maternal circulation. There are clear and distinct differences in the response of oestradiol and progesterone to the appearance of hCG . At least with respect to progesterone, corpus-luteum rescue is a short-lived event. These data also suggest that placental steroid production is well under way and the corpus luteum relatively redundant from about 3 weeks after the first missed menses.

References

Batzer FR (1980) Hormonal evaluation of early pregnancy. Fertil Steril 34: 1–13

Batzer FR, Schlaff S, Goldfard AF, Corson SL (1981) Serial subunit human chorionic gonadotrophin doubling time as a prognosticator of pregnancy outcome in an infertile population. Fertil Steril 35: 307–312

Boyko WL (1979) Determination of serum hCG levels by radioreceptor assay in the clinical laboratory. Am J Med Technol 45: 797–805

Boyko WL, Russell HT (1979) Evaluation and clinical application of the quantitative radioreceptor assay for serum hCG. Obstet Gynecol 54: 737–745

Catt KJ, Dufau ML, Vaitukaitis JL (1975) Appearance of hCG in pregnancy following the initiation of implantation of the blastocyst. J Clin Endocrinol Metab 40: 537–540

Chartier M, Roger M, Barrat J, Michelon B (1979) Measurement of plasma human chorionic gonadotrophin (hCG) and β-hCG activities in the late luteal phase: evidence of the occurrence of spontaneous menstrual abortions in infertile women. Fertil Steril 31: 134

Csapo AI, Pulkkinen MO, Ruttner B, Sauvage JP, Wiest WG (1972) The significance of the human corpus luteum in pregnancy maintenance. Am J Obstet Gynecol 112: 1061–1067

Fishel SB, Edwards RG, Evans CJ (1984) Human chorionic gonadotropin secreted by preimplantation embryos cultured in vitro. Science 223: 816–818

Goodman AL, Hodgen GD (1979) Corpus luteum–conceptus–follicle relationships during the fertile cycle in rhesus monkeys: pregnancy maintenance despite early luteal removal. J Clin Endocrinol Metab 49: 469–471

Hertig AT (1975) Implantation of the human ovum. In: Behrman SJ, Kistner RW (eds) Progress in infertility. Little, Brown & Co, Boston, p 411

Lenton EA, Adams M, Cooke ID (1978) Plasma steroid and gonadotrophin profiles in ovulatory but infertile women. Clin Endocrinol 8: 241–255

Lenton EA, Grudzinskas JG, Gordon YB, Chard T, Cooke ID (1981a) Pregnancy specific β_1

glycoprotein and chorionic gonadotrophin in early human pregnancy. Acta Obstet Gynecol Scand 60: 489–492
Lenton EA, Grudzinskas JG, Neal LM, Chard T, Cooke ID (1981b) Chorionic gonadotrophin concentration in early human pregnancy: comparison of specific and non-specific assays. Fertil Steril 35: 40–45
Lenton EA, Cripps K, Sulaiman R, Sobowale O, Ryle M, Cooke ID (1982a) Plasma prolactin concentrations during conception and the first ten weeks of human pregnancy. Acta Endocrinol 100: 295–300
Lenton EA, Neal LM, Sulaiman R (1982b) Plasma concentrations of human chorionic gonadotrophin from the time of implantation until the second week of pregnancy. Fertil Steril 37: 773–778
Lenton EA, Sulaiman R, Sobowale O, Cooke ID (1982c) The human menstrual cycle: plasma concentrations of prolactin, LH, FSH, oestradiol and progesterone in conceiving and non-conceiving women. J Reprod Fertil 65: 131–139
Manganiello PD, Nazian SJ, Ellegood JO, McDonough P, Mahesh VB (1981) Serum progesterone, 17-hydroxyprogesterone, human chorionic gonadotropin, and prolactin in early pregnancy and a case of spontaneous abortion. Fertil Steril 36: 55–60
Marshall JR, Hammond CB, Ross GT, Jacobson A, Rayford P, Odell WD (1968) Plasma and urinary chorionic gonadotropin during early pregnancy. Obstet Gynecol 32: 760
Neill JD, Johansson EDB, Knobil E (1969) Patterns of circulating progesterone concentratio during the fertile menstrual cycle and the remainder of gestation in the rhesus monkey. Endocrinology 84: 45–48
Yoshimi T, Strott CA, Marshall JR, Lipsett MB (1969) Corpus luteum function in early pregnancy. J Clin Endocrinol 29: 225–230

Section III
PLACENTAL HORMONES AND PROTEINS

3. Recognition of Early Pregnancy: Human Chorionic Gonadotrophin

P. G. Whittaker

Introduction

The issues discussed in this chapter are those associated with the recognition of pregnancy by the clinician. How soon after conception can pregnancy be detected by measurement of human chorionic gonadotrophin (hCG)? Can pregnancy be detected in this way before it is manifest clinically? Can hCG levels predict impending failure in early pregnancy?

Assay of HCG

The first difficulty is how to be confident of discerning true from false positive hCG values. Hussa et al. (1985) pointed out that some hCG assays will detect up to 30 mU/ml in occasional serum samples from normal non-pregnant women when other assays will not detect hCG in the same samples. They also remarked, as did Whittaker et al. (1983a) and Wilcox et al. (1985), that occasional non-pregnant patients have persistent though low levels of the hormone. The suggested reasons for these variations were: the detection of free subunits, modification of hCG saccharide content and the presence of large-molecular weight forms. Whereas the hCG used as the international radioimmunoassay (RIA) standard is defined as uniform, purified preparations of intact hCG for use as iodinated tracers, which are homogeneous by gel filtration, show striking differences when analysed by gradient gel electrophoresis. This is reflected in differing assay performance (Whittaker, unpublished observations). There is also ectopic production of hCG-like substances in non-pregnant healthy women. Other confounding factors include variations in polyclonal antibody specificity for hCG and β-hCG, cross-reactions with elevated LH levels and unidentified non-specific serum factors. The combined effect of these differences means that detectable low levels of hCG can be expected in about 3% of samples.

The suggested solutions for minimising aberrant results include the use of two hCG assay systems, one of which should be an immunoradiometric assay (IRMA). LH assays to determine any cross-reactivity and parallelism on sample dilution also help to confirm the hCG values. In repeated samples during early normal pregnancy the concentrations of hCG should double every 1 to 2 days.

An important issue is whether blood or urine should be the sample of choice. Urine collection is more acceptable to patients, but is assay of serum more sensitive, or that of urine more prone to false positive results? Human chorionic gonadotrophin can be first detected in serum on average 9 days after ovulation (Lenton et al, 1982) and at 13 days in urine (Armstrong et al. 1984); both groups used an RIA with sensitivities of 5 mU/ml, but IRMA sensitive to 0.5 mU/ml detected hCG in urine 9 days after ovulation (Armstrong et al. 1984). It would be interesting to see the results of IRMA techniques with serum samples: there are no published concurrent studies on both serum and urine over conception or during very early normal pregnancy using these techniques. Marshall et al. (1968) suggested that concentrations of hCG in urine and plasma over conception were similar, whereas Wehmann and Nisula (1981) showed that hCG concentrations in urine are directly proportional to plasma levels. Norman et al. (1985) found that in early pregnancy concentrations of intact hCG in serum were higher than in urine, though related, while β-hCG subunit levels were much higher in urine and did not correlate with serum β-hCG levels. Serum β-hCG levels have been estimated as 16% of total hCG, 4–6 weeks after the last menstrual period (LMP) (Cole et al. 1984), though Norman et al. (1985) found them to be less than 1% at 6 weeks.

HCG and Unsuspected Pregnancy

In deciding whether conception has occurred before pregnancy becomes clinically obvious, two considerations are important. Firstly, non-specific binding and LH cross-reactivity must be allowed for. This should be based on a knowledge of LH cross-reaction in the assay used and on repeated measurements of samples from non-pregnant women. We have undertaken measurements on a variety of serum samples from men and from women during the follicular phase of the menstrual cycle. None had values over 4 mU/ml. Cross-reaction with LH was 2.5%, and we therefore chose a pregnancy diagnosis level of 16 mU/ml. Ideally every hCG assay would be combined with a concomitant assay of LH, although exclusion of hCG cross-reaction in LH assays has only recently become possible. Secondly, to signify true early pregnancy, hCG should be apparent on more than 1 day in any given menstrual cycle, and 8 days or more after estimated ovulation, since hCG is unlikely to be detected before implantation.

Early studies of women using an intrauterine contraceptive device (IUCD) suggested that hCG was detectable during the luteal phase of regular cycles and that the effect of an IUCD was to interfere with implantation (Beling et al. 1976; Landesman et al, 1976; Seppälä et al, 1978). Other reports claimed that such hCG coincided with LH peaks (Klein and Mishell 1977; Sharp et al. 1977; Orloff et al. 1979). In women who had had a tubal ligation occasional positive hCG results were obtained when daily serum samples were assayed for both LH and hCG (Segal et al. 1985). Thus if true hCG cannot be reliably detected before implantation, then it may always remain unclear whether IUCDs are true contraceptive devices or interfere with implantation after conception.

The rate of subclinical early pregnancy loss reported among apparently normal women trying to conceive is influenced not only by assay method but also by subject selection and sampling protocol. Miller at al. (1980) estimated unsuspected pregnancy loss as 33% (as a percentage of total conceptions). However, they used a β-hCG tracer, which may overestimate hCG content (Tyrey and Hammond 1976), and they did not indicate how many of the positive urinary hCG values occurred more than once in a cycle. Our own study (Whittaker et al. 1983a), using intact hCG label and measuring weekly serum samples, showed an 8% rate of preclinical pregnancy loss. At the other extreme Edmonds et al. (1982) using a β-hCG tracer, found a rate of 57%. Only 48% of these positives results were detected more than once in a cycle – i.e., if unsuspected pregnancies were diagnosed on the basis of two or more hCG values per cycle, then the loss rate was 27% of all pregnancies, a rate similar to that reported by Chartier et al. (1979) and Sharp et al. (1986) in infertile women. Though Edmonds et al. (1982) chose a high cut-off limit (50 mU/ml), derived from ovulatory cycles of sterilised women, they were able to detect pregnancy surprisingly soon after ovulation. Subclinical pregnancy was identified on average 8–9 days after ovulation (earlier than successful pregnancy), though how ovulation was dated is not clear.

Wilcox et al. (1985) took daily early morning urine samples from 30 women trying to become pregnant and assayed them with RIA and IRMA for hCG as well as for LH. They picked up four early pregnancy losses out of 21 total conceptions (hCG having been maintained over 4 or more days). This yielded a rate of 19%. Three of these (14%) were detected only by the IRMA for hCG (not by the RIA), and the IRMA also detected eight 1-day spikes of hCG in six of 68 other study cycles. While LH peaks were not always clearly identifiable in urine, two of the early losses were identified late in the menstrual cycle, occurring within the menses.

HCG and Pregnancy Failure

The changes in serum hCG concentrations during early normal pregnancy are both rapid and predictable. Various authors (Whittaker et al. 1983b; Lagrew et al. 1984; Ahmed et al. 1984) have shown that hCG levels can be used to predict the length of gestation up to about 60 days post LMP, with an error of 3 to 4 days. Although an exponential regression of serum hCG against length of gestation can be constructed, the changes are probably more complex, with a slowing rate of increase (Pittaway et al. 1985a) yielding a smooth parabolic curve.

Some reports have suggested that maternal serum hCG levels will predict subsequent failure of an established pregnancy, but sensitivity is low since time-to-time variation may be 25% (Owens et al. 1981). Braunstein et al. (1978) found normal hCG levels in 14 of 33 women subsequently going on to abort and who had had two or more blood samples after day 28 from the LMP. Examining the rate of hCG increase might remove some of the ambiguity in assessing patients against the normal range (Batzer et al. 1981). However, Pittaway et al. (1985b) found that 9 out of 25 women subsequently aborting and 3 out of 8 women with ectopic pregnancy had a normal hCG doubling time of about 2 days between days 35 and 42 post LMP. Our own work with early pregnancy failure compared serial determinations in 25 women whose pregnancies ended in spontaneous abortion (blighted ovum) with 72 normal pregnancies (Aspillaga et al. 1986). Though the mean hCG concentrations in the abortion group were significantly lower from 8 weeks of pregnancy onwards, the hCG levels showed a normal pattern of change. In particular, comparison of within-patient changes in hCG during the 4 to 6 weeks post LMP showed that normal and aborting women were not statistically different. It thus appears that in women whose pregnancies could never succeed – i.e., those having a blighted ovum – early rates of increase in hCG are not different to those with a successful ongoing pregnancy.

HCG and In-vitro Fertilisation

The increased use of in-vitro fertilisation (IVF) has provided opportunities for investigating some of the physiological changes during early pregnancy. Englert et al. (1984) have suggested that IVF pregnancies show delayed appearance of hCG, but this may be a quirk of their regression analysis, which extrapolated beyond their actual data. The average day of first hCG detection in their IVF conceptions does not appear different, and initial input of hCG may be increased in the first few days after implantation (Lenton et al. 1982; Hay 1985). Hay (1985), using monoclonal RIAs, showed that in successful IVF pregnancies intact hCG is first detectable in serum 9 days after oocyte retrieval (10 days post hCG stimulation) but free β-hCG subunit levels are detectable on day 6, though declining to less than 5% of the total by day 22. The terms "biochemical pregnancy" and "preclinical abortion" have been applied to women having two hCG values greater than 10 mU/ml but going on to apparently normal menstruation (Jones et al. 1983); it is not clear in practice whether a delay in menses distinguishes the two terms. Such biochemical pregnancies were first detected at 12 days and were judged to be of predominantly β-hCG secretion (Hay 1985). The summation of outcomes in recent studies yields a total of 566 pregnancies, 94 (17%) of which were preclinical abortions, 93 (16%) were clinical abortions and 19 (3%) ectopic pregnancies (Jones et al, 1983; Deutinger et al. 1986; Confino et al, 1986a; Okamoto et al. 1987). This suggests that once implantation has occurred, subsequent pregnancy loss due to abortion is similar to normal. A multicentre study (Confino et al. 1986a) demonstrated that multiple transfer of embryos resulting in a singleton pregnancy sometimes have higher serum hCG levels than single-transfer pregnancies. It was suggested that the cyclical pattern of hCG found in early pregnancy reflected embryo loss, though this did not explain why it was also seen after single transfer. There has been some enthusiasm for the use of hCG assays in early prediction of IVF outcome (Deutinger et al. 1986; Confino et al, 1986b). This has been due in part to the assumption that improved knowledge of the gestational age of normal and abnormal IVF pregnancies will remove some of the overlap between them in hCG levels, but the reported rates of change in hCG do not support this view. Tarlatzis et al. (1986) suggest that while initial detection of hCG may be delayed by 3 days, the subsequent exponential rise is not attenuated in women destined to abort. Deutinger et al. (1986) showed that the two groups were not different before 17 days (post hCG stimulus). In a large-scale study Okamoto et al. (1987) showed that hCG values 2 weeks after oocyte retrieval diagnose 100% of ectopic pregnancies but only 64% of spontaneous abortions (predictive value was less good in both cases). Whereas the serum concentration of hCG in the ectopic pregnancies was significantly lower, the mean rate of increase in hCG from 2 to 4 weeks after oocyte retrieval was the same as normal. This contrasts with another study which showed that the sensitivity of an abnormal hCG slope in detecting ectopic pregnancy is 90% (Romero et al. 1986), though the stages of gestation were not described.

Conclusion

The application of sensitive IRMAs will clearly have an important influence on the early recognition of pregnancy and the assessment of preclinical abortion. It is also evident that unsuspected pregnancy loss, though adding to the overall reproductive failure rate in the first trimester of pregnancy, does not constitute the major proportion of spontaneous abortions. In the period of early pregnancy, before ultrasound gives reliable diagnosis of fetal viability, great confidence should not be placed on hCG changes as an indication of outcome. The intensive monitoring and research effort that is a part of IVF studies may shed further light on these questions.

Acknowledgements. We thank Professor Tom Lind for fruitful discussion and the Medical Research Council for financial support.

References

Ahmed AG, Klopper A, Francesco D (1984) Determination of the stage of gestation by the assay of hCG and Schwangerschafts protein 1. Br J Obstet Gynaecol 91: 1234–1239

Armstrong EG, Ehrlich PH, Birken S et al. (1984) Use of a highly sensitive and specific immunoradiometric assay for detection of hCG in urine of normal, non-pregnant and pregnant individuals. J Clin Endocrinol Metab 59: 867–874

Aspillaga M, Whittaker P, Lind T (1986) Placental hormones and early pregnancy failure. Placenta 7: 458–459

Batzer FR, Schlaff S, Goldfarb AF, Corson SL (1981) Serial β-subunit hCG doubling time as a prognosticator of pregnancy outcome in a infertile population. Fertil Steril 35: 307–312

Beling CG, Cederqvist LL, Fuchs F (1976) Demonstration of gonadotropin during the second half of the cycle in women using intrauterine contraception. Am J Obstet Gynecol 125: 855–858

Braunstein GD, Karow WG, Gentry WC, Rasor J, Wade ME (1978) First trimester hCG measurements as an aid in the diagnosis of early pregnancy disorders. Am J Obstet Gynecol 131: 25–32

Chartier M, Roger M, Barrat J, Michelon B (1979) Measurement of plasma hCG and β-hCG activities in the late luteal phase: evidence of the occurrence of spontaneous menstrual abortions in infertile women. Fertil Steril 31: 134–137

Cole LA, Krole TG, Ruddon RW, Hussa RO (1984) Differential occurrence of free beta and free alpha subunits of hCG in pregnancy sera. J Clin Endocrinol Metab 58: 1200–1202
Confino E, Demir RH, Friberg J, Gleicher N (1986a) The predictive value of hCG beta subunit levels in pregnancies achieved by in vitro fertilization and embryo transfer: an international collaborative study. Fertil Steril 45: 526–531
Confino E, Demir RH, Friberg J, Gleicher N (1986b) Does cyclic hCG secretion indicate embryo loss in in vitro fertilization? Fertil Steril 46: 897–902
Deutinger J, Neumark J, Reinthaller A et al. (1986) Pregnancy specific parameters in early pregnancies after in vitro fertilization: prediction of the course of pregnancy. Fertil Steril 46: 77–80
Edmonds DK, Lindsay KS, Miller JF, Williamson E, Wood PJ (1982) Early embryonic mortality in women. Fertil Steril 38: 447–453
Englert Y, Roger M, Belaisch-Alart J, Jondet M, Frydman R, Testart J (1984) Delayed appearance of plasmatic hCG in pregnancies after in vitro fertilization and embryo transfer. Fertil Steril 42: 835–838
Hay DL (1985) Discordant and variable production of hCG and its free alpha and beta subunits in early pregnancy. J Clin Endocrinol Metab 61: 1195–1200
Hussa RO, Rinke ML, Schwester PG (1985) Discordant hCG results: causes and solutions. Obstet Gynecol 65: 211–219
Jones HW, Acosta AA, Andres MC et al. (1983) What is a pregnancy? A question for programs of in vitro fertilization. Fertil Steril 40: 728–733
Klein TA, Mishell DR (1977) Absence of circulating hCG in wearers of intrauterine contraceptive devices. Am J Obstet Gynecol 129: 626–628
Lagrew DC, Wilson EA, Fried AM (1984) Accuracy of serum hCG concentrations and ultrasonic fetal measurements in determining gestational age. Am J Obstet Gynecol 149: 165–168
Landesman R, Continho EM, Saxena BB (1976) Detection of hCG in blood of regularly bleeding women using copper intrauterine contraceptive devices. Fertil Steril 27: 1062–1066
Lenton EA, Neal LM, Sulaiman R (1982) Plasma concentrations of hCG from the time of implantation until the second week of pregnancy. Fertil Steril 37: 773–778
Marshall JR, Hammond CB, Ross GT, Jacobsen A, Rayford P, Odell WD (1968) Plasma and urinary hCG during early human pregnancy. Obstet Gynecol 32: 760–764
Miller JF, Williamson E, Glue J, Gordon YB, Grudzinskas JG, Sykes A (1980) Fetal loss after implantation: a prospective study. Lancet ii: 554–556
Norman RJ, Poulton T, Gard T, Chard T (1985) Monoclonal antibodies to hCG: implications for antigenic mapping, immunoradiometric assays and clinical applications. J Clin Endocrinol Metab 61: 1031–1038
Okamoto SH, Healy DL, Morrow LM, Rogers PAW, Trounson AO, Wood EC (1987) Predictive value of plasma hCG beta-subunit in diagnosing ectopic pregnancy after in vitro fertilisation and embryo transfer. Br Med J 294: 667–670
Orloff VS, Yamamoto S, Greenwood FC, Bryant-Greenwood GD (1979) hCG beta-subunit-like immunoreactive material in the plasma of women wearing an intrauterine progesterone contraceptive system. Am J Obstet Gynecol 134: 632–637
Owens MO, Ryan KJ, Tulchinsky D (1981) Episodic secretion of hCG in early pregnancy. J Clin Endocrinol Metab 53: 1307–1309
Pittaway DE, Reish RL, Wentz AC (1985a) Doubling times of hCG increase in early viable intrauterine pregnancies. Am J Obstet Gynecol 152: 299–302
Pittaway DE, Wentz AC, Maxson WS, Herbert C, Daniell J, Fleischer AC (1985b) The efficacy of early pregnancy monitoring with serial hCG determinations and real-time sonography in an infertility population. Fertil Steril 44: 190–194
Romero R, Kadar N, Copel JA, Jeanty P, De Cherney AH, Hobbins JC (1986) The value of serial hCG testing as a diagnostic tool in ectopic pregnancy. Am J Obstet Gynecol 155; 392–394
Segal SJ, Alvarez-Sanchez F, Adejuwon CA, Brache de Mejia V, Leon P, Fawndes A (1985) Absence of hCG in sera of women who use intrauterine devices. Fertil Steril 44: 214–218
Seppälä M, Rutanen EM, Jalanko H, Lehtovirta P, Stenman UH, Engvall E (1978) Pregnancy-specific beta-glycoprotein and hCG-like immunoreactivity during the latter half of the cycle in women using intrauterine contraception. J Clin Endocrinol Metab 47: 1216–1219
Sharp NC, Anthony F, Miller JF, Masson GM (1986) Early conceptual loss in subfertile patients. Br J Obstet Gynaecol 93: 1072–1077
Sharp RM, Wrixon W, Hobson BM, Corker CS, McLean HA, Short RV (1977) Absence of hCG-like activity in the blood of women fitted with intrauterine contraceptive devices. J Clin Endocrinol Metab 45: 496–499

Tarlatzis BC, De Cherney AH, MacLusky N, Fakih H, Barnea ER, Naftolin F (1986) Embryo–maternal interaction in conceptions after in vitro fertilization and embryo transfer. J In Vitro Fertil Embryo Transf 3: 196–197

Tyrey L, Hammond CB (1976) The hCG beta subunit radioimmunoassay: potential error in hCG measurement related to choice of labelled antigen. Am J Obstet Gynecol 125: 160–165

Wehmann RE, Nisula BC (1981) Metabolic and renal clearance rates of purified hCG. J Clin Invest 68: 184–194

Whittaker PG, Taylor A, Lind T (1983a) Unsuspected pregnancy loss in healthy women. Lancet i: 1126–1127

Whittaker PG, Aspillaga MO, Lind T (1983b) Accurate assessment of early gestational age in normal and diabetic women by serum hPL concentration. Lancet ii: 394–396

Wilcox AJ, Weinberg CR, Wehman RE, Armstrong EG, Canfield RE, Nisula BC (1985) Measuring early pregnancy loss: laboratory and field methods. Fertil Steril 44: 366–374

4. Recognition of Early Pregnancy: Human Placental Lactogen and Schwangerschaftsprotein 1

T. Chard

The human placenta produces a wide variety of "specific" materials which have been evaluated as markers of early pregnancy. These materials have been divided into three groups (Chard 1986): group 1 includes the "classical" trophoblast products (e.g., human placental lactogen [hPL], human chorionic gonadotrophin [hCG], Schwangerschaftsprotein 1 [SP1], placental steroids); group 2 includes placental protein 5 (PP5) and pregnancy associated plasma protein-A (PAPP-A); group 3 includes the endometrial/decidual proteins (e.g., plasma proteins 12, 14 [PP12, PP14]). hCG was the first marker to be described, and its measurement is currently the most widely used in clinical practice. The question addressed here is whether any other material of this class – notably hPL or SP1 – might replace or supplement the clinical use of hCG.

There are three areas of clinical application of biochemical tests in early pregnancy: detection of pregnancy; estimation of the stage of gestation; and evaluation of fetal viability.

Detection of Early Pregnancy

Although the trophoblast probably synthesises specific proteins at the blastocyst stage, these do not enter the maternal circulation in significant quantities until intimate contact is established between the fetal and maternal tissues at the time of implantation (about 7 days after conception). Thereafter, the level of trophoblast products shows a progressive increase in maternal blood. The term "increase" is used, since it is likely that there are small amounts of all these materials in non-pregnant blood – i.e., the levels are never zero.

The time of first detection of the increase of a trophoblast product will depend upon a variety of factors, most of which relate to the sensitivity, specificity and precision of the system of measurement. In the case of hPL, there have been no substantial attempts to develop highly sensitive methods. Consequently, the earliest reported detection is at 6 weeks' gestation (4 weeks post-conception), and current technology would not, therefore, be competitive with hCG for the detection of early pregnancy. There is no reason to believe that earlier detection would not be possible; but it is not available at the present time and, if it were, it would be unlikely to be superior to hCG. In the case of SP1, highly sensitive methods are available and have been used for the detection of early pregnancy (Grudzinskas et al. 1977). However, although SP1 may be equivalent to hCG in this respect, there is no evidence that it is superior. The only situation in which it might replace hCG measurement is in patients who are receiving hCG for therapeutic purposes, though current studies on this possibility are not encouraging (R.J.S. Howell, personal communication).

Estimation of Stage of Gestation

A single determination of blood hPL, hCG or SP1 can be applied to a formula that relates the result to the day of gestation (Lagrew et al. 1983; Whittaker et al. 1983; Westergaard et al. 1983, 1985; Ahmed and Klopper 1986). The same principle can also be applied to predict the day of spontaneous delivery (Westergaard et al. 1985). The accuracy of these estimations is similar to that which can be achieved with the best ultrasound technology. Up to 60 days of pregnancy (from the last menstrual period) there appears to be little to choose between hCG and SP1 (hPL cannot be used at this time because routine assays do not have appropriate sensitivity). Consequently, hCG estimation would be the method of choice because of its widespread availability. After 60 days hCG cannot be used because the increase ceases to be linear. At this time either hPL or SP1 would be suitable, with availability of the method favouring hPL.

Evaluation of Fetal Viability

There is an extensive literature on the use of biochemical tests in prognosis of threatened abortion in early pregnancy (see Grudzinskas et al. 1986). Most of this literature concerns hCG or hPL, with rather more limited information on SP1 and other products. Three general conclusions can be drawn: (1) all group 1 products give the same result – i.e., the levels are generally low in pregnancies in which the outcome is unfavourable (Salem et al. 1984; Fig. 4.1); (2) none of the tests are of much value before 7–8 weeks of pregnancy (though serial levels might be of value at an earlier stage – an attractive hypothesis which has yet to be adequately documented); and (3) essentially the same clinical information (i.e., whether the fetus is present, or alive or dead) can be obtained by ultrasound.

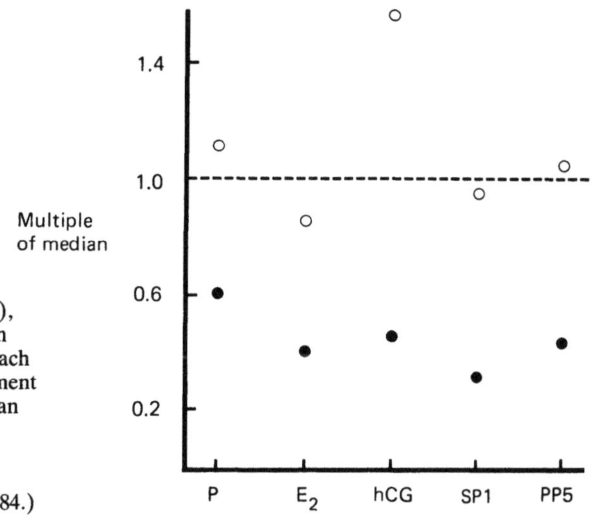

Fig. 4.1. Levels of progesterone (P), oestradiol (E_2), hCG, SP1 and PP5 in patients with threatened abortion. Each symbol represents the mean measurement of each test as a multiple of the median of the control group. ○, pregnancy continued beyond 27 weeks; ●, spontaneous abortion at less than 18 weeks. (Modified from Salem et al. 1984.)

Conclusions

The clinical applications of hPL and SP1 in early pregnancy are limited – not because they are inefficient in this respect but because for each potential application there is a more widely available test (hCG or ultrasound).

References

Ahmed AG, Klopper A (1986) Estimation of gestational age by last menstrual period, by ultrasound scan and by SP1 concentration: a comparison with date of delivery. Br J Obstet Gynaecol 93: 122–127

Chard T (1986) Placental synthesis. In: Chard T (ed) Clinics in obstetrics and gynaecology, vol 13. WB Saunders, London, pp 447–467

Grudzinskas JG, Gordon YB, Jeffrey D, Chard T (1977) Specific and sensitive determination of pregnancy-specific beta-1 glycoprotein by radioimmunoassay. Lancet i: 333–335

Grudzinskas JG, Westergaard JG, Teisner B (1986) Biochemical assessment of placental function: early pregnancy. In: Chard T (ed) Clinics in obstetrics and gynaecology, vol 13. WB Saunders, London, pp 553–569

Lagrew DC, Wilson EA, Jawad MJ (1983) Determination of gestational age by serum concentrations of human chorionic gonadotrophin. Obstet Gynecol 62: 37–42

Salem HT, Ghaneimah SA, Shaaban MM, Chard T (1984) Prognostic value of biochemical tests in the assessment of fetal outcome in threatened abortion. Br J Obstet Gynaecol 91: 382–385

Westergaard JG, Teisner B, Grudzinskas JG, Chard T (1983) Accurate assessment of early gestational age by measuring serum hCG and SP1. Lancet ii: 567

Westergaard JG, Teisner B, Grudzinskas JG, Chard T (1985) Single measurements of chorionic gonadotrophin and schwangerschaftsprotein 1 for assessing gestational age and predicting the day of delivery. J Reprod Med 30: 57–60

Whittaker PG, Aspillaga MO, Lind T (1983) Accurate assessment of early gestational age in normal and diabetic women using serum human placental lactogen (hPL). Lancet ii: 304–306

5. Pregnancy-Associated Plasma Protein-A: Fact, Fiction and Future

M.J. Sinosich

Introduction

Sixty years ago a gonadotrophic component, termed chorionic gonadotrophin (hCG), was detected in the urine of pregnant women. Since then, many other biologically active pregnancy associated proteins have been identified including hormones (placental lactogen), enzymes (heat stable alkaline phosphatase), enzyme inhibitors (urokinase inhibitor), immunosuppressive agents (early pregnancy factor) and thromboregulators (embryo derived platelet activating factor). During the past two decades a large series of "pregnancy specific" or pregnancy associated proteins have been described, resulting in a plethora of synonyms. Since the function of this latter group of proteins, detected with immunological techniques, remains to be elucidated, they were collectively termed "new generation" placental proteins (Klopper 1979).

However, the immunological approach has resulted in confusion surrounding the new generation placental proteins. Synonyms and misnomers abound. Pregnancy associated plasma protein-A (PAPP-A) is the most recent of the "new generation" placental proteins to have landed in this immunological quagmire. The present review evaluates the literature on PAPP-A, attempts to separate fact from fiction, and suggests avenues for future research.

Isolation and Purification

Early attempts to purify PAPP-A from human pregnancy serum were based on classical physicochemical approaches and proved technically cumbersome (Lin et al. 1974). Subsequent purification schedules included affinity chromatography on concanavalin-A Sepharose (Bischof 1979) and positive immuno-affinity matrices (Folkersen et al. 1979, 1981). However, because of physicochemical similarities to α_2-macroglobulin (α_2M) and pregnancy zone protein (PZP), these glycoproteins were frequently copurified with PAPP-A (Folkersen et al. 1979; Gore and Sutcliffe 1984). Furthermore, the chaotropic eluants required to dissociate antigen-antibody complexes may irreversibly denature PAPP-A (Sutcliffe et al. 1979, 1980a).

Qualitative physicochemical analysis demonstrated many similarities between PAPP-A and α_2M, including α_2 electrophoretic mobility (net surface charge), dual absorption maxima at 214 nm and 280 nm (Sinosich 1985; Sinosich et al. 1984a), metal chelate (Sinosich et al. 1983a) and lectin interactions (Table 5.1). Whereas immunoelectrophoretic analysis after incubation with β-glucuronidase virtually eliminated the PAPP-A immunoprecipitate, this glycolase had no effect on the electrophoretic mobility or immunoprecipitate morphology of α_2M (Table 5.1). This difference was substantiated by crossed affinity immunoelectrophoresis (CAIE), which demonstrated a specific interaction between PAPP-A and heparin, a highly charged anionic glycosylaminoglycan (Sinosich et al. 1981).

Table 5.1. Comparative carbohydrate analysis of PAPP-A and α_2-macroglobulin (α_2M) by immunoelectrophoresis

	Specificity	Mitogenic activity	Interaction[a]	
			PAPP-A	α_2M
Lectin				
Bandeiraea simplicifolia 1	α-D-GalNAC α-gluNAC		±	±
Bandeiraea simplicifolia 2	α-NAC		–	–
Concanavalin A	α-D-man; α-D-glc	Yes	+	+
Helix pomatia	D-gal NAC		+	+
Lens culinaris	α-D-man; α-D-glc	Yes	+	+
Limulus polyphemus	NeuNAC		+	+
Peanut agglutinin	B-D-gal (1–3)-D-gal NAC	Yes	–	–
Phytohaemagglutinin	D-galNAC	Yes	+	+
Pokeweed mitogen	(D-glc NAC)$_3$	Yes	–	–
Ricinus communis	D-galNAC; B-D-gal		+	+
Soya bean	D-galNAC	Yes	–	–
Ulex europeus	α-L-fuc		–	–
Wheatgerm	D-glcNAC; NeuNAC	Yes	+	±
Glycolase				
β-Glucuronidase	Glucuronic acid		+	–

[a]+, complete elimination of immunoprecipitate; ±, altered immunoprecipitate; –, unaltered immunoprecipitate.

The heparin-PAPP-A interaction was confirmed by chromatography on heparin-Sepharose. Quantitative desorption of heparin-Sepharose bound PAPP-A was achieved with protamine sulphate (a heparin antagonist), decreased buffer pH or increased buffer ionic strength. Sample application in 50-mmol Tris-HCl, pH 7.8, containing 0.3 mol/l NaCl, eliminated weak non-specific heparin interactions, and the resultant PAPPA-A preparation was free of PZP, α_2M and SPl (Sinosich et al, 1987a).

PAPP-A was purified in three stages (Table 5.2). Firstly, positive affinity chromatography on heparin-Sepharose to concentrate PAPP-A and simul-

Table 5.2. PAPP-A purification procedure

	Fractionation mechanism	PAPP-A (mg)	Recovery (%)	Yield (%)	Purification factor
Pooled pregnancy serum	Starting material	4.85	100	100	1.0
Heparin–Sepharose	Positive affinity	1.90	103.5	39.1	485.8
Ultrogel AcA 34	Size	1.84	98.8	37.9	826.9
Antibody matrix	Negative immunoaffinity	1.08 (both steps)	58.7 (both steps)	22.2 (both steps)	1482.8 (both steps)
Heparin–Sepharose	Positive affinity				

taneously eliminate α_2M and PZP. Secondly, heparin-Sepharose eluted PAPP-A was applied, in high ionic strength buffer, to gel filtration on Ultrogel AcA34 to remove any leached heparin and small heparin-binding proteins. The third stage was balanced negative immunoaffinity chromatography (Sinosich et al. 1987a). The overall yield of pure PAPP-A was 22.2% (1.1 mg) with a purification factor of 1483 (Table 5.2). Thus for the first time large quantities of pure PAPP-A could be rapidly prepared with mild chromatographic procedures. Although this procedure selects high-heparin-affinity PAPP-A, as yet no physicochemical, immunological or in-vitro functional differences have been demonstrated between PAPP-A of low and high heparin affinity.

Homogeneity and purity were confirmed with various techniques, including polyacrylamide electrophoresis (conventional and autoradiographic), isoelectric focusing (visualised or assayed by radioimmunoassay), gel filtration, radioimmunoassay (RIA), and injection into rabbits with subsequent demonstration of monospecific antiserum. This evaluation was far more extensive than previous studies (Bischof 1979; Folkersen et al. 1979).

Characterisation

Earlier studies reported that PAPP-A was a large (750–820 kd) homotetrameric glycoprotein (Bischof 1979, 1980) with an isoelectric point (pI) of 4.4 (Lin et al. 1974). Qualitative studies confirmed that the molecular conformation of PAPP-A was labile to trypsin, elastase, chaotropic ions, extremes of pH and chromatography on metal-chelate Sepharose, but inert to nuclease activity and lipid (Sudan Black or Oil Red O) stains (Sinosich 1985; Sinosich et al. 1983a, 1984a; Sutcliffe et al. 1979). In addition, PAPP-A was degraded by incubation with chondroitinase AC, neuraminidase, α-glucosidase and β-glucuronidase, but not β glucosidase (Table 5.1). By contrast, α_2M was inert to β glucuronidase and chondroitinase AC, as well as β-glucosidase (M.J. Sinosich, unpublished data). This confirmed the presence of glucuronic acid, as chondroitin sulphate

Table 5.3. Comparison of PAPP-A and α_2-macroglobulin (α_2M) molecular conformation

	PAPP-A	α_2-Macroglobulin
Molecular weight (d)	820 000	725 000
Molecular conformation	Homotetrameric	Homotetrameric
Carbohydrate (%)	19.4	8.6
Carbohydrate (d)	159 080	62 350
Protide (d)	660 920	662 650

moieties, in the carbohydrate component of PAPP-A and suggested a marked difference in post-translational processing of the two proteins. Hence, PAPP-A is a proteoglycan, whereas α_2M is a glycoprotein.

Despite differences in carbohydrate content and composition, the peptide cores of both antigens were of similar molecular size – approximately 661 000 d (Table 5.3). Qualitative carbohydrate composition of PAPP-A includes glucuronic acid, glucose, galactose, fucose, sialic acid and mannose (Fig. 5.1). The native homotetrameric PAPP-A consists of two dimeric subunits bonded electrostatically by van der Waal's forces (Fig. 5.1). Each dimer is composed of two monomeric subunits covalently bound by disulphide bridges.

Chondroitin moieties are highly charged glycans, like heparin, with many potent biological actions including protease inhibition, binding of metal ions and functioning as an adhesive for extracellular matrix. Thus, the heparin-PAPP-A interaction may be considered a glycan-glycan interaction, and indeed the heterogeneity of this interaction may be due to variability in the chondroitin sulphate content of PAPP-A. However, potential variation in post-translational composition of PAPP-A was not apparent as variability in molecular size. By gel

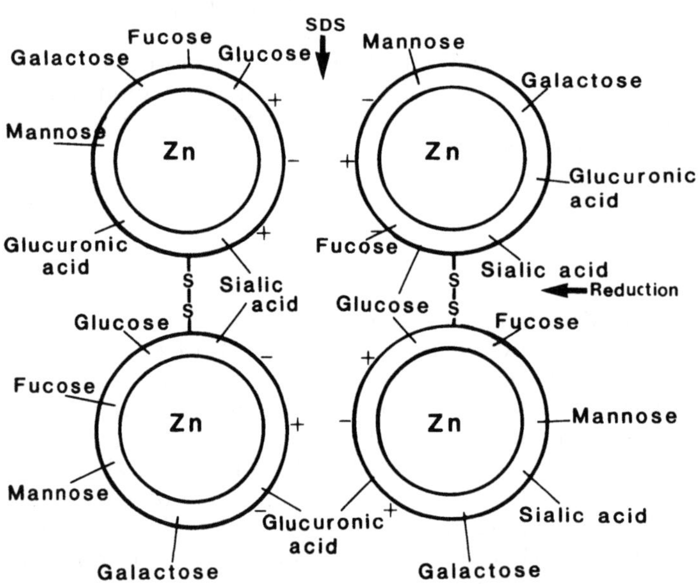

Fig. 5.1. Schematic representation of PAPP-A molecular conformation.

filtration or SDS-PAGE, the molecular size of pure PAPP-A has been consistently reported at 750–820 kd (Lin et al. 1974; Bischof 1979; Sutcliffe et al. 1980b; Davey et al. 1983). Indeed, size fractionation on Ultrogel AcA34 of pregnancy serum, pregnancy heparinised plasma, pregnancy EDTA plasma, placental extract, preovulatory follicular fluid, seminal plasma, vesicular fluid from hydatidiform molar tissue, radioiodinated (^{125}I) PAPP-A and culture media from granulosa or trophoblast cells consistently eluted one single Gaussian peak of PAPP-A immunoreactivity of molecular size (Mr) 820 kd (Table 5.4) (Sinosich et al. 1982b, 1984b, 1985g, 1987d). By contrast, the PAPP-A RIA as reported by Bischof et al. (1981) was based on an immunoreactive tracer of Mr 200 kd.

Table 5.4. Physicochemical and immunological characterisation of trophoblastic and non-trophoblastic derived PAPP-A

	Pregnancy serum	Placenta	Follicular fluid	Seminal plasma	Granulosa cell culture media
Slope of logit-log regression line (m)	−2.267	−2.288	−2.271	−2.271	−2.369
Molecular weight (kd)	820	820	820	820	820
Affinity interactions[a]					
Heparin–Sepharose	+	+	+	+	+
Zinc chelate-Sepharose	+	+	+	+	+
Thrombin–Sepharose	−	−	−	−	−

[a] +, reversible interaction; −, no interaction.

Function

Pregnancy-associated antigens fall into two major categories – those detected by monitoring bioactivity (hormones, enzymes, thromboregulators) and the "new generation" placental proteins for which biological actions remain to be elucidated. It is widely accepted that the luteotrophic action of human chorionic gonadotrophin (hCG) was to "rescue" the corpus luteum (Short 1969), but an immunosuppressive role has also been suggested (Adcock et al. 1973; Teasdale et al. 1973). However, pure hCG does not suppress mitogen induced lymphocyte transformation (Caldwell et al. 1975), and this activity has been attributed to a contaminant, early pregnancy factor (Rolfe et al. 1983).

Mechanisms by which the fetal allograft survives or escapes maternal immune rejection have preoccupied researchers for decades. Of the many hypotheses, three possibilities remain: (1) the uterus as a privileged site, (2) maternal immunosuppression, and (3) the presence of an immunologically inert barrier at the fetomaternal interface. Pregnancy serum is a more potent inhibitor of in-vitro immune responses (Anderson and Yunis 1985; Davies and Browne 1985a, b; Nicholas and Panayi 1985). However, tests such as mitogen induced lymphoblastogenesis are non-specific (Nelson and Gatti 1976; Miyasaka 1985) and are susceptible to lectin–antigen interaction and to variations in dose and antigen purity. Yet by such assays many varied antigens have been implicated as maternal immunomodulators (Table 5.5), including decidualised stromal cells of the endometrium (Kearns and Lala 1985), suppressor T lymphocytes (Vanderbeeken et al. 1984) and newly identified factors (Chaouat et al. 1985; McIntyre and Page Faulk 1979; Daya and Clark 1986; Culouscou et al. 1986). However, the tenet of systemic immunosuppression is not compatible with maternal wellbeing, and few, if any, of these antigens (Table 5.5) are now considered immunosuppressive (see reviews in Table 5.5).

Table 5.5. Antigens thought to suppress maternal immune responses

Antigen	Molecular size (d)	Reference
Alphafetoprotein	70 000	Murgita and Tomasi (1975)
Androgens	–	See reviews[a]
α_1-Acid glycoprotein	–	See reviews[a]
α_1-Antitrypsin	54 000	See reviews[a]
α_2-Macroglobulin	725 000	See reviews[a]
Caeruloplasmin	–	See reviews[a]
Chorionic gonadotrophin	47 000	Kaye and Jones (1971)
Complement factor 3	150 000	See reviews[a]
Cortisol	–	Munroe (1971)
C-reactive protein	118 000	See reviews[a]
Early pregnancy factor	20 000–250 000	Noonan et al. (1979)
Fibrinogen degradation products	–	See reviews[a]
Lipoproteins	–	See reviews[a]
Oestrogens	–	Ablin et al. (1974)
Placental lactogen	20 000	Contractor and Davies (1973)
Placental protein 14	42 000	Bolton et al. (1986)
Placental protein 15	31 000	Bohn et al. (1980)
Pregnancy associated plasma protein-A	820 000	Martin-du-Pan et al. (1983)
Pregnancy specific β1-glycoprotein	110 000	Horne et al. (1976)
Pregnancy zone protein	389 000	von Schoultz et al. (1973)
Progesterone	314	Mori et al. (1977)
Prostaglandins	–	Clark (1985)

[a] Rocklin et al. (1979); Murgita and Wigzell (1981); Gill (1984); Jacoby et al. (1984); Keller and Calvanico (1984).

PAPP-A was added to this ever-increasing list of alleged immunosuppressive factors simply because diluted seminal plasma suppressed mitogen induced lymphoblastogenesis (Martin-du-Pan et al. 1983). However, subsequent studies showed no correlation between circulating PAPP-A concentrations and in vitro immunosuppression (Bischof et al. 1983b), and an immunosuppressive role for PAPP-A was unanimously rejected (Stimson 1980, 1981; McIntyre et al. 1981; Nicholas and Panayi 1986; Sinosich et al. 1985e). The apparent immunosuppressive activity of seminal plasma was attributed to the cytotoxic property of oxidised spermine (Allen and Roberts 1986).

Many other in-vitro functions have been attributed to PAPP-A (Table 5.6). Bischof (1979) suggested that PAPP-A inhibited complement mediated red cell haemolysis. However, the protease inhibitory effects attributed to PAPP-A by Bischof and his colleagues (Table 5.6) were due to contaminants, including α_2M (Gore and Sutcliffe 1984) and heparin. Although in the absence of appropriate

Table 5.6. Reported in-vitro actions of PAPP-A

Inhibitory	
Plasmin	Bischof (1979)
Urokinase activation of plasminogen	Bischof (1979)
Complement system	Bischof (1979)
Leucocyte elastase	Sinosich et al. (1982a)
Lectin induced lymphoblastogenesis	Martin-du-Pan et al. (1983)
Thrombin	Bischof et al. (1983c)
Activator	
Acrosome reaction	Bischof et al. (1987)
Transport	
Zinc carrier protein	Sinosich et al. (1983a)
Biological action	
Barrier against phagocytic–proteolytic defences	Sinosich et al. (1984c)

controls (column blanks) induction of the acrosome reaction may be attributed to heparin (Delgado et al, 1980, 1982), leached from affinity matrix during PAPP-A purification, follicular proteoglycans do enhance activation of proacrosin (Lenz et al. 1982).

By contrast, pure PAPP-A did not inhibit plasmin, urokinase activation of plasminogen, trypsin, collagenase, chymotrypsin (Sinosich et al, 1982a, 1985f) or complement activation (Gore and Sutcliffe 1984; R. Simm, personal communication) but specifically and non-competitively inhibited granulocyte elastase (Sinosich et al, 1982a). Controls to determine elastase inhibition kinetics and zinc content of PAPP-A included column buffer blanks, elution buffers and pooled male serum fractionated in parallel with pregnancy serum. Although glycan moieties have affinity for divalent metal ions via ionic interactions with sulphate groups (Parish and Fair 1981; Casu 1985), this is not the explanation for Zn^{2+}– PAPP-A interaction. Firstly, PAPP-A specifically interacted with chelated Zn^{2+} but not with Ca^{2+}, Cu^{2+}, Mg^{2+} or Mn^{2+}. Secondly, the 214 nm absorption maximum of PAPP-A and α_2M (Sinosich et al. 1984a) is characteristic of Zn^{2+}–sulphide bond interactions in zinc carrier proteins (Vallee and Galdes 1984).

Measurement

Quantitation of PAPP-A includes isotopic and non-isotopic immunoassays. Non-isotopic immunodiffusion and immunoelectrophoretic techniques were widely applied because only an antiserum was required to perform the assay. These techniques were insensitive (detection limit 10 mg/l), and PAPP-A measurements were therefore limited to later stages of pregnancy. The first radioimmunoassay was reported in 1980 (Sinosich et al. 1980). At a detection limit of 3.0 μg/l, (487 μg = 1 IU WHO 78/610) PAPP-A was detected in maternal peripheral blood 28 days after conception (Sinosich et al. 1982b).

Table 5.7. Comparative characteristics of radioimmunoassays for PAPP-A

Reference	Antigen prepared by	Antibody prepared by	Iodination	Specific activity (Ci/g)	Tracer size (d)	Separation phase	Diluent	Assay sensitivity	Plasma or serum	Concentration in blood of normal non-pregnant adults
Sinosich et al. (1980, 1982b)	Bischof	Teisner	B–H	14	820 000	2ndAb-PEG[e]	Male serum	2.9 μg/l	Serum	<2.0 μg/l[a]
Bischof et al. (1981)	Bischof	Bischof	Lacto	60–100	200 000	solid phase 2nd Ab	Horse serum	32.5 μg/l	EDTA	112 μg/l
Anthony et al. (1983)	Bischof	Bischof	ChT	20	⩾200 000	2ndAb-PEG	Male serum	20 μg/l	Serum	<20 μg/l
Pinto-Furtado et al. (1984)	Pinto-Furtado	Dako	ChT	NR	NR	2ndAb-PEG	NR	5 μg/l	Serum	<5 μg/l
Sinosich et al. (1984b)	Sinosich	Dako	ChT	10–650	820 000	2ndAb-PEG	Buffer	12.2 mIU/l	Serum	<12.2 mIU/l[a]
Sjoberg et al. (1984b, 1985)	Sinosich	Dako	ChT	NR	820 000	2ndAb-PEG	Buffer	3 μg/l	Serum	<3 μg/l

B–H, Bolton–Hunter reagent; Lacto, lactoperoxidase; ChT, chloramine T; EDTA, ethylenediaminetotetra-acetic acid; NR, not reported. 2nd Ab-PEG, second antibody-polyethylene glycol.
[a] 1 IU PAPP-A WHO 78/610=487 μg PAPP-A.

However, the description of a second RIA for PAPP-A (Bischof et al, 1981) led to a number of controversies.

PAPP-A immunoassays are susceptible to interference by anticoagulants, and plasma should be avoided (Toop and Klopper 1983; Pinto-Furtado et al. 1984; Sinosich 1985). Addition of EDTA, at levels sufficient to prevent coagulation, suppressed antibody-tracer binding and/or precipitation of antibody bound tracer, thereby doubling apparent PAPP-A levels. This was eliminated by addition of a neutralising dose of Ca^{2+}. By contrast, heparin increased precipitated radioactivity, and resulted in apparent lower PAPP-A levels (Sinosich 1985). The heparin interference was eliminated by neutralisation with protamine sulphate or establishing a PAPP-A standard curve in heparinised plasma.

An International Reference Preparation (WHO 78/610) has been available for almost a decade and it is unfortunate that only four research centres (Geneva, London, Odense, Sydney) provide conversion factors to facilitate comparative studies.

There are fundamental differences between the various PAPP-A RIAs (Table 5.7), particularly the assay of Bischof et al. (1981). The use of horse serum as assay diluent is not recommended because of batch-to-batch variation and variable effects on tracer precipitation. Recommended diluents include assay buffer and/or pooled serum from normal, healthy males. However, when buffer diluents are used the protein concentration in the assay tube must be restored for optimal precipitation of antibody bound tracer.

The vascular system may contain circulating PAPP-A, albeit at levels below current assay detection limits (3 mIU/l or 1.5 µg/l). Such is the case with other pregnancy associated antigens (hCG, schwangerschaftsprotein 1 [SP1]). By contrast to other groups, Bischof et al. (1981) reported PAPP-A levels of 112 µg/l in the blood of non pregnant adults, though Anthony et al. (1983) could not confirm this (Bischof et al. 1981) using the same primary reagents. Furthermore, a combination of Bischof anti-PAPP-A with pure PAPP-A from another source (Sinosich et al. 1982b) did not detect circulating PAPP-A in normal individuals (Table 5.7).

Bischof et al. (1981) demonstrated multiple ($\geqslant$4) peaks of radioactivity after gel filtration of their radioiodination mixture. Maximal PAPP-A immunoreactivity eluted with Mr 200 kd, as a shoulder of a major peak (Mr 70 kd), whereas in other studies immunoreactive ^{125}I-PAPP-A consistently eluted with Mr 820 kd (Sinosich et al. 1984b).

Tracer characterisation by isoelectric focusing confirmed additional differences between the Bischof and Sinosich PAPP-A tracers (Fig. 5.2). After radioiodination the isoelectric point (pI) of (Sinosich) serum-derived ^{125}I-PAPP-A increased by 0.25 units to 4.75. The pI of Bischof's labelled antigen was 5.1.

The earlier RIAs for PAPP-A used Bischof's rabbit anti-PAPP-A antiserum as primary antibody (Table 5.7). More recently, a rabbit anti-PAPP-A antiserum was prepared (Fig. 5.3) with pure PAPP-A as immunogen. The antibody titre achieved far surpassed other available antisera. Analysis by RIA, immunoelectrophoresis, autoradiography and immunohistochemistry confirmed monospecificity, before and after absorption with male serum.

All four current anti-PAPP-A antisera have been evaluated with multidimensional immunoelectrophoresis, RIA, autoradiography and immunohistochemistry. Sinosich (Sydney), Teisner (Odense) and Dakopatts antisera were untreated. The Bischof (Geneva) antiserum (prepared in Aberdeen) was

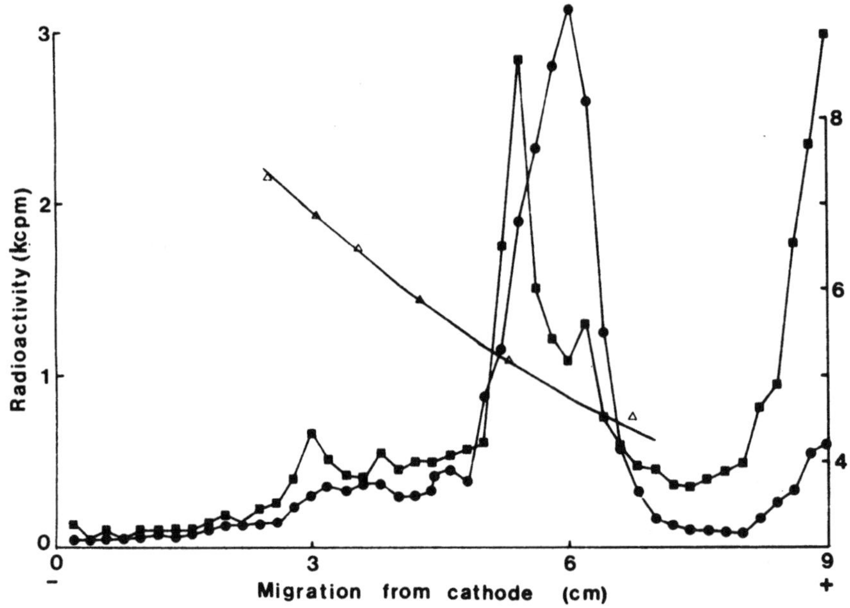

Fig. 5.2. Isoelectric focusing profile of Bischof [■] and Sinosich (●) radioiodinated ^{125}I-PAPP-A. The pH gradient was determined with Pharmacia standards (△).

analysed before and after absorption with pooled male serum. Each antiserum was diluted 1/200 and separately incubated with each tracer: PAPP-A (Sinosich et al. 1987c), SPl (Dr. E. Engvall), α_2M (Boehringer-Mannheim), PZP (Dr. J. Folkersen), antithrombin 111 (AT111; Boehringer-Mannheim) and rabbit IgG (Miles). All five antisera bound PAPP-A tracer, with the Odense preparation showing the lowest binding (Fig. 5.4). Neither the Sydney nor the Odense antibodies reacted with any of the five remaining tracers. The Dakopatts preparation bound SP1 (58.6%) and PAPP-A (56.7%) tracers equally, but no significant interactions were detected with α_2M, PZP, AT111 or rabbit IgG (negative control). Bischof unabsorbed preparation significantly interacted with all tracers, except the negative control. A single absorption with male serum (1 : 1) removed specificities against AT111 and significantly reduced binding of α_2M and PZP, but reactivity against PAPP-A and SP1 remained unchanged.

The Bischof anti-PAPP-A antiserum is polyspecific, but RIA specificity is maintained if pure PAPP-A is used as tracer (Sinosich et al. 1982b) – in which case no PAPP-A (above assay detection limit 12.2 mIU/l) is detected in non-pregnant adults. Clearly, this antiserum is unsuitable for direct application to positive immunoaffinity chromatography, immunohistochemistry and in-vitro functional studies (Bischof et al. 1983c, 1984). Thus, some controversies may be attributed to application of impure primary reagents; and, indeed, an empirical relationship relating specificities of anti-PAPP-A antisera may prove useful when comparing independent PAPP-A data (Fig. 5.5). This relationship states that monospecific PAPP-A antisera (Sydney, Odense) can only identify PAPP-A (as recognised by the Bischof antiserum) when combined with at least six other specificities including α_2M, AT111, SP1, PZP and β-lipoprotein.

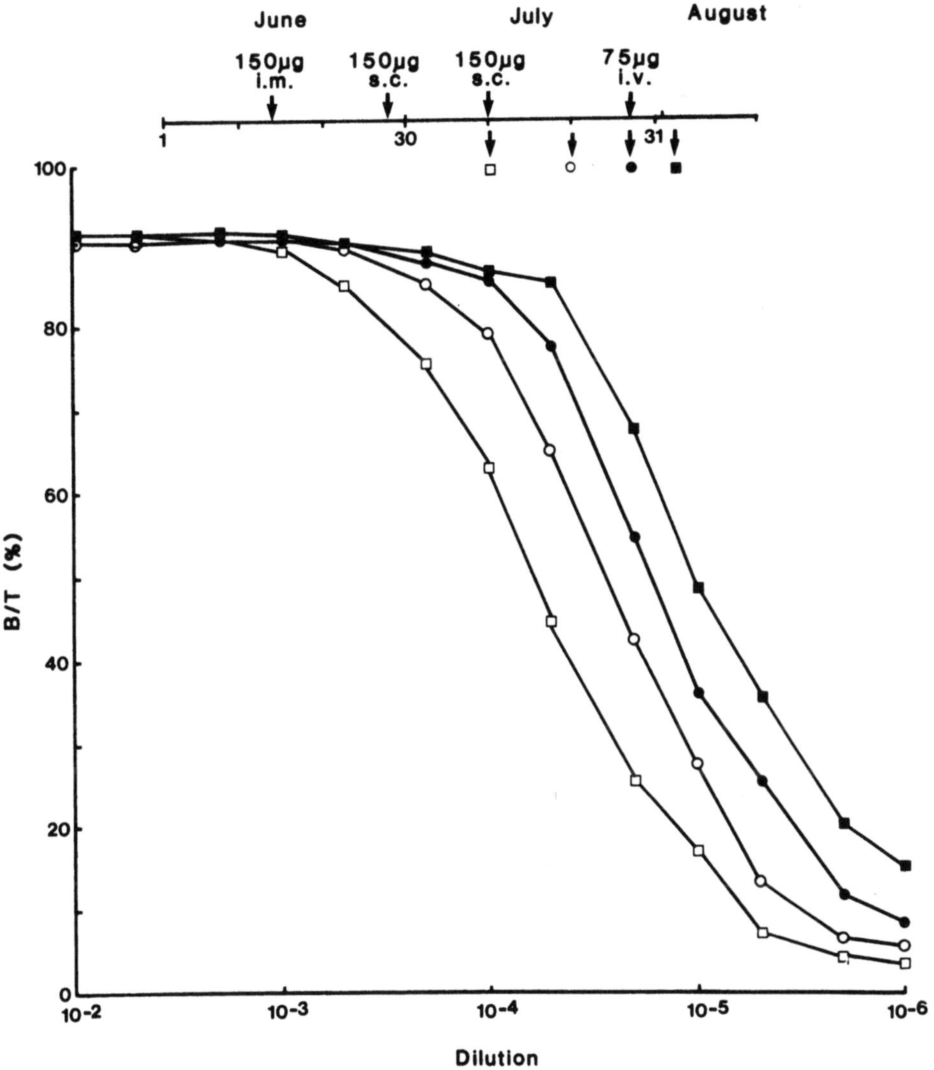

Fig. 5.3. Rabbit immunisation (i.m., intramuscular; s.c., subcutaneous; i.v., intravenous) and bleeding (□, ○, ■, ●) schedule for preparation of (Sydney) anti-PAPP-A antiserum. Portions of each bleed were simultaneously titrated against (Sydney) ^{125}I-PAPP-A tracer.

Site of Production

Defining sites of production of pregnancy associated antigens has proved a daunting task. Methodologies commonly applied included quantitative analysis of body fluids and tissues, demonstration of concentration gradients between afferent and efferent blood flow, immunohistochemistry and cell cultures. Although such studies refer to production sites, they actually indicate storage

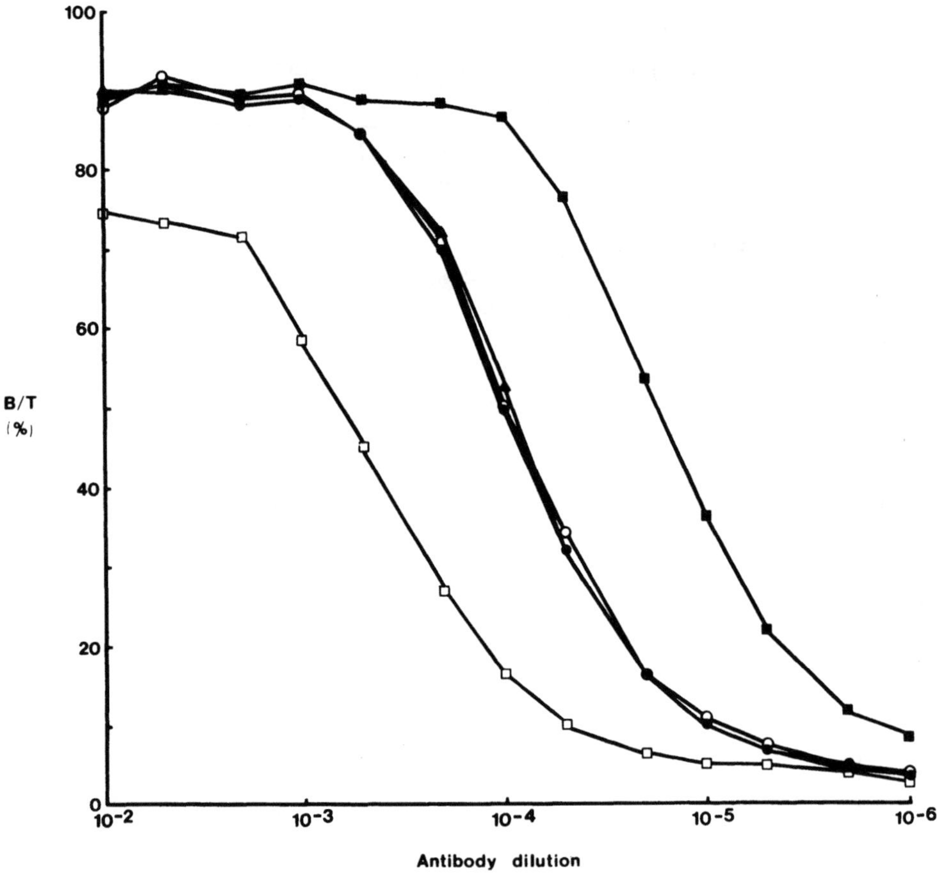

Fig. 5.4. Titration curves of Bischof neat (○) and absorbed (●), Dakopatts (▲), Sinosich (Sydney, ■) and Teisner (Odense, □) anti-PAPP-A antisera against Sydney radioiodinated ^{125}I-PAPP-A.

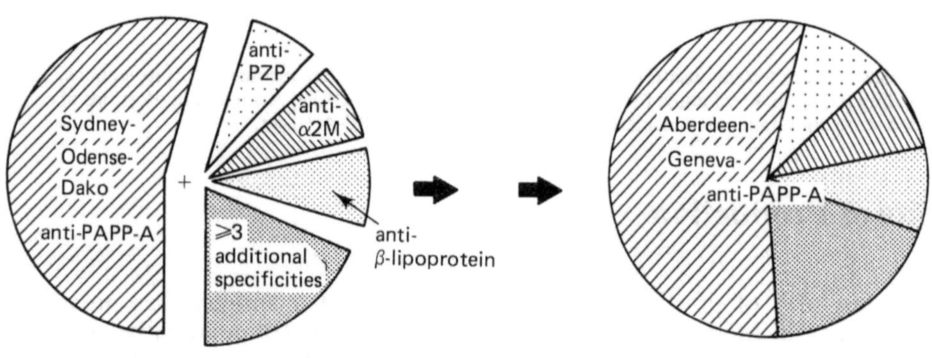

Fig. 5.5. Empirical relationship demonstrating specificity differences between Sydney–Odense–Dakopatts and Bischof (Aberdeen–Geneva) anti-PAPP-A antisera.

and/or secretory sites. The reservation about cell culture technology rests with the ability of isolated cells to undergo spontaneous transformation or regression. Unless appropriate controls, as markers of cellular status or dedifferentiation, are included in such studies, then one must be cautious when extrapolating in-vitro findings to in-vivo situations.

Tissue Distribution

Venous blood trophoblastic and uterine decidual tissues were obtained from 10 women undergoing surgical removal of tubal ectopic pregnancies, 13 undergoing elective termination during early pregnancy and five being delivered of normal live babies at term. Endometrial tissues were collected from six volunteers during the luteal phase of their ovulatory cycle. After sectioning for histology or culture, tissues were rinsed in 50 mmol/l phosphate (pH 7.4) containing 0.15 mol/l phosphate buffered saline (PBS), patted dry, weighed and transferred to a Ten Broeck (Selby Scientific, Sydney) tissue grinder, to which was added an equal volume of PBS. After homogenisation and extraction, supernatants were assayed for albumin (to correct for contamination by maternal blood), hCG (hCG-B, Serono) and PAPP-A (Sinosich et al. 1984b).

Both PAPP-A and hCG were detected in maternal and placental tissues (Table 5.8). There were no significant differences between PAPP-A and hCG content of endometrial or decidual tissues obtained from women with intra- or extra-uterine pregnancies. By contrast, in all cases the content of either antigen was significantly greater ($P<0.001$) in trophoblastic than in endometrial/decidual tissue. Furthermore, PAPP-A ($P<0.05$) and hCG ($P<0.001$) content were depressed in tubally implanted trophoblast when compared to eutopically implanted placentas. Although placental PAPP-A concentration remained constant from first trimester to term, total placental PAPP-A content increased with advancing gestation.

As yet, endometrial PAPP-A immunoreactivity remains to be characterised and confirmed. It may well be that glycan moieties from the extracellular matrix inhibit either antibody binding of tracer or precipitation of antibody bound tracer, and thereby register as PAPP-A positive. There is also the possibility that some tissue glycan moieties cross-react with the glycan component of PAPP-A, as indicated by selective absorption of anti-PAPP-A antisera to remove specificities against connective tissue (Chemnitz et al. 1986).

In-vitro Cell Cultures

Whereas none of the pregnancy associated antigens (hCG, PAPP-A, SP1, placental protein 5 [PP5]) were detected in decidual cell culture media, 45%–50% of the incorporated radioactivity was immunoreactive prolactin. By contrast, after 24 h incubation PAPP-A, hCG, SP1 and PP5 were readily detected in culture media of trophoblast cells (Fig. 5.6). Concentrations of both PAPP-A and SP1 increased with prolonged incubation, whereas hCG and PP5 secretion decreased. Unlike PP5/hCG ratios, PAPP-A/hCG and SP1/hCG ratios mimicked those calculated from serum concentrations measured in women

Table 5.8. Mean (± SEM) PAPP-A and chorionic gonadotrophin concentrations in pregnancy and non-pregnancy derived tissues and blood

	PAPP-A			Chorionic gonadotrophin		
	Serum (mIU/l)	Endometrium (mIU/g)	Trophoblast (mIU/g)	Serum (IU/l)	Endometrium (IU/g)	Trophoblast (IU/g)
Non-pregnant (Luteal phase)	<12.2	0.209(±0.085)	–	5.3(±1.676)	0.171(±0.099)	–
Tubal pregnancy (6–9 weeks)	15.7(±2.706)	0.294(±0.094)	0.946(±0.166)	3200(±1228)	0.406(±0.140)	3.731(±0.876)
Intrauterine pregnancy (6–12 weeks)	1309(±597)	0.248(±0.078)	1.860 (±0.208)	19 917(±3460)	0.789(±0.269)	33.928(±5.505)
Term pregnancy (>38 weeks)	72 300(±15 800)	0.248(±0.091)	1.988(±0.308)	12 419(±3020)	0.678(±0.286)	18.538(±3.971)

Table 5.9. Immunohistochemical evaluation of anti-PAPP-A antisera

Site	Bischof (unabs)[a]	Bischof (abs)[b]	Teisner[c]	Dakopatts	Sinosich	Lin[d]	SP1	hCG
Endometrium	+	–	–	–	–	NR	–	–
Placenta								
Syncytiotrophoblast	+	+	+	+	+	+	+	+
Cytotrophoblast	+	–	–	–	–	NR	–	–
Connective tissue	+	–	–	–	–	–	–	–
Fetal vasculature	+	–	–	–	–	–	–	–

+, positive reaction; –, negative reaction; NR, not reported.
[a] Bischof (unabs), neat antiserum; [b] Bischof (abs), antiserum absorbed (1 : 1) with male serum; [c] Teisner, Chemnitz et al. (1986); [d] Lin, Lin and Halbert (1976).

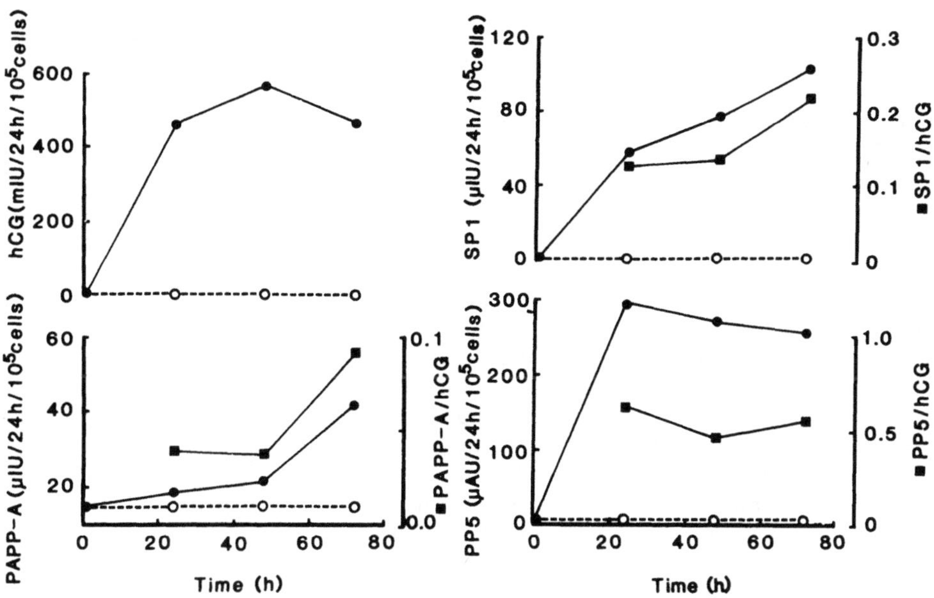

Fig. 5.6. In-vitro production of pregnancy associated proteins (hCG, PAPP-A, PP5, SP1) by trophoblast (●) and decidual (○) cells. Molar ratios (■) of each protein to hCG were used as an indicator of in-vitro genetic expression.

throughout the first trimester of pregnancy. Thus, the integrity of protein synthesis was maintained during the culture period.

Immunohistochemical Distribution

Tissue localisation of PAPP-A by direct or indirect immunohistochemistry has proved a rich source of controversy. Tissue processing should be a standardised procedure, but this ideal is seldom achieved. Reported procedures range from unspecified (Wahlstrom et al. 1981) to detailed descriptions (Tornehave et al. 1984). Fixatives included Zamboni's solution (Dobashi et al. 1984), Bouin's fluid (Schindler et al. 1984), at 10% (Sjoberg et al. 1984a) and 4% (v/v) phosphate buffered formalin (Chemnitz et al. 1986).

Deparaffinised sections may (Tornehave et al. 1984) or may not (Sjoberg et al. 1984a) be pretreated with trypsin. Although trypsinisation is thought to improve resolution by unmasking antigenic determinants, duration of proteolysis must be empirically determined for each lot of enzyme. Prolonged exposure will hydrolyse tissue proteins. Little consideration is given to the immunological stability of PAPP-A in paraffin embedded tissues, bearing in mind that immunoreactivity is inversely related to duration of storage. Thus, because of the potential for false negative findings, one must be wary of retrospective studies.

Methods for visualising tissue antigen–antibody reactions are varied and include direct and indirect immunofluorescence (Lin and Halbert 1976; Sinosich and Saunders 1987), peroxidase conjugated second antibody (Schindler et al, 1984; Tornehave et al. 1984) and the biotin–avidin immunoperoxidase method

(Sjoberg et al. 1984b). The results may vary widely between different studies. For example, using Dakopatts rabbit anti-PAPP-A antiserum Schindler and Bischof (1984) demonstrated PAPP-A in the epididymis and Leydig cells of the human testis in 17 of 17 cases, whereas Sjoberg et al. (1985) demonstrated PAPP-A only in the distal portions of the male reproductive tract, but not the testis or epididymis.

The greatest controversies arise with the use of different antisera. Three anti-PAPP-A antisera (Dakopatts, Sinosich, Bischof [1979], neat and absorbed), anti-SP1 and anti-hCG antisera (Dakopatts) were evaluated with the enzyme bridge immunoperoxidase technique on trypsin treated sections. The Bischof anti-PAPP-A reacted positively with endometrial and placental tissue. After one absorption with male serum (Bischof absorbed), only placental syncytiotrophoblast cells reacted positively (Table 5.9). By contrast, absorption of Bischof antiserum with pure PAPP-A marginally reduced the staining intensity of syncytiotrophoblast cells only. Thus, the apparently ubiquitous distribution of PAPP-A reported by Bischof may be an artefact. Most studies have confirmed the original immunofluorescent studies of Lin and Halbert (1976): PAPP-A, like SP1 and hCG, is localised to placental syncytiotrophoblast cells.

Table 5.10. Physiological distribution of PAPP-A in the circulation, fluids and secretions of normal non-pregnant adults

	PAPP-A (mIU/l)	Positive (%)	Sample size (*n*)
Circulation			
Male serum	<12.2[a]	0	17
Female serum			
Follicular phase	<12.2	0	17
Luteal phase	<12.2	0	17
Fluids			
Seminal	<12.2–735	91.5	47
Ovarian follicular	<12.2–1977	99.6	275
Cervical mucus	58–3015	100	7
Peritoneal	<12.2–21.5	16.3	43
Culture media			
Human embryo	<12.2	0	39

[a] Assay detection limit.

Physiological Distribution

Normal Non-pregnant Adults

Even at enhanced assay sensitivity (3 mIU/l), no PAPP-A was detected in serum of normal male volunteers. Similarly, PAPP-A was not detected in serum samples obtained from 17 fertile female volunteers at mid-follicular and mid-luteal phases of their ovulatory cycles, and no PAPP-A was detected in culture media of viable and non-viable human embryos (Table 5.10). Low PAPP-A concentrations were detected in 16.3% of peritoneal fluids. Seminal and follicular fluids contain PAPP-A which is immunologically and physicochemically indistinguishable from pregnancy derived PAPP-A (Table 5.4).

Oncological Disorders

Studies evaluating PAPP-A measurement in oncological disorders are sparse (Tsakok et al. 1983; Bischof et al. 1982) and generally consider PAPP-A as inferior to hCG. In the present study, PAPP-A, hCG, SP1 and alphafetoprotein (AFP) were measured in serum samples collected from 85 patients with various oncological and non-oncological disorders (Table 5.11).

Table 5.11. Proportion (%) of samples testing positive for pregnancy associated proteins

Category	*n*	PAPP-A (>12.2 mIU/l)[a]	hCG (>5 IU/l)[a]	SP1 (>4/mIU/l)[a]	AFP (>10 UG/l)[a]
Testicular					
Teratoma	8	0	75.0	75.0	87.5
Teratoma + seminoma	2	0	0	50.0	100
Seminoma	3	0	33.3	33.3	33.3
Embryonal	5	0	100	40.0	60.0
Gestational trophoblastic disease (GTD)	11	100	100	100	–
Hydatidiform mole	3	0	100	100	–
Invasive mole	4	50.0	100	100	–
Choriocarcinoma	1	100	100	100	100.0
Transitional mole + pregnancy	3	33.3	100	100	
Persistent GTD					
Non-specific					
Ovarian	5	0	40.0	40.0	60.0
Hepatoma	19	0	0	15.8	100.0
Fatty liver disease	1	0	0	100.0	0
Infectious liver disease	1	0	0	0	0
Adenocarcinoma	9	0	44.4	22.2	0
Lymphoma	1	0	0	0	0
Myeloma	1	0	0	100	0
Melanoma	2	0	0	100	0
Pinealoma	1	0	0	0	100
Undefined metastatic tumour	5	0	0	20.0	0
Total	85				

[a] Assay detection limit.

AFP, as a marker of yolk-sac-derived cell lines, was consistently present in gonadal, hepatic and neural tumours, but not in gestational trophoblastic diseases. By contrast, hCG was produced by all trophoblastic tumours. Maximum hCG concentration (1 786 000 IU/l) was detected in a patient with disseminated choriocarcinoma who died 10 days after admission. Although this tumour also secreted SP1, no PAPP-A was detected in peripheral blood. Tissue was not available for examination.

Whereas SP1 was detected in trophoblastic and non-trophoblastic tumours, PAPP-A was present only in patients with gestational trophoblastic tumours. Expression of PAPP-A was inversely related to disease aggression. Thus, the progression from benign to malignant disease state is associated with an apparent suppression of PAPP-A relative to hCG.

Ovarian Follicle

The biochemical composition of human preovulatory follicular fluid (hFF) is qualitatively similar to peripheral blood (Sinosich 1987). Small solutes such as divalent zinc cations exist in equimolar amount, across the follicular basement membrane, whereas follicular prostaglandin ($PGF_2\alpha$) and cyclic adenosine monophosphate (cAMP) concentrations far exceed circulating levels. The total protein concentration of hFF is about half the serum level. Follicular proteins include enzymes (galactosyl transferase), complement factors, immunoglobulins (IgG), hormones (inhibin), transport proteins (albumin), acute phase reactants (orosomucoid), protease inhibitors (α_2M, α_1-antitrypsin), glycosylaminoglycans and proteins with unknown function (PP5, α-follicular fluid antigen [Sinosich 1987]).

Protease inhibitors (PIs) constitute a significant proportion of the intrafollicular protein environment (Table 5.12).

Follicular PI concentration is inversely related to molecular size:

$$[\text{Protease inhibitor}]_{\%} = \frac{2.578 \times 10^5}{(\text{Mr of PI})^{0.72}} \qquad (r = 0.918)$$

where protease inhibitor is PI concentration given as percentage of peripheral blood (Table 5.12) and (Mr of PI) the respective apparent molecular weight (in daltons). This relationship suggests that many follicular PIs originate from blood and enter the follicle by passive diffusion. Whereas α_1-antitrypsin was equally distributed across the follicular basement membrane, larger PIs (α_2M) were higher in blood than in the ovary. PAPP-A was not detected in blood, indicating a local origin for follicular PAPP-A.

In aspirates obtained from spontaneously ovulating women, PAPP-A concentrations were positively related to folliculogenesis with maximum PAPP-A levels achieved immediately before ovulation (Sinosich et al. 1985h). The positive relationship between intrafollicular PAPP-A and progesterone/oestradiol concentrations suggested a relationship between granulosa/theca steroidogenesis and PAPP-A production (Westergaard et al. 1985), but the distinction between cause and coincidence is not always apparent. In aspirates obtained from women undergoing ovarian hyperstimulation before in-vitro fertilisation/embryo transfer (IVF-ET), PAPP-A concentrations were positively related to granulosa cell maturity and progesterone production, but not to oocyte quality (Sinosich et al. 1987c). This supports a causal relationship between granulosa cell luteinisation and production of PAPP-A and progesterone. Quantitative follicular biochemistry cannot predict clinical outcome of harvested oocytes, after IVF ET (Sinosich 1987; Sinosich et al. 1985f, 1987c). Yet the question remains, why do

Table 5.12. Quantitative analysis of protease inhibitor distribution across the human blood-ovarian barrier

Protease inhibitor	M_r (kd)	Circulation (AU/l)[a]	Follicular fluid (AU/l)	Specificity
α_1-antitrypsin	54	100	123	Elastase
Antithrombin 111	65	100	77	Thrombin, acrosin
α_1-antichymotrypsin	68	100	53	Chymotrypsin
α_2-antiplasmin	70	100	90	Plasmin
C1-esterase inhibitor	104	100	54	Kallikrein
Inter-α-trypsin inhibitor	160	100	77	Trypsin, acrosin
α_1-thiolproteinase inhibitor	165	100	69	Cathepsin B_1
Fibronectin	350	100	28	Collagenase
Pregnancy zone protein	360	100	25	Elastase
α_2-macroglobulin	725	100	13	Non-specific
Pregnancy associated plasma protein-A	820	ND	0.57[b]	Elastase

ND, not detected.
[a] Pooled normal male serum.
[b] Percentage of WHO International Reference 78/610(100IU/l).

maturing follicles produce increasing quantities of this proteoglycan elastase inhibitor, PAPP-A?

Seminal Plasma

Seminal plasma contains PAPP-A immunoreactivity (Martin-du-Pan et al. 1983) and this PAPP-A exists in the same form as that in other sites. (The physicochemical and immunological characteristics of trophoblastic and non-trophoblastic, in-vitro and in-vivo derived PAPP-A are summarised in Table 5.4)

PAPP-A is one of many protease inhibitors in seminal plasma (Sinosich and Saunders 1987) and is presumably synthesised in the reproductive tract. Although Martin-du-Pan et al. (1983) suggested a relationship between seminal plasma PAPP-A concentrations and number of spermatozoa, others have shown that seminal plasma PAPP-A concentrations do not correlate with any clinical parameter of sperm function (morphology, number, motility) (Sinosich et al. 1985e; Sjoberg et al. 1985). Furthermore, as seminal plasma PAPP-A concentrations are the same in normospermic and azospermic/vasectomised males, a testicular origin for PAPP-A is excluded (Sinosich et al. 1985g).

Ejaculates (n=5) were split into three components, each three contractions. All three components readily coagulated, and the first component liquefied most

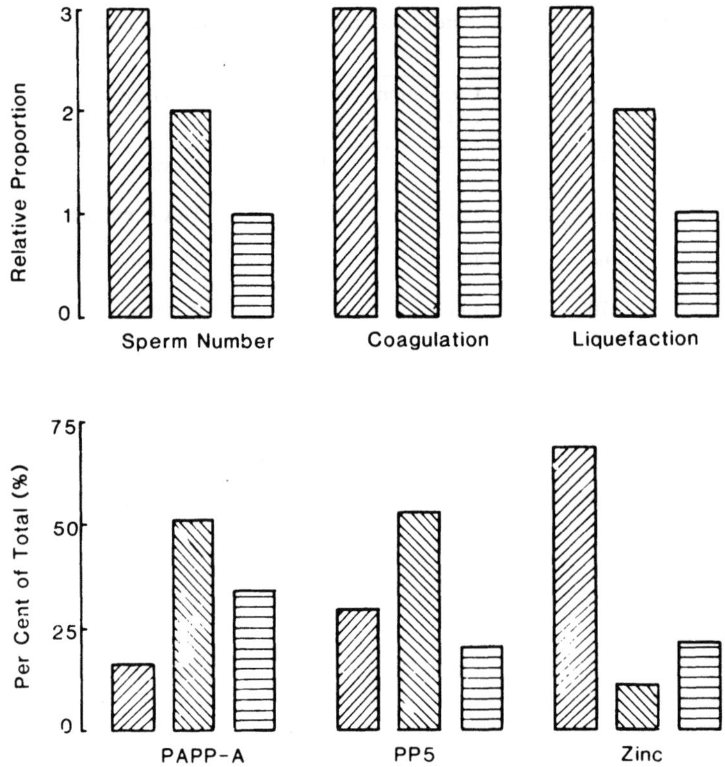

Fig. 5.7. Quantitative analysis of human ejaculates split into first (▨), second (▧) and third (▤) components.

rapidly. The major portion of seminal zinc and spermatozoa were concentrated in the first contractions (Fig. 5.7), indicating a testicular origin for the constituents of the first orgasmic contractions. By contrast, the greatest proportion of seminal PAPP-A and PP5 was found in the later contractions, which are predominantly accessory gland (prostate gland, seminal vesicles) secretions. The hypothesis of a non-testicular origin was further supported by immunohistochemical localisation of PAPP-A to the accessory glands in the distal portions of the reproductive tract (Sjoberg et al. 1985).

After liquefaction, freshly ejaculated spermatozoa were washed and incubated for 0 and 6 h at 37 °C in Ham's F10 medium supplemented with 10% fetal bovine serum. Under these conditions, used to induce capacitation and acrosome reaction before IVF ET (Rogers 1978), there was no significant increase in sperm bound radiolabelled (^{125}I-) anti-PAPP-A immunoglobulins. By contrast, ^{125}I-antifibronectin and ^{125}I-placental-protein-14 binding increased after capacitation (Table 5.13). However, Bischof et al. (1987) demonstrated, with autoradiography, apparent ^{125}I-PAPP-A bound to sperm heads and suggested that PAPP-A induced the acrosome reaction.

PAPP-A was localised on heads of about 2% of total spermatozoa by direct immunofluorescence. This sheath of PAPP-A was thought to be a protective

Table 5.13. Influence[a] of radiolabelled (^{125}I) antibodies and antigens on capacitation and acrosome reaction of human spermatozoa

	Initial (0 h)			Final (6 h)		
	1 μl[b]	10 μl[b]	100 μl[b]	1 μl[b]	10 μl[b]	100 μl[b]
Antisera to						
PAPP-A						
Rat immunoglobulin G	1.4	0.5	0.5	0.4	0.7	0.3
Immunoglobulin G	0.2	0.5	0.4	0.6	0.6	0.4
SP1	1.5	1.7	2.1	3.3	3.1	2.7
α_2-macroglobulin	0.2	0.6	0.6	0.5	0.6	0.5
Fibronectin	1.4	1.5	1.1	0.9	1.1	1.4
	3.4	3.4	3.6	5.6	4.8	6.2
Antigens						
PAPP-A						
Placental protein 14	0	0.4	0.3	0	0	0
Rat follitropin	1.3	0.7	0.9	1.4	1.8	2.1
	0.1	0.1	0.3	0.3	0.3	0.5

[a] Expressed as percentage radioactivity bound to spermatozoa.
[b] Volume of tracer added, activity range 4700–470 000 cpm.

barrier against localised phagocytic–proteolytic degradation within the female reproductive tract, a host defence activated within 2 h of coitus and insemination (Sinosich and Saunders 1987). Thus, PAPP-A may be part of a protease inhibitor protective mechanism that ensures sperm survival.

Pregnancy

The development of sensitive and specific RIAs for PAPP-A allows quantitation during the first trimester of pregnancy. For this purpose IVF ET and infertility clinics have proved a useful source of patients. Two hundred and eleven serial blood samples were collected during the first half of pregnancy from 13 volunteers who underwent successful IVF ET. Similarly, 323 serial blood samples were collected from women who conceived in vivo after treatment/insemination at Royal North Shore Hospital. Circulating PAPP-A levels and normal ranges (80% confidence interval) for in-vitro and in-vivo singleton conceptions are shown in Table 5.14. There were no significant differences according to mode of conception, and PAPP-A values were pooled for the normal range.

After initial detection (6 weeks), serum PAPP-A concentrations increased exponentially (Table 5.15) with a doubling time of 5.9 days during the first trimester (Table 5.16). Similar exponential growth curves were also determined for hCG, SP1 and PP5 (Table 5.15). Regression equations for hCG and SP1 were similar; the curves for PAPP-A and PP5 were virtually superimposable. Bischof et al. (1981) also reported the first significant increase in PAPP-A at 6 weeks' gestation. However, the doubling time (as calculated from Fig. 1 and Table 1 of Bischof et al. 1981) was 3.6 weeks.

Table 5.14. Circulating levels of PAPP-A (mIU/1) after implantation of embryos fertilized in vivo and in vitro

Gestation (weeks)	In vitro[a]				In vivo[a]			
	n	10th	Median	90th	*n*	10th	Median	90th
6.0	19	<12.2	<12.2	28	11	<12.2	17	69
6.5	15	<12.2	14.4	47				
7.0	14	16	69	153	11	28	52	120
8.0	14	67	138	473	22	66	160	810
8.5	21	119	272	1000				
9.0	15	204	551	1508	26	150	360	1100
10.0	14	311	781	2325	27	290	810	1900
10.5	5	360	1378	2116				
11.0	7	659	1400	2703	29	810	1500	3500
12.0	7	998	2425	3385	16	1100	2100	4500
13.0	4	1500	3097	10 500	16	1500	4000	7600
	135				158			
Regression[b]								
Slope (*a*)		−4.370	−4.695	−3.984		−4.243	−4.209	−2.927
Intercept (*b*)		2.972	3.278	3.984		2.926	3.062	2.682
Correlation (*r*)		0.990	0.982	0.965		0.995	0.998	0.982

[a] No significant differences across each gestation.
[b] $\log [\text{PAPP-A}]_{\text{mIU/1}} = a + b \ln (\text{gestation})_{\text{wks.}}$

Table 5.15. Production and disappearance rates of pregnancy associated proteins

Production rates		
$[\text{PAPP-A}]_{\text{mIU/l}}$	= 0.1635 exp [0.820 × (gestation) wks]	(r = 0.962)
$[\text{hCG}]_{\text{IU/l}}$	= 3.512 exp [1.226 × (gestation) wks]	(r = 0.969)
$[\text{SP1}]_{\text{mIU/l}}$	= 1.123 exp [0.360 × (gestation) wks]	(r = 0.882)
$[\text{PP5}]_{\text{mAU/l}}$	= 0.142 exp [0.616 × (gestation) wks]	(r = 0.889)
Post-partum clearance rates		
$[\text{PAPP-A})_{\text{mIU/l}}$	= 170.11 × 10^3 exp (−0.0068t)	(r = 0.887)
$[\text{hCG}]_{\text{IU/1}}$	= 8.49 × 10^3 exp (−0.0454t)	(r = 0.932)
$[\text{SPl}]_{\text{mIU/1}}$	= 169.79 × 10^3 exp (−0.0181t)	(r = 0.964)

Table 5.16. Comparative physiological kinetics of pregnancy associated proteins during first trimester of pregnancy

	HCG (IU/l)	SP1 (mIU/l)	PAPP-A (mIU/l)	PP5 (mAU/l)
Calibration	2nd IRP 61/6[a]	WHO 78/610	WHO 78/610	Internal[b]
Initial detection (weeks LMP)	~4	>4.5	6.1	7.6
Extrapolated initial detection (days LMP)	16	22	41	49
Maximum concentration (weeks LMP)	8–9	Term	Term	Term
Concentration at term	2 × 10^4	10^5	10^5	10^3
Doubling rate during first trimester (days)	4.0[c]	2.6	5.9	7.9
Post-partum decay ($t_{1/2}$)	15.3 h	38.3 h	4.2 d	10–20 min[d]

LMP, last menstrual period.
[a] 1 mIU 2nd International Reference Preparation (IRP) 61/6 = 2.2 mIU 1st International Reference Preparation 75/537.
[b] Pooled late pregnancy serum (n>1000).
[c] Determined from 4.5 to 8.5 weeks' gestation.
[d] Nisbet et al. (1981).

Maximum levels of PAPP-A, SP1 and PP5 were attained at term (Table 5.16). At this time the mean vascular content of PAPP-A approximates 250 mg, far exceeding (25–50-fold) the placental PAPP-A pool (Sinosich 1985). Traces of PAPP-A were also detected in amniotic fluid, colostrum and fetal blood. After parturition, PP5 was rapidly cleared from peripheral blood. By contrast, the clearance rate of PAPP-A ($t_{1/2}$=4.2 days) was greatly prolonged (Tables 5.15 and 5.16), as shown in earlier studies (Lin et al. 1976).

Clinical Application

Biochemical diagnosis of pregnancy depends on detection of hCG in maternal blood or urine; with sensitive immunoassays this may be achieved almost at the time of implantation. It is unlikely that PAPP-A could serve as an early pregnancy test.

Recurrent Abortion

Serial measurement of pregnancy associated proteins (hCG, SP1 and PAPP-A) demonstrated no PAPP-A throughout two unsuccessful conception cycles in which hCG and SP1 were present. In the third conception cycle circulating kinetics of all three glycoproteins were normal, and this pregnancy progressed uneventfully.

Anembryonic Pregnancy

Depressed mean serum levels of hCG, PAPP-A and steroid hormones may be detected as early as 4.5 weeks' gestation (Yovich et al, 1986; Sinosich et al. 1987b). Examination of individual patient profiles suggested two situations in this disorder: (1) apparently normal placental physiology retained in absence of a fetus, as evidenced by normal levels of hCG, SP1 and PAPP-A; and (2) depressed serum levels of placental products suggesting abnormal placental function culminating in fetal (and subsequently, pregnancy) demise (Sinosich et al. 1987b). These findings confirm that low levels of placental proteins are suggestive of failed pregnancy, at a time before ultrasonography can provide a definitive diagnosis.

Tubal Pregnancy

Application of rapid and sensitive hCG immunoassays has revolutionised patient management and significantly reduced the rate of clinical complications associated with extra-uterine pregnancies. If hCG is detected (>25 IU/l), this alerts the clinician to the possibility of a pregnancy-related disorder. The diagnostic value of hCG and SP1 measurements are similar, whereas PAPP-A levels are more severely compromised in this condition (Sinosich et al. 1985a).

Biochemical assessment is often carried out at a time when clinical symptoms are readily apparent. Under these conditions circulating levels of pregnancy associated antigens are usually depressed (Grudzinskas et al. 1985). However, serial measurements from time of conception to surgical intervention provide a biochemical window into the physiology of an extrauterine implantation (Fig. 5.8). Development of tubally implanted trophoblast is not compromised during the first 6 weeks of pregnancy (Sinosich et al, 1985a). After 6 weeks, the ectopic pregnancy cannot be adequately sustained by the fallopian tube, and clinical symptoms become manifest in association with depressed hCG production.

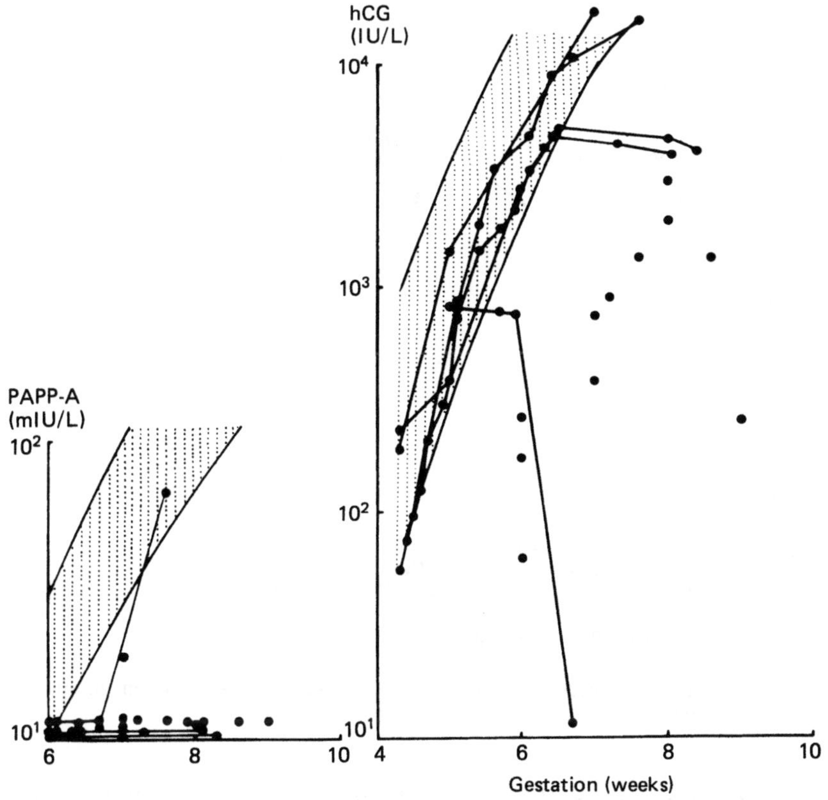

Fig. 5.8. Circulating levels of PAPP-A and hCG in women with histologically confirmed tubal pregnancies.

Examination of individual hCG profiles (Fig. 5.8) reveals at least three clinical categories: (1) sharp reduction or absence of circulating hCG, suggestive of spontaneous regression; (2) clinical symptoms and, at the time of surgery, depressed serum hCG levels; and (3) normal serum hCG concentrations, even at the time of surgical intervention. Generally, PAPP-A was not detected in blood of category 1, seldom detected in category 2, and readily detected in category 3 patients.

The hCG and PAPP-A contents of eutopically implanted placentas were positively correlated (Fig. 5.9). By contrast, hCG and PAPP-A in tubally implanted trophoblasts were negatively related.

Spontaneous Abortion

The frequency of postimplantation pregnancy wastage has been estimated at 25%–35%. Early pregnancy factor or embryo derived platelet activation factor have been used as markers of preimplantation embryo viability, and antigens of fetal, placental and maternal origin have been used to predict the outcome of

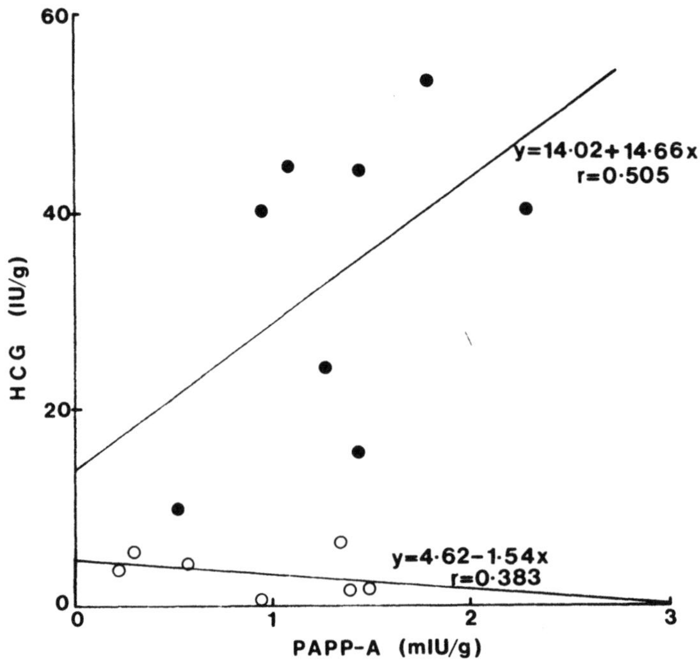

Fig. 5.9. Regression analysis in placental content of hCG and PAPP-A between eutopically (intrauterine, •) and ectopically (tubal, ○) implanted pregnancies.

compromised early pregnancies (Morton et al. 1977; O'Neill et al. 1985; Niven et al. 1972; Braunstein et al. 1978). Non-invasive imaging technology has revolutionised management of patients with vaginal bleeding and rendered some of these tests obsolete (Jouppila et al. 1980a, b). Nevertheless, a small proportion of patients with ultrasonically confirmed viable pregnancies will spontaneously miscarry.

In the absence of fetal heart action – i.e., a non-viable pregnancy – maternal serum concentrations of fetal, maternal and placental antigens are consistently depressed. However, if abortion occurs after detection of fetal heart action, maternal blood levels of hCG, SP1, hPL, oestradiol, AFP and progesterone are generally within normal limits, whereas PAPP-A levels are depressed below the 10th percentile. Conversely, positive ultrasonographic findings in conjunction with normal serum PAPP-A levels suggest a better than 98% chance of successful pregnancy outcome (Westergaard et al. 1983b, 1985b; Grudzinskas et al. 1986; Sinosich et al. 1983b, 1985b, 1985c, 1987b).

Multiple Pregnancy

The presence of multiple fetuses was frequently associated with increased circulating levels of PAPP-A (Lin et al, 1977; Sinosich et al. 1982b; Westergaard et al. 1985a). However, the doubling rate of PAPP-A is unchanged. In women with a "vanishing twin" (Landy et al. 1986) (ultrasonographically identified viable multiple pregnancies with subsequent resorption of all but one fetus) PAPP-A levels were normal. Serum hCG concentrations could not distinguish between uncomplicated multiple pregnancy and a "vanishing twin". Serum PAPP-A levels provide a test capable of diagnosing multiple pregnancy some 3 weeks after conception.

Fetal Congenital Abnormality

Levels of pregnancy associated proteins in pregnancies complicated by fetal congenital abnormality are usually normal. Recently, however, two cases with Cornelia de Lange syndrome were reported (Westergaard et al. 1983a), in both of which no PAPP-A could be detected with immunohistochemistry.

Biological Role of PAPP-A

The ability of PAPP-A to inhibit neutrophil or granulocyte elastase (HGE) activity has many biological implications. HGE is one of many potent serine proteases contained in the azurophilic granules of neutrophils (Ohlsson and Olsson 1974). This lysosomal enzyme is associated with the pathology of rheumatoid arthritis and emphysema, involving destruction of connective tissue structural proteins. Its substrate specificity includes elastin, proteoglycans and collagen types III, IV and V, structural proteins of basement membranes and extracellular matrix. Lysosomal proteases are secreted in response to an antigenic challenge. Phagocytosis–proteolysis constitutes one of the earliest evolved and most mobile host defence mechanisms.

PAPP-A is readily detected in reproductive tract fluids and secretions and can be localised to cellular interfaces which pose an antigenic challenge to maternal defence systems (Fig. 5.10). Following coitus and insemination, an acute localised leucocytic reaction can be detected in cervical mucus (Sinosich and Saunders 1987), an inflammatory reaction not unique to man (Brown et al. 1985). Seminal immunomodulators do not adequately explain how 2%–5% of ejaculated spermatozoa survive this localised phagocytic–proteolytic response, to reach the oocyte within the oviduct (Pandya and Cohen 1985). Furthermore, the apparent immunosuppressive property of human seminal plasma has been attributed to the cytotoxic effect of spermine oxidation by amine oxidase (Allen and Roberts 1986).

By direct immunofluorescence, PAPP-A has been localised on the heads of 2% of ejaculated spermatozoa (Sinosich and Saunders 1987). Thus, PAPP-A may contribute to sperm survival by forming a barrier against phagocytic–proteolytic attack by maternal leucocytes. This hypothesis may be extended to include fertilisation and survival of the preimplantation embryo, at which stage embryo derived immunosuppressive factors may be secreted in preparation for implantation. After implantation, PAPP-A is secreted into the maternal circulation, where it may exert its protease inhibiting effect systemically and locally at the utero-placental interface.

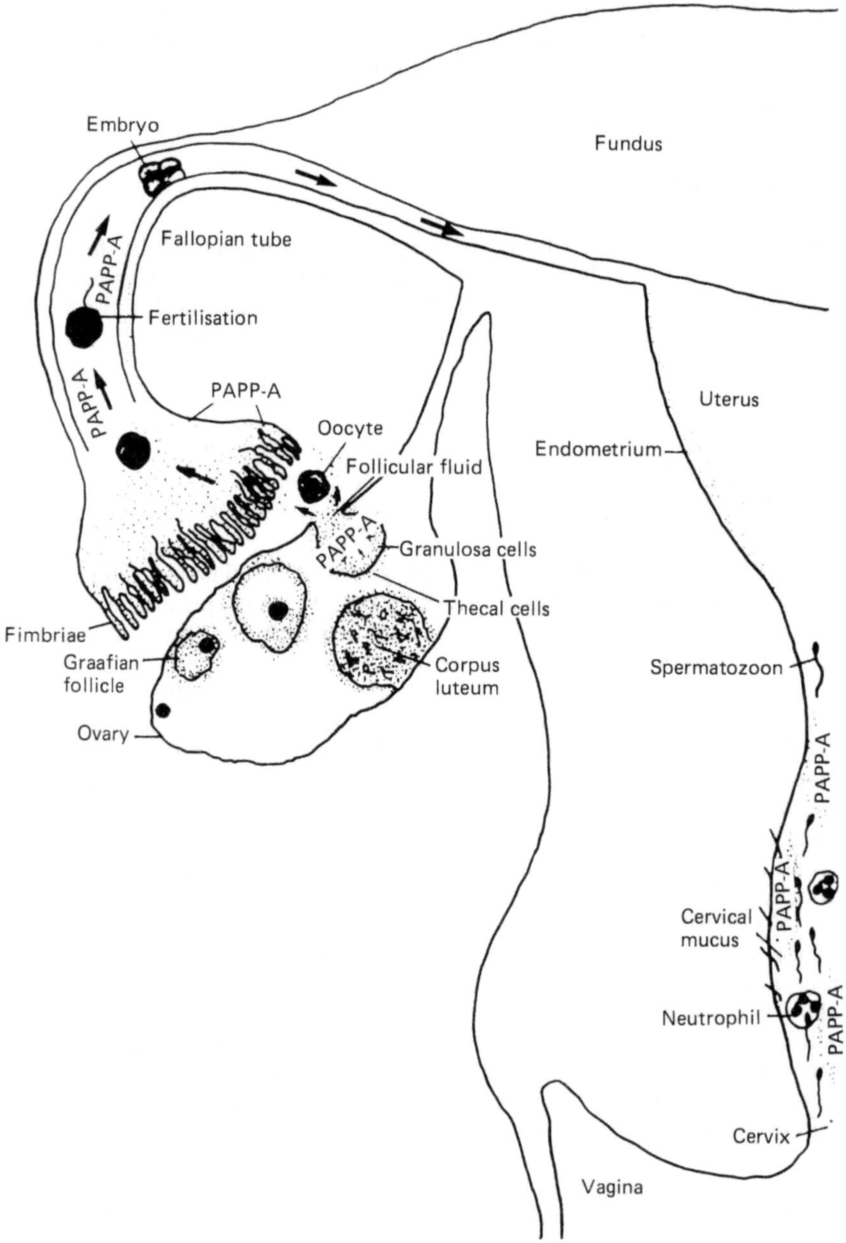

Fig. 5.10. Proposed biological actions and distribution of PAPP-A within the female reproductive tract.

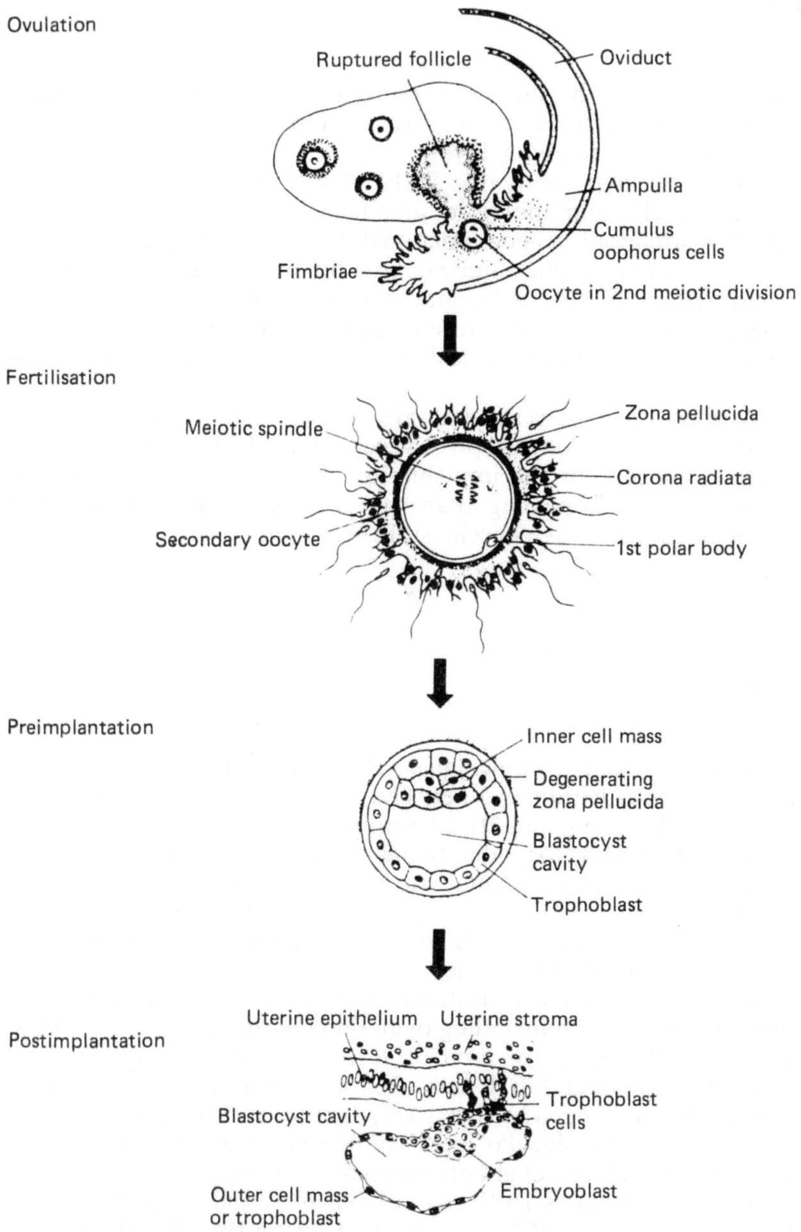

Fig. 5.11. Potential sites of action of immunological contraception using PAPP-A or pregnancy associated elastase inhibitor (PAEI) as target antigen.

Future of PAPP-A

PAPP-A has proved one of the most interesting of the pregnancy associated antigens. It appears to be involved at each crucial step of human reproduction from folliculogenesis onwards (Fig. 5.11).

Future research on the hypothesis that PAPP-A is a molecular condom will depend on animal models. Immunological analogues of human PAPP-A have been found in Old World monkeys and functional PAPP-A analogues in rodents (Sinosich 1985). Evolution of placental heparin binding elastase inhibitor is closely associated with evolution of haemochorial placentation. Heterologous models based on human PAPP-A offer little prospect for research. Two homologous animal models have been developed – in the cynomolgus monkey and guinea-pig.

Since PAPP-A does not cross-react with any known serum protein, it is an ideal target antigen for active contraceptive vaccination. With the potential for inhibiting ovulation, conception and implantation (Fig. 5.11), vaccination with PAPP-A is ethically more appealing than the peri-implantation mechanism associated with hCG contraceptive vaccines (Talwar et al, 1983).

Conclusions

Since the initial reports by Lin and colleagues (1976) the PAPP-A literature has become contradictory both within and between research centres. Contentious issues include in-vitro function, quality of primary reagents, the selection of biological fluids (serum/plasma) for radioimmunoassay, apparent concentrations in blood of normal non-pregnant adults, site of production and physiological kinetics during early pregnancy (Table 5.17). A critical review suggests that the apparent ubiquitous distribution of PAPP-A as an immunosuppressive factor is based on relatively non-specific techniques.

Clinical and immunohistochemical studies support a vital role for PAPP-A at all stages of human reproduction including ovulation (Westergaard et al. 1985a; Sinosich et al. 1985d, 1987c), sperm survival, fertilisation and embryo survival before (Sinosich and Saunders 1987) and after implantation (Sinosich et al. 1984c). Within the female reproductive tract, before and after conception, PAPP-A forms a protective molecular interface between maternal phagocytic–proteolytic defences and cellular elements with paternal antigens.

Absence of placental products (hPL, SP1) from maternal blood in normal term pregnancies (Nielsen et al. 1979; Borody and Carlton 1981; Grudzinskas et al. 1979) raises doubts as to their function in the maintenance of the fetoplacental unit. Westergaard et al. (1983a) have reported immunoelectrophoretic evidence for absence of PAPP-A in pregnancies afflicted with gross fetal congenital deformities, such as Cornelia de Lange syndrome. However, with radioimmunoassay PAPP-A immunoreactivity was readily detected in maternal serum, and levels remained at about 2 IU/l throughout the latter half of gestation (M.J. Sinosich, unpublished data). Levels of hPL and SP1 were

Table 5.17. Summary of confirmed and contradictory aspects of PAPP-A physiology and biochemistry

	Confirmed	Contradictory
Molecular	Physicochemistry, heparin interaction	Size heterogeneity
Functional	Elastase inhibition	Broad spectrum immunosuppressive
Radioimmunoassay	^{125}I as label	Serum/plasma, tracer size, reagent quality, unitage
Site of production	Trophoblast, granulosa cell	Endometrium/decidua
Localisation	Syncytiotrophoblast, male accessory glands, molar tissue	Endometrium/decidua, testicle
Distribution	Pregnancy, seminal plasma, follicular fluid	Ubiquitous
Physiology	Diagnostic at 6 weeks	Doubling rate, prognostic indices, non-pregnant levels, disappearance rate

normal. It is possible this influx of maternal polymorphonuclear leucocytes, across a partially disrupted PAPP-A barrier, into the fetoplacental unit during embryonic organogenesis may produce these characteristic deformities. It is equally feasible that Cornelia de Lange syndrome may be a chromosomal aberration, expressing a PAPP-A variant with little functional and immunological homology to native PAPP-A. It is also possible that suppressed PAPP-A levels may reduce the availability of zinc ions to the fetus during organogenesis, a condition known to produce a wide spectrum of congenital malformations (Hurley 1981; Dreosti et al. 1985; Dura Trave et al. 1986).

There is no doubt that, of the "new generation" of pregnancy associated antigens, PAPP-A is functionally and physiologically one of the most interesting. Many of the remaining questions will be addressed by new techniques of molecular biology, but the crucial question remains: is PAPP-A or heparin binding pregnancy associated elastase inhibitor (PAEI) essential for mammalian reproduction? The answer to this question must await the development of homologous animal models actively immunised with PAPP-A or PAEI.

Acknowledgements. The data reviewed here is the culmination of several years of excellent technical support provided by Miss D Laverack, Miss C. O'Brien, Mrs. V. Carrillo, Mr. M.D. Bonifacio and Mr. A. Zakher. My gratitude also to Mrs. E.A. Sinosich for support and preparing this manuscript. My thanks also to Professor B. Teisner and Dr. P. Bischof for making their reagents available for evaluation.

References

Ablin RM, Bruns GR, Guinan P (1974) The effect of estrogen on the incorporation of ^{3}H-thymidine by PHA-stimulated human peripheral blood. J Immunol 113: 705–707

Adock EW, Teasdale F, August CS et al. (1973) Human chorionic gonadotropin: its possible role in maternal lymphocyte suppression. Science 181: 845–847

Allen RD, Roberts TK (1986) The relationship between the immunosuppressive and cytotoxic effects of human seminal plasma. Am J Reprod Immunol Microbiol 11: 59–64

Anderson DJ, Yunis EJ (1985) The elusive immunosuppressive factors of pregnancy. Am J Reprod Immunol Microbiol 9: 91–92

Anthony F, Masson GM, Wood PJ (1983) Development of radioimmunoassay for pregnancy-associated plasma protein-A and establishment of normal levels in the first trimester of pregnancy. Ann Clin Biochem 20: 26–30

Bischof P (1979) Purification and characterization of pregnancy-associated plasma protein-A (PAPP-A). Arch Gynecol 227: 315–326

Bischof P (1980) Observations on the isolation of macromolecules of placental origin. In: Klopper A, Genazzani A, Crosignani PG (eds) The human placenta. Academic Press, London, pp 47–55

Bischof P, Tseng L (1986) In vitro release of pregnancy-associated plasma protein-A (PAPP-A) by human endometrial cells. Am J Reprod Immunol Microbiol 10: 139–142

Bischof P, Haenggeli L, Sizonenko MT, Herrmann W, Sizonenko PC (1981) Radioimmunoassay for the measurement of pregnancy-associated plasma protein-A (PAPP-A) in humans. Biol Reprod 24: 1076–1081

Bischof P, Rapin CH, Weil A, Herrmann W (1982) Is pregnancy-associated plasma protein-A a tumor marker? Am J Obstet Gynecol 143: 378–381

Bischof P, DuBerg S, Herrmann W, Sizonenko P (1983a) Pregnancy-associated plasma protein-A (PAPP-A) and hCG in early pregnancy. Br J Obstet Gynecol 88: 973–975

Bischof P, Lauber K, Girard JP, Herrmann WL, Sizonenko PC (1983b) Circulating levels of pregnancy proteins and depression of lymphoblastogenesis during pregnancy. J. Clin Lab Immunol 12: 93–96

Bischof P, Meisser A, Haengeli L, et al. (1983c) Pregnancy-associated plasma protein-A (PAPP-A) inhibits thrombin-induced coagulation of plasma. Thromb Res 32: 45–55

Bischof P, Geinoz A, Herrmann WL, Sizonenko PC (1984) Pregnancy-associated plasma protein-A (PAPP-A) specifically inhibits the third component of human complement (C3). Placenta 5: 1–8

Bischof P, Bilat M, Herrmann WL (1987) Pregnancy-associated plasma protein-A (PAPP-A) induces the acrosomal reaction in human spermatozoa. In: Proceedings of the 34th Annual Meeting of the Society for Gynecologic Investigation, p 118

Bohn H, Johannsen R, Kraus W (1980) New placental protein (PP15) with immunosuppressive properties. Arch Gynecol 230: 167–172

Bolton AE, Clough KJ, Bohn H, Chapman MG (1986) Placental protein 14 (PP14) in the human reproductive tract. In: Hau J (ed) Pregnancy proteins in animals. Walter de Gruyter, Berlin, pp 165–171

Borody IB, Carlton MA (1981) Isolated defect in human placental lactogen synthesis in a normal pregnancy. Br J Obstet Gynecol 88: 447–449

Braunstein GD, Karow WG, Gentry WC, Rasor I, Wade MM (1978) First trimester chorionic gonadotropin measurements as an aid in the diagnosis of early pregnancy disorders. Am J Obstet Gynecol 143: 25–32

Brown AE, Hansen PJ, Asbury AC (1985) Opsonization of bacteria by uterine secretions of cyclic mares. Am J Reprod Immunol Microbiol 9: 119–123

Caldwell JL, Stiles DP, Fudenberg HH (1975) Human chorionic gonadotropin: effects of crude and purified preparations on lymphocyte responses to phytohaemagglutinin and allogeneic stimulation. J Immunol 115: 1249–1253

Casu B (1985) Structure and biological activity of heparin. Adv Carbohyd Chem Biochem 43: 51–134

Chaouat G, Kalb JP, Riviere M, Lankar D, Provost MA (1985) Immunoactive products of placenta. VIII. Translation products of messenger RNA extracted from murine placenta or A6B91C5 teratocarcinoma are immunosuppressive in vitro. Am J Reprod Immunol Microbiol 9: 48–51

Chemnitz J, Folkersen J, Teisner B et al. (1986) Comparison of different antibody preparations against pregnancy-associated plasma protein-A (PAPP-A) for use in localization and immunoassay studies. Br J Obstet Gynecol 93: 916–923

Clark DA (1985) Prostaglandins and immunoregulation during pregnancy. Am J Reprod Immunol

Microbiol 9: 111–112
Contractor SF, Davies H (1973) Effect of human chorionic somatomammotropin and hCG on phytohaemagglutinin-induced lymphocyte transformation. Nature 243: 284
Culouscou JM, Remacle-Bonnet MM, Pommier G, Rance RJ, Depieds RC (1986) Immunosuppressive properties of human placenta: study of supernatants from short-term syncytiotrophoblast cultures. J Reprod Immunol 9: 33–47
Davey MW, Teisner B, Sinosich MJ, Grudzinskas JG (1983) Interaction between heparin and human pregnancy-associated plasma protein-A (PAPP-A): a simple purification procedure. Anal Biochem 131: 18–24
Davies M, Brown CM (1985a) Pregnancy-associated nonspecific immunosuppression: kinetics of the generation and identification of the active factors. Am J Reprod Immunol Microbiol 9: 77–83
Davies M, Brown CM (1985b) Pregnancy-associated nonspecific immunosuppression: mechanism for the activation of the immunosuppressive factors. Am J Reprod Immunol Microbiol 9: 84–90
Daya S, Clark DA (1986) Production of immunosuppressor factor(s) by preimplantation human embryos. Am J Reprod Immunol Microbiol 11: 98–101
Delgado NM, Huacija L, Merchant H, Reyes R, Rosado A (1980) Species specific decondensation of human spermatozoa nuclei by heparin. Arch Androl 4: 305–313
Delgado NM, Reyes R, Huacija L, Merchant H, Rosado A (1982) Heparin binding sites in the human spermatozoa membrane. Arch Androl 8: 87–95
Dobashi K, Ajika K, Ohkawa T, Okano H, Okinaga S, Arai K (1984) Immunohistochemical localization of pregnancy-associated plasma protein-A (PAPP-A) in placentae from normal and pre-eclamptic pregnancies. Placenta 5: 205–212
Dreosti IE, Record IR, Manuel SJ (1985) Zinc deficiency and the developing embryo. Biol Trace Elem Res 7: 103–122
Dura Trave T, Puig Abuli M, Monreal I, Willa Elizaga I (1986) Plasmatic zinc concentrations during pregnancy in zinc deficient rats in relation to neonatal outcome. Gynecol Obstet Invest 22: 134–139
Folkersen J, Westergaard JG, Hindersson P, Teisner B (1979) Affinity chromatographic purification of a new high molecular weight pregnancy-specific protein SP4. In: Lehmann FG (ed) Carcinoembryonic proteins, vol 2. Elsevier, Amsterdam, pp 503–508
Folkersen J, Grudzinskas JG, Hindersson P, Teisner B, Westergaard JG (1981) Purification of pregnancy-associated plasma protein-A by a two step affinity chromatographic procedure. Placenta 2: 11–18
Gill TJ (1984) Immunity and pregnancy. CRC Critical Rev Immunol 5: 201–227
Gore CH, Sutcliffe RG (1984) Pregnancy-associated plasma protein-A: purification under mild conditions, peptide mapping, and tests for possible interactions with trypsin, plasmin and complement. Placenta 5: 293–314
Grudzinskas JG, Gordon YB, Davies Humphreys J, Brudenell M, Chard T (1979) Circulating levels of pregnancy-specific β_1-glycoprotein in pregnancies complicated by diabetes mellitus. Br J Obstet Gynecol 86: 978–980
Grudzinskas JG, Westergaard JG, Teisner B (1985) Pregnancy-associated plasma protein-A in normal and abnormal pregnancies. In: Bischof P, Klopper A (eds) Proteins of the placenta. Karger, Basel, pp 184–197
Grudzinskas JG, Westergaard JG, Teisner B (1986) Biochemical assessment of placental function: early pregnancy. Clin Obstet Gynaecol 13: 553–569
Horne C, Towler C, Pugh-Humphreys R, Thomson A, Bohn H (1976) Pregnancy-specific β_1-glycoprotein product of syncytiotrophoblast. Experientia 32: 1197–1199
Hurley LS (1981) Teratogenic aspects of manganese, zinc and copper nutrition. Physiol Rev 61: 249–295
Jacoby DR, Olding LB, Oldstone MBA (1984) Immunologic regulation of fetal–maternal balance. Adv Immunol 35: 157–208
Jouppila P, Huhtaniemi I, Tapanainen J (1980a) Early pregnancy failure: study by ultrasonic and hormonal methods. Obstet Gynecol 55: 42–47
Jouppila P, Seppala M, Chard T (1980b) Pregnancy-specific β_1-glycoprotein in complications of early pregnancy. Lancet i: 667–668
Kaye MD, Jones WR (1971) Effect of hCG in vitro lymphocyte transformation. Am J Obstet Gynecol 109: 1029–1031
Kearns M, Lala PK (1985) Radioautographic analysis of surface markers on decidual cells shared by cells of the lymphomyeloid tissues. Am J Reprod Immunol Microbiol 9: 39–47
Keller RH, Calvanico NJ (1984) Suppressor macromolecules. CRC Crit Rev Immunol 5: 149–199
Klopper A (1979) The new placental proteins. Biol Med 1: 89–104

Landy HL, Weiner S, Corson SL, Batzer FR, Bolognese RJ (1986) The "vanishing twin": ultrasonographic assessment of fetal disappearance in the first trimester. Am J Obstet Gynecol 155: 14–19

Lenz RW, Ax RL, Grimek HJ, First NL (1982); Proteoglycan from bovine follicular fluid enhances an acrosome reaction in bovine spermatozoa. Biochem Biophys Res Comm 106: 1092–1098

Lin TM, Halbert SP (1976) Placental localization of human pregnancy-associated plasma proteins. Science 193: 1249–1252

Lin TM, Halbert SP, Kiefer D, Spellacy WN, Gall S (1974) Characterisation of four pregnancy-associated plasma proteins. Am J Obstet Gynecol 118: 223–236

Lin TM, Halbert SP, Spellacy WN, Gall S (1976) Human pregnancy-associated plasma proteins during the postpartum period. Am J Obstet Gynecol 124: 382–387

Lin TM, Halbert SP, Spellacy WN, Berne BH (1977) Plasma concentrations of four pregnancy proteins in complications of pregnancy. Am J Obstet Gynecol 128: 808–810

McIntyre JA, Page Faulk W (1979) Trophoblast modulation of maternal allogeneic recognition. Proc Natl Acad Sci 76: 4029–4032

McIntyre JA, Hsi B, Page Faulk W, Klopper A, Thomson A (1981) Immunological studies of the human placenta: functional and morphological analysis of pregnancy-associated plasma protein-A (PAPP-A). Immunology 44: 577–583

Martin-du-Pan RC, Bischof P, Bourrit B, Lauber K, Girard JP, Herrmann WL (1983) Immunosuppressive activity of seminal plasma and pregnancy-associated plasma protein-A (PAPP-A) in men. Arch Androl 10: 185–188

Merz WE, Schmidt W, Hilgenfeldt U, Schackert K, Lenhard V (1976) Isolation of substances from crude preparation of human chorionic gonadotropin (hCG) which strongly inhibit the transformation of lymphocytes. Z Immunitaetsforsch 152: 286–289

Miyasaka M (1985) Is the mother really immunosuppressed? In: Ellendorf F, Koch E (eds) Early pregnancy factors. Perinatology Press, New York, pp 75–87

Mori T, Kobayashi H, Nishimoto H, Suzuki A, Nishimura T, Mori T (1977) Inhibitory effect of progesterone and 20α-hydroxypregn-4-en-3-one on the phytohaemagglutinin induced transformation of human lymphocytes. Am J Obstet Gynecol 127: 151–157

Morton H, Rolfe B, Clunie GJA, Anderson MJ, Morrison J (1977) An early pregnancy factor detected in human serum by the rosette inhibition test. Lancet i: 394–397

Munroe JS (1971) Progesteroids as immunosuppressive agents. J Reticuloendothel Soc 9: 361–375

Murgita RA, Tomasi TB (1975) Suppression of the immune response by α-fetoprotein. I. The effect of mouse α-fetoprotein on the primary and secondary antibody response. J Exp Med 141: 269–286

Murgita RA, Wigzell H (1981) Regulation of immune functions in the fetus and newborn. Prog Allergy 29: 54–133

Nelson DS, Gatti RA (1976) Humoral factors influencing lymphocyte transformation. Prog Allergy 21: 261–341

Nicholas NS, Panayi GS (1985) Inhibition of interleukin-2 production by retroplacental sera: a possible mechanism for human fetal allograft survival. Am J Reprod Immunol Microbiol 9: 6–11

Nicholas NS, Panayi GS (1986) Immunosuppressive properties of pregnancy serum on the mixed lymphocyte reaction. Br J Obstet Gynecol 93: 1251–1255

Nielsen PV, Pedersen H, Kampmann EM (1979) Absence of human placental lactogen in an otherwise uneventful pregnancy. Am J Obstet Gynecol 135: 322–326

Nisbet AD, Bremner RD, Herriot R, Jandial V, Horne CHW, Bohn H (1981) Placental protein 5 (PP5): development of a radioimmunoassay and measurement of circulating levels in normal pregnancy. Br J Obstet Gynecol 88: 484–491

Niven PAR, Landon J, Chard T (1972) Placental lactogen levels as a guide to outcome of threatened abortion. Br Med J iii: 799–801

Noonan FP, Holliday WJ, Mcrton H, Clunie GJA (1979) Early pregnancy factor is immunosuppressive. Nature 278: 649–650

Ohlsson K, Olsson I (1974) The neutral proteases of human granulocytes: isolation and partial characterization of granulocyte elastases. Eur J Biochem 42: 519–527

O'Neill C, Gidley-Baird AA, Pike IL, Porter RN, Sinosich MJ, Saunders DM (1985) Maternal blood platelet physiology and luteal-phase endocrinology as a means of monitoring pre- and post-implantation embryo viability following in vitro fertilization. JIVFET 2: 59–65

Pandya IJ, Cohen J (1985) The leucocyte reaction of the human uterine cervix to spermatozoa. Fertil Steril 43: 417–421

Parish RF, Fair WR (1981) Selective binding of zinc ions to heparin rather than to other glycosylaminoglycans. Biochem J 193: 407–410

Pinto-Furtado LG, Bolton AE, Grudzinskas JG, Chapman MG, Sinosich MJ, Sharma V (1984) The development and validation of a radioimmunoassay for human pregnancy-associated plasma protein-A (PAPP-A). Arch Gynecol 236: 83–91

Rocklin RE, Kitzmiller JL, Kaye MD (1979) Immunobiology of the maternal–fetal relationship. Ann Rev Med 30: 375–404

Rogers BJ (1978) Mammalian sperm capacitation and fertilization in vitro: a critique of methodology. Gamete Res 1: 165–223

Rolfe BE, Morton H, Clarke FM (1983) Early pregnancy factor is an immunosuppressive contaminant of commercial preparations of human chorionic gonadotropin. Clin Exp Immunol 51: 45–52

Schindler AM, Bischof P (1984) Histochemical localization of pregnancy-associated plasma protein-A in fetal, infant and adult organs and comparison between antisera. Gynecol Obstet Invest 18: 88–94

Schindler AM, Bordignon P, Bischof P (1984) Immunohistochemical localization of pregnancy-associated plasma protein-A in decidua and trophoblast: comparison with human chorionic gonadotropin and fibrin. Placenta 5: 227–236

Short RV (1969) Implantation and the maternal recognition of pregnancy. In: Wolstenholme GEW, O'Connor M (eds) Ciba foundation symposium on fetal autonomy. Churchill, London, pp 2–26

Sinosich MJ (1985) Biological role of pregnancy-associated plasma protein-A (PAPP-A) in human reproduction. In: Bischof P, Klopper A (eds) Proteins of the placenta. Karger, Basel, pp 158–183

Sinosich MJ (1987) Human ovarian follicular antigens. In: Feichtinger W, Kemeter P (eds) Future aspects in human in vitro fertilization. Springer-Verlag, Berlin Heidelberg New York, pp 64–76

Sinosich MJ, Saunders DM (1987) Potential role of pregnancy-associated plasma protein-A in human reproduction. J Reprod Immunol 10: 55–65

Sinosich MJ, Teisner B, Grudzinskas JG, Saunders DM (1980) Radioimmunoassay for pregnancy-associated plasma protein-A (PAPP-A). In: Proceedings of the Australian Society for Research into Obstetrics and Gynecology, Armidale, NSW

Sinosich MJ, Teisner B, Davey M, Grudzinskas JG (1981) Pregnancy-associated plasma protein-A; interaction with heparin in crossed affinity immunoelectrophoresis. Aust NZ J Med 11: 429–433

Sinosich MJ, Davey MW, Ghosh P, Grudzinskas JG (1982a) Specific inhibition of human granulocyte elastase by human pregnancy-associated plasma protein-A. Biochem Intern 5: 777–786

Sinosich MJ, Teisner B, Folkersen J, Saunders DM, Grudzinskas JG (1982b) Radioimmunoassay for pregnancy-associated plasma protein-A. Clin Chem 28: 50–53

Sinosich MJ, Davey MW, Teisner B, Grudzinskas JG (1983a) Comparative studies of pregnancy-associated plasma protein-A and α_2-macroglobulin using metal chelate chromatography. Biochem Int 7: 33–42

Sinosich MJ, Smith DH, Grudzinskas JG, Saunders DM, Westergaard JG, Teisner B (1983b) The prediction of pregnancy failure by measurements of PAPP-A following in vitro fertilization and embryo transfer. Fertil Steril 40: 539–541

Sinosich MJ, Davey MW, Grudzinskas JG (1984a) Interaction of pregnancy-associated plasma protein-A (PAPP-A) and alpha 2 macroglobulin (α_2M) on metal chelate. Prot Biol Fluids 31: 19–22

Sinosich MJ, Porter P, Sloss P, Bonifacio M, Saunders DM (1984b) Pregnancy-associated plasma protein-A in human ovarian follicular fluid. J Clin Endocrinol Metab 58: 500–504

Sinosich MJ, Saunders DM, Grudzinskas JG (1984c) Pregnancy-associated plasma protein-A: a barrier to maternal proteolytic attack. In: Harrison RF, Bonnar J, Thompson W (eds) In vitro fertilization, embryo transfer and early pregnancy. MTP Press, Lancaster, pp 209–212

Sinosich MJ, Ferrier A, Porter RN et al. (1985a) Circulating and tissue concentration of pregnancy-associated plasma protein-A (PAPP-A) in tubal ectopic gestation. J. Clin Reprod Fertil 3: 311–317

Sinosich MJ, Ferrier A, Saunders DM (1985b) Monitoring of postimplantation embryo viability following successful in vitro fertilization and embryo transfer by measurement of placental protein. Fertil Steril 44: 70–74

Sinosich MJ, Ferrier A, Saunders DM (1985c) Monitoring of post implantation embryo viability by measurement of pregnancy-associated plasma protein-A, SP1 and beta-hCG. Ann NY Acad Sci 442: 440–444

Sinosich MJ, Grudzinskas JG, Saunders DM (1985d) Placental proteins in the diagnosis and evaluation of the "elusive" early pregnancy. Obstet Gynecol Survey 40: 273–282

Sinosich MJ, King M, Bonifacio MD, Nelson D, Saunders DM (1985e) Pregnancy-associated plasma protein-A in human seminal plasma. Prot Biol Fluids 32: 289–292

Sinosich MJ, Saunders DM, Grudzinskas JG (1985f) Pregnancy-associated plasma protein-A and placental protein 5 in human ovarian follicular fluid. Ann NY Acad Sci 442: 269–275

Sinosich MJ, Seppala M, Saunders DM, Grudzinskas JG (1985g) Pregnancy-associated plasma protein-A and placental protein 5 in human seminal plasma. Ann NY Acad Sci 442: 287–292

Sinosich MJ, Westergaard L, Teisner B, Westergaard JG, Saunders DM, Grudzinskas JG (1985h) Pregnancy-associated plasma protein-A in human ovarian follicular fluid. Prot Biol Fluids 32: 259–262

Sinosich MJ, Bonifacio MD, Hodgen GD (1987a) Affinity immuno-electrophoresis and chromatography for isolation and characterization of a placental granulocyte elastase inhibitor. In: Burgess R (ed) Protein purification: micro to macro. Alan R Liss, New York

Sinosich MJ, Bonifacio MD, Saunders DM (1987b) Biochemical diagnosis of early pregnancy failure after IVF-ET. In: Proceedings of the 43rd annual meeting of the American Fertility Society, p 107

Sinosich MJ, Danforth DR, Veeck LL, Hodgen GD (1987c) Pregnancy-associated plasma protein-A (PAPP-A): a bio-chemical marker of granulosa cell status. In: Proceedings of the 43rd annual meeting of the American Fertility Society, p 9

Sinosich MJ, Dodd J, Bonifacio MD et al. (1987d) Production of pregnancy associated plasma protein-A (PAPP-A) by cultured tumour granulosa cells. Gynecol Obstet Invest 23: 60–66

Sjoberg J. Wahlstrom T, Seppala M (1984a) Pregnancy-associated plasma protein-A in the human endometrium is dependent on the effect of progesterone. J Clin Endocrinol Metab 58: 359–362

Sjoberg J, Wahlstrom T, Seppala M et al. (1984b) Hyperstimulated human preovulatory follicular fluid, luteinized cells of unruptured follicles and corpus luteum contain pregnancy-associated plasma protein-A (PAPP-A). Fertil Steril 41: 551–557

Sjoberg J, Wahlstrom T, Rutanen EM et al. (1985) Seminal plasma levels of PAPP-A in normospermic and oligospermic men and tissue localisation of PAPP-A in the male genital tract. Arch Androl 14: 253–261

Stimson WH (1980) Are pregnancy-associated serum proteins responsible for the inhibition of lymphocyte transformation by pregnancy serum? Clin Exp Immunol 40: 157–160

Stimson WH (1981) Pregnancy-associated serum proteins and hormones. Do they contribute to the regulation of the maternal immune response in pregnancy. Clin Immunol Newsl 2: 7–11

Sutcliffe RG, Kukulska BM, Nicholson LVB, Paterson WF (1979) The use of antibody affinity chromatography and other methods in the study of pregnancy-associated proteins. In: Klopper A, Chard T (eds) Placental proteins. Springer-Verlag, Berlin Heidelberg New York, pp 55–70

Sutcliffe RG, Hunter JB, Gibb S, MacLean AB (1980a) Studies on human pregnancy-associated plasma protein-A. In: Klopper A, Genazzani A, Crosignani PG (eds) The human placenta. Academic Press, London, pp 57–66

Sutcliffe RG, Kukulska-Langlands BM, Coggins JR, Hunter JB, Gore CH (1980b) Studies on human pregnancy-associated plasma protein-A. Biochem J 191: 799–809

Talwar GP, Mullick A, Ramakrishnan S et al. (1983) Immunointerception of fertility. In: Wegmann TG, Gill TJ (eds) Immunobiology of reproduction. Oxford University Press, New York, pp 449–467

Teasdale F, Adcock EW, August CS, Cox S, Battaglia FC, Naughton MA (1973) Human chorionic gonadotropin: Inhibitory effect on mixed lymphocyte cultures. Gynecol Invest 4: 263–269

Toop K, Klopper A (1983) The effect of anticoagulants on the measurement of pregnancy-associated plasma protein-A (PAPP-A). Br J Obstet Gynecol 90: 150–155

Tornehave D, Chemnitz J, Teisner B, Folkersen J, Westergaard JG (1984) Immunohistochemical demonstration of pregnancy-associated plasma protein-A (PAPP-A) in the syncytiotrophoblast of the normal placenta at different gestational ages. Placenta 5: 427–432

Tsakok FTM, Koh MS, Ratnamm SS et al. (1983) Pregnancy-associated proteins in trophoblastic disease. Br J Obstet Gynecol 90: 483–486

Vallee BL, Galdes A (1984) The metallobiochemistry of zinc enzymes. Adv Enzymol 56: 283–430

Vanderbeeken Y, Vlieghe MP, Duchateau J, Delespesse G (1984) Suppressor T-lymphocytes in pregnancy. Am J Reprod Immunol 5: 20–24

von Schoultz B, Stigbrand T, Tarnvik A (1973) Inhibition of PHA-induced lymphocyte stimulation of the pregnancy zone protein. FEBS Lett 38: 23–26

Wahlstrom T, Teisner B, Folkersen J (1981) Tissue localization of pregnancy-associated plasma protein-A (PAPP-A) in normal placenta. Placenta 2: 253–258

Westergaard JG, Chemnitz J, Teisner B et al. (1983a) Pregnancy-associated plasma protein-A: a possible marker in the classification of prenatal diagnosis of Cornelia de Lange syndrome. Prenatal Diagnosis 3: 225–232

Westergaard JG, Sinosich MJ, Bugge M, Madsen LT, Teisner B, Grudzinskas JG (1983b) Pregnancy-associated plasma protein-A in the prediction of early pregnancy failure. Am J Obstet Gynecol

145: 67–69
Westergaard JG, Teisner B, Hau J, Grudzinskas JG (1985a) Placental protein and hormone measurements in twin pregnancy. Br J Obstet Gynecol 92: 72–76
Westergaard JG, Teisner B, Sinosich MJ, Madsen LT, Grudzinskas JG (1985b) Does ultrasonic examination render biochemical tests obsolete in the prediction of early pregnancy failure? Br J Obstet Gynecol 92: 77–83
Westergaard L, Sinosich MJ, Grudzinskas JG et al. (1985) Pregnancy-associated plasma protein-A (PAPP-A) in preovulatory, nonovulatory healthy and atretic human ovarian follicles during the natural cycle. Ann NY Acad Sci 442: 205–211
Yovich JL, Willcox DL, Grudzinskas JG, Bolton AE (1986) The prognostic value of β-hCG, PAPP/A, estradiol and progesterone in early human pregnancy. Aust NZ J Obstet Gynecol 26: 59–64

6. Embryo-Derived Platelet Activating Factor

C. O'Neill and N. Spinks

Introduction

Three factors with physiologically important effects have been claimed to be secreted by the preimplantation embryo. These are:

1. Embryo-derived platelet activating factor (PAF) (O'Neill and Saunders 1984; O'Neill 1985b,c)
2. Ovum factor (Cavanagh et al. 1982)
3. Immunosuppressive factor (Daya and Clarke 1986)

This chapter reviews each of these factors, offers evidence that they are all the same, and shows that production of embryo-derived PAF is essential for the establishment of pregnancy.

Embryo-Derived Platelet Activating Factor

The discovery of embryo-derived PAF (O'Neill 1985 a,b) resulted from the observation in mice that, after conception, there was an embryo-dependent retardation in the hormonally induced increase in spleen weight normally associated with the early luteal phase. On day 2 of pregnancy (day 1 = day of mating), the spleen was only 40%–50% of the weight of pseudopregnant control mice. By day 4 of pregnancy, the weight was again comparable in the two groups. This retardation was due to contraction of the spleen after conception, resulting in a dramatic (50%) reduction in the platelet count of the spleen. It was apparent that this was due to enhanced vascular consumption of platelets, since a significant reduction in the peripheral platelet count also occurred by 10.30 h on day of mating (some 6–12 h after the expected time of mating). This reached a maximum by 14.30 h on day 1 (20%) and persisted throughout the peri-implantation phase of pregnancy.

This early pregnancy associated thrombocytopenia depended on the presence of the zygote. It did not occur in female mice mated with vasectomised males nor in bilaterally ligated females mated with intact males. Thrombocytopenia was apparent within 3 h of transfer of zygotes to pseudopregnant recipients, and there was a significant correlation between the extent of thrombocytopenia and the number of embryos in the reproductive tract. Early pregnancy associated thrombocytopenia was also observed in the common marmoset and in man, although the extent of thrombocytopenia was variable (O'Neill 1987). This variability means that monitoring the peripheral platelet count has little practical value other than to confirm the existence of early pregnancy associated platelet activation.

This early pregnancy associated platelet activation was caused by the secretion by the zygote of a platelet activating factor. Mouse embryos cultured for 24 h in simple defined media (HTF) with bovine serum albumin as the only protein source, rapidly caused thrombocytopenia when injected into splenectomised mice (O'Neill 1985 b,c). A drop of up to 30% occurred within only 10 min of injection. The responses were dose dependent and did not occur with unconditioned media or media in which unfertilised oocytes had been cultured for 24 h. This embryo derived PAF was also produced by human embryos following in-vitro fertilisation (O'Neill and Saunders 1984). Culture of single pronuclear stage embryos for 24 h (to the 2–4 cell stage) in defined medium with 10% maternal serum resulted in production of detectable levels of PAF by about 40% of embryos. All embryos which resulted in ongoing pregnancies produced detectable levels of PAF in the splenectomised mouse bioassay, whereas all embryos which failed to produce PAF failed to result in pregnancy (O'Neill et al. 1985). However, many embryos that produced PAF also failed to establish pregnancy.

Embryo derived PAF was homologous to the platelet activating factor, 1-o-alkyl-2-acetyl-sn-glyceryl-3-phosphocholine (PAF-acether), in terms of its structural characteristics, dose and time response and pharmacology (O'Neill 1985c, 1987). Murine embryo derived PAF is a phospholipid that co-migrates on thin layer chromatography with PAF-acether. Its activity is inhibited by PAF-acether specific antagonists, and antiserum raised against a synthetic PAF-acether neutralises the biological activity of PAF.

The platelet activating action of this molecule represents the first demonstration of a biologically active phospholipid. It is also a potent hypotensive agent (Muirhead et al. 1981) and elicits a vast number of pathophysiological effects, including airway and pulmonary oedema, bronchial hyperresponsiveness of asthma, cardiac arrhythmias and anaphylaxis. It is a mediator of endotoxin-included hypotension and shock and a potent vasodilator, and it enhances vascular permeability (for review see O'Neill 1987).

PAF is extremely potent, being active at concentrations as low as 10^{-11} to 10^{-14} mol/l, has a very short half-life in serum, owing to its inactivation by specific acetylhydrolases, and its effects are receptor mediated. Thus PAF appears to be an important autacoid, its actions being determined by its site of production.

Studies of the production of PAF by mouse embryos in vitro using the splenectomised mouse assay show considerable variability in its production

(Spinks et al. 1986). One of three cultures of 2-cell embryos displayed activity after 1h of culture, whereas all cultures displayed activity after 4h. However, after longer periods in culture production was reduced. After 24h, eight of 10 cultures had detectable levels of PAF. For 2-cell embryos cultured to the blastocyst stage over 72h (with transfer to fresh tubes every 24h) only five of 10 were PAF positive. By contrast, all four mouse blastocysts collected fresh from the uterus and cultured for 12h were PAF positive in culture. These results suggest that prolonged culture of mouse embryos in vitro reduced their PAF production.

Ovum Factor

Ovum factor is produced by mice and sheep embryos. (Morton 1984; Smart et al. 1981). Several authors have claimed that the ovum factor can be detected in human embryo culture medium following in-vitro fertilisation, but other groups have been unable to verify this. Ovum factor was detected by its ability to cause female animals to express the putative immunosuppressive early pregnancy factor (EPF).

In the mouse, EPF has two components, A and B. Component A is a product of the oviduct and is not pregnancy specific. In the presence of a pituitary hormone, thought to be prolactin, ovum factor promotes the production of component B by the ovary. Together, components A and B express EPF activity (Cavanagh et al. 1982).

Nancarrow et al. (1980) proposed a different scheme for the sheep. In mono-ovulatory sheep, ovum factor stimulated the secretion of a second messenger from the oviduct which resulted in the release of EPF activity from the ipsilateral ovary. No activity was produced by the contralateral ovary after infusion of the factor from fertilised eggs into the contralateral oviduct. This suggested that the oviduct message arrives at the ovary by local transport and that the ovary must have ovulated for EPF activity to be produced.

The putative EPF has been described on the basis of an alteration in lymphocyte reactivity in the rosette inhibition test. This test was devised as a means of assessing, in vitro, the immunosuppressive characteristics of antilymphocyte serum. Thymus-derived (T) lymphocytes from many species will spontaneously bind xenogeneic erythrocytes – e.g., human T lymphocytes bind sheep erythrocytes. This is termed rosette formation. The addition of antiserum directed against lymphocytes (antilymphocyte serum) may cause inhibition of rosette formation. The inhibition of rosette formation by antilymphocyte serum was augmented when the lymphocytes forming the rosettes were from the spleen of pregnant mice. This effect could be acquired by spleen cells from non-pregnant animals following incubation with serum from pregnant animals.

In a rigorous examination of the procedure in pigs, Koch and Ellendorff (1985) found an interassay coefficient of variation of 23.8% for "pregnancy" sera and 6.5% for "non-pregnancy" sera. False positives were found in 8.6% and false negatives in 12.2% of assays, and sera were consistently positive for a short time after mating with vasectomised males. The authors concluded that the cumbersome, time-consuming and easily disturbed assay was suitable only for limited investigations under well-defined experimental conditions.

Since antilymphocyte sera that inhibit rosette formation are immunosuppressive, the observation that serum from pregnant animals augments rosette inhibition led Morton et al. (1976) to suggest that "pregnant" serum also has an immunosuppressive function. The validity of this claim was tested (Noonan et al. 1979) by examining the effects of the putative EPF on the adoptive transfer of contact hypersensitivity to trinitrochlorobenzene.

The delayed type hypersensitivity response is a measure of T-cell mediated immune responses and is thought to involve the same mechanisms as skin and organ graft rejection (Loveland et al. 1981). The ability of EPF activity to provide protection from such a response suggests that this activity may be important in the suppression of the maternal immune response to the genetically alien fetus. The results also suggest that the immunosuppression is neither antigen specific nor species specific (Noonan et al. 1979). This is clearly not so in early pregnancy. It was subsequently shown (Morton 1984) that upon binding to a small number of lymphocytes, EPF mediates the release of soluble factors that induce the alterations to the rosette inhibition test. This complicates characterisation of ovum factors and EPF even further.

The method of detection of ovum factor or EPF is by monitoring the changes in rosette inhibition test results, the new observation (Morton 1984) of secondary suppressive factors predicts that EPF characterised to date may well be these secondary factors. Since such release products are probably normal components of lymphocytes, they are most unlikely to be pregnancy specific, hence the difficulty in the development of an antiserum to the factor which can distinguish between pregnancy and non-pregnancy. Apart from this disturbing implication, there is some question as to the physiological significance of putative immunosuppression in early pregnancy. Indeed some studies suggest the requirement for enhanced maternal immunological recognition of the early conceptus for the establishment of pregnancy.

There is now good evidence that ovum factor is in fact embryo derived PAF. Apart from the great similarities in the kinetics of production, physicochemistry and biology of these two factors (O'Neill 1986), it has now been shown (Orozco et al. 1986) that PAF induces the expression of EPF in mice within 1 h of injection. In animals this occurs only during dioestrus, pro-oestrus or oestrus. It does not occur in males or immature females.

Immunosuppressive Factor

Whereas the immunosuppressive actions of ovum factor are exerted indirectly via the generation of EPF, the human zygote produces a factor in vitro which is directly immunosuppressive (Daya and Clark 1986). The factor, obtained from the culture media of human embryos following in-vitro fertilisation, suppressed mitogen (concanavalin A) induced proliferation of peripheral lymphocytes. Such culture media caused suppression in 43% (12/28) of embryos tested (mean suppression 11.2± 3%). Chromatography showed the suppressive factors to exist in two fractions, both of which had an apparent molecular weight of less then 10 000 daltons. It was suggested that suppressive activity was important, since only embryos producing the activity resulted in pregnancy. The results were said to be consistent with the hypothesis that successful implantation depends on prior maternal immunosuppression.

It is likely that this immunosuppression, like ovum factor, is another manifestation of embryo-derived PAF. PAF was also produced by 43% of human embryos following in-vitro fertilisation (O'Neill and Saunders 1984), and embryos failing to produce PAF failed to result in pregnancy. PAF binds to carriers, so activity can be found at a number of apparent molecular weights following chromatography. Most importantly, PAF is a potent (10^{-8} to 10^{-6} mol/l) inhibitor of mitogen (phytohaemagglutinin) induced proliferation of human peripheral lymphocytes (Rola-Pleszczynski et al. 1986). This inhibition was reversible by PAF antagonists. It will be important to determine whether the putative immunosuppressive factor derived from the embryo can be inhibited by PAF antagonists and is extractable by the Bligh and Dyer (1959) technique.

Role of Embryo-Derived PAF in the Establishment of Pregnancy

The apparent correlation between the production of human embryo-derived PAF and the pregnancy potential of embryos suggests that its production may serve a fundamental role in the establishment of pregnancy. There is some circumstantial evidence to support this. Gasic and Gasic (1970) showed, in mice, that depletion of blood platelets following intravenous injection of antiplatelet antiserum on days 4 and 7 of pregnancy resulted in a 50% reduction in the number of successful pregnancies. Such treatment did not result in platelet activation or the accompanying release of important biologically active molecules but led to their destruction by the reticuloendothelial system.

PAF did not appear to exert major effects on preimplantation development. PAF-acether did not enhance cell division rates (Fig. 6.1) of 2-cell embryo development to the blastocyst stage in vitro. PAF specific antagonists also had no adverse effects on 2-cell mouse embryo development. However, small concentrations did enhance the utilisation of sodium lactate by 2-cell and 4-cell embryos (Fig. 6.2). This suggests that while PAF directly enhances metabolism by preimplantation embryos it does not have an essential role.

The use of inhibitors of PAF activity in vivo, however, provided conclusive evidence for the essential role of embryo-derived PAF in the establishment of pregnancy (Spinks and O'Neill 1987).

A broad-specificity inhibitor, Iloprost (an analogue of prostacyclin) and a PAF-acether specific antagonist, SRI 63–441, were used. Both were potent inhibitors of implantation in mice. The drugs were prepared in saline and given as an intraperitoneal injection at various doses over the first four days of pregnancy. Animals were killed on day 8 of pregnancy. A total dose of 12 μg of Iloprost or 80 μg of SRI 63 441 over this time caused a 70% reduction in the implantation rate. Animals treated similarly but killed on day 4 had their reproductive tract flushed to collect embryos. The inhibitors did not affect the

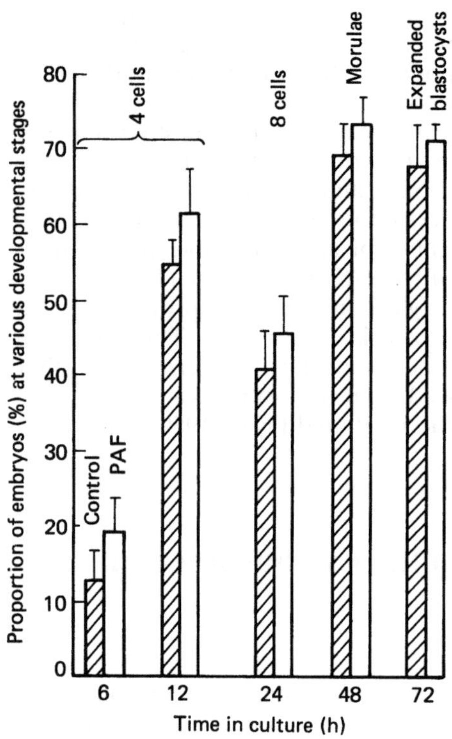

Fig. 6.1. Effect of PAF on embryo development. Embryos (2-cell) were cultured in HTFM (3 mg BSA/ml) ± 1 μg/ml PAF. The developmental stage at various times was recorded. Over the 72 h culture period there was no advantage in culturing in PAF-containing media. The result was the mean ± SEM of three replicates.

survival of embryos in utero and, as noted above, the drugs had no effect on the development of 2-cell embryos through to the blastocyst stage in vitro. Thus the contraceptive action of the drugs was not a result of direct toxic effects on the embryo.

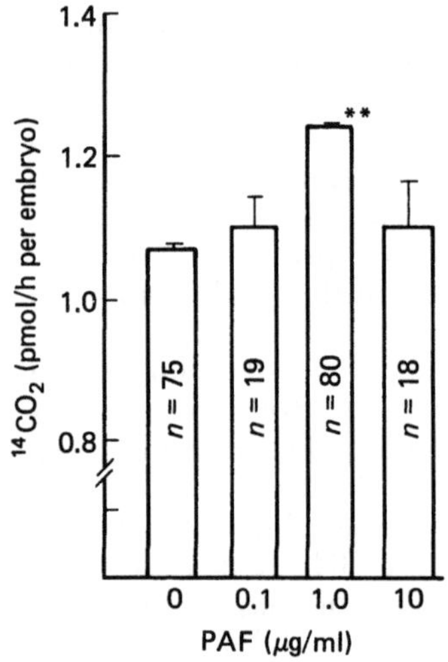

Fig. 6.2. Effect of PAF on lactate utilisation by 2-cell mouse embryo. The generation of $^{14}CO_2$ from ^{14}C-lactate by single mouse embryos cultured in various concentrations of PAF. There was a quadratic effect with maximum utilisation occurring at 1.0 μg/ml.

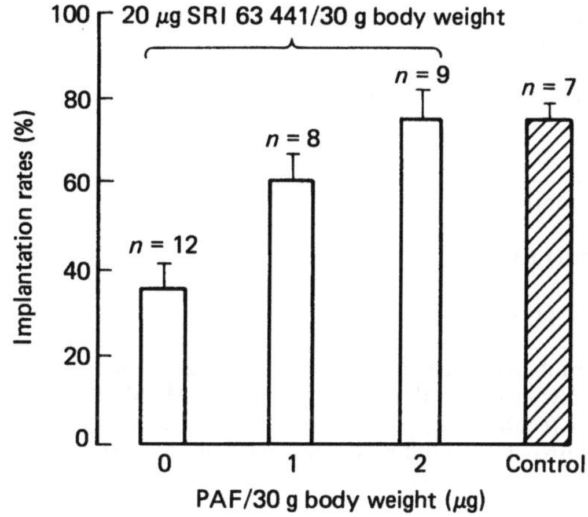

Fig. 6.3. Reversibility of contraceptive action of PAF antagonist. Implantation rates inhibited by SRI 63 441 can be re-established to control levels by administration of PAF.

$$\text{Implantation rate} = \frac{\text{no. of implantations per uterine horn}}{\text{no. of corpora lutea per ovary}} \times 100$$

Doses of PAF-acether were given concomitantly with both Iloprost and SRI 63 441. At doses of Iloprost (2 μg) and SRI 63 441 (10 μg) that gave significant inhibition of implantation, increasing doses of PAF-acether caused the re-establishment of pregnancy such that a dose of 1.0 μg of PAF-acether gave implantation rates not significantly different ($P>0.05$) from the control (Fig. 6.3). Thus, the activity of embryo-derived PAF was essential for the implantation of the embryo in mice. The mechanism by which it exerts this effect is probably multifactorial.

PAF antagonists do not have a significant effect on progesterone production during early pregnancy, but PAF has luteotrophic effects on human granulosa cells in culture (O'Neill 1987). Thus luteolysis is not the cause of implantation failure.

Blastocysts transferred from SRI 63 441 treated donors to untreated recipients have a similar implantation rate as control transfer. This suggests that embryonic exposure to the antagonists is in itself not sufficient to prevent implantation but that maternal exposure is also required. By contrast, PAF antagonists can abolish the ability of mouse blastocyst to undergo blastocyst outgrowth (the equivalent of implantation) in vitro (Table 6.1), suggesting a direct role for PAF in implantation.

Blastocyst outgrowth in vitro, however, requires serum supplementation. It has been shown (O'Neill 1986) that the platelet-dependent serum factors provide the stimulus for blastocyst activation and outgrowth in vitro. Mouse blastocysts were cultured for 72 h in culture medium supplemented with various doses (1%–10% v/v) of either platelet rich or platelet poor serum. Whereas all concentrations of platelet rich serum supported blastocyst outgrowth in vitro, platelet poor serum was incapable of supporting such differentiation. This suggests that a role for PAF in early pregnancy is the provision of platelet derived trophic factors necessary for the initiation of implantation. Implantation may therefore require both the direct effect of PAF on the blastocyst and the

Table 6.1. Effect of PAF and PAF antagonists on mouse blastocyst outgrowth in vitro[a]

Treatment	Blastocyst outgrowth (%)	
	Without PAF-acether	With PAF-acether
Control	60 ± 11.3	53 ± 5.6
SRI 63 441	8 ± 7.5	25 ± 12.3
Iloprost	15 ± 12.3	61 ± 16

Embryos were grown from 2-cell stage to blastocyst (72 h) in medium with 1 μg/ml PAF-acether then transferred to MEM 1% FBS ± inhibitors. Values are percentage outgrowth after 72 h. PAF antagonist prevented blastocyst outgrowth in vitro. This could be at least partially overcome by the presence of PAF.
[a] Results are mean ± SEM.

PAF-dependent production of platelet-derived embryonic trophic factors, in the presence of a suitable maternal environment.

Acknowledgements. This work was supported by grants 860550 and 870322 from the Australian National Health and Medical Research Council. NRS is supported by an RNSH postgraduate scholarship.

References

Bligh EG, Dyer WJ (1959) A rapid method of total lipid extraction and purification. Can J Biochem 258: 911–918.

Cavanagh AC, Morton H, Rolfe BE, Gidley-Baird AA (1982) Ovum factor: a first signal of pregnancy? Am J Reprod Immunol 2: 97–101

Daya S, Clarke DA (1986) Immunosupressive factor produced by human embryos in vitro. N Engl J Med 315: 1551–1552

Gasic GJ, Gasic TB (1970) Total suppression of pregnancy in mice by post-coital administration of neuraminidase. Proc Natl Acad Sci 67: 793–798

Koch E. Ellendorff F (1985) Prospects and limitations of the rosette inhibition test to detect activity of early pregnancy factor in the pig. J Reprod Fertil 74: 29–38

Loveland B, Ceredig R, Hogarth M, McKenzie I (1981) The key role of Lyt-1^+ cells in skin graft rejection in the mouse. Transplant Proc 13: 1079–1081

Morton H (1984) Early pregnancy factor; a link between fertilisation and immunomodulation. Aust J Biol Sci 37: 393–407

Morton H, Hegh V, Clunie GJA (1976) Studies of the rosette inhibition test in pregnant mice: evidence of immunosuppression. Proc R Soc Lond B 193: 413–419

Muirhead EE, Byers LW, Desiderio D, Smith KA, Prewitt RL, Brooks B (1981) Alkyl ether analogs of phosphatidylcholine are orally active in hypertensive rabbits. Hypertension 3: 107–113

Nancarrow CD, Quinn PJ, Rigby NW, Wallace ALC, Grewal AS (1980) Involvement of the zygote, oviduct and ovary in the production of an early pregnancy factor in sheep. Proc Soc Study Reprod 13

Noonan JP, Halliday WJ, Morton H, Clunie GJA. (1979) Early pregnancy factor is immunosuppressive. Nature 278: 649–650

O'Neill C (1985a) Thrombocytopenia is an initial maternal response to fertilization in the mouse. J Reprod Fertil 73: 559–566
O'Neill C (1985b) Examination of the cause of early pregnancy associated thrombocytopenia. J Reprod Fertil 73: 567–577
O'Neill C (1985c) Partial characterization of the embryo-derived platelet activating factor in mice. J Reprod Fert 75: 375–389
O'Neill (1985d) Embryo-derived platelet activating factor. In: Early pregnancy factors. Ellendorff F, Koch E (eds) Promethean Press, New York, pp 261–267
O'Neill C (1986) The role of blood platelets in the establishment of pregnancy. In: Hau J (ed) Pregnancy proteins in animals. Walter de Gruyter, Berlin, pp 225–234
O'Neill C (1987) Embryo-derived platelet activating factor: a pre-implantation embryo mediator of maternal recognition of pregnancy. Domestic Animal Endocrinol 4: 69–85
O'Neill C, Saunders DM (1984) Assessment of embryo quality. Lancet ii: 1035
O'Neill C, Gidley-Baird AA, Pike IL, Porter RN, Sinosich MJ, Saunders DM (1985) Maternal blood platelet physiology and luteal phase endocrinology as a means of monitoring pre and post implantation embryo viability following in-vitro fertilisation. J In Vitro Fertil Embryo Transf 2: 59–65
Orozco C, Perkins F, Clarke FM (1986) Platelet activating factor induces the expression of early pregnancy factor activity in female mice. J Reprod Fertil 78: 549–555
Rola-Pleszczynski M, Pignol B, Michel F, Braquet P (1986) Effect of PAF on the production of interleukin 2 by lymphocytes; differential activities of enantiomers and PAF antagonists. Proc 2nd int conference on platelet activating factor, Tennessee, Abs 73
Smart YC, Cripps AL, Clancy RL, Roberts TK, Lopata A, Shutt D (1981) Detection of an immunosuppressive factor in human preimplantation embryo cultures. Med J Aust i: 78–79
Spinks RN, O'Neill C (1987) Embryo-derived PAF activity is essential for the establishment of pregnancy in the mouse. Lancet i: 106–107
Spinks RN, Ammit AJ, Saunders DM, O'Neill C (1986) Embryo-derived platelet activating factor: kinetics of its production (in vitro). Proc Aust Soc Reprod Biol 18:1

Section IV
ENDOMETRIAL/DECIDUAL PROTEINS

7. Synthesis and Secretion of Proteins by the Endometrium and Decidua

S.C. Bell

Introduction

The human endometrium, a complex tissue containing many cell types, undergoes hormone-dependent proliferation and differentiation (Robertson 1981). This differentiation presumably fulfils functions during the implantation process and is involved in trophoblast–endometrial interactions, since implantation of the embryo within the uterus occurs during the period of endometrial differentiation. The endometrium may fulfil different functions during pregnancy, because of its differential anatomical relationship with the fetoplacental unit. As with other tissues, the functions of the human endometrium may be mediated by proteins synthesised and secreted in vivo. In this respect the endometrium, with its intimate and changing relationship with various populations of trophoblasts during implantation and pregnancy, presents an intriguing possibility that such proteins act in a paracrine manner, affecting genetically non-identical tissues (Bell and Smith 1987). Certainly research on the differentiated endometrium in experimental rodents has linked a number of functions with secretory protein products, including endocrinological and immunological functions and protection from trophoblast invasion (Bell 1983, 1985). The identification and characterisation of secretory protein products of the endometrium, especially if they represent products of individual cell types of this tissue, would facilitate developments in a number of areas. Given the value of the hormone-dependent endometrial secretory product of the rabbit, uteroglobin, in research into aspects of steroid hormone control of gene expression, the identification of such products in man could facilitate clinical research in these areas (Bell 1985, 1986a). They would be of great value in assessing the impact of endometrial cellular dysfunction on implantation and pregnancy. The current knowledge about the range of proteins synthesised and secreted by the endometrium will be reviewed in this chapter. Detailed consideration of the potential function and clinical applications of the measurement of a number of these proteins will be discussed.

Histologically Defined Endometrial Differentiation

Certain features of endometrial differentiation during implantation and pregnancy are unique to man, and these features will affect the interpretation of in-vitro studies on the endometrial tissue available for study and the potential function of secretory proteins. Although it is becoming apparent that a range of cell types are present in the endometrium, including those of the immune system (Bulmer and Sunderland 1984), the most obvious differentiation at the histological level occurs in the glandular epithelium and stromal cells.

During the luteal phase two periods of differentiation are observed, the first occurring during the first postovulatory week and characterised by differentiation of the glandular epithelium, which acquires secretory activity and is maximal on the 7th post-ovulatory day. Since implantation would occur during this period in a conceptual cycle, secretions from this cell type might be expected to be intimately involved with implantation. The second phase of differentiation occurs on postovulatory days 9–10, when the stromal cell population differentiates to form predecidual cells in restricted areas of the endometrium, principally around spiral arteries and arterioles defining the late secretory endometrium. As the late luteal phase progresses, this spreads through the upper layer of the endometrium, the zona compacta, the glandular epithelium of this layer tending to become non-secretory and inconspicuous. Glands still actively secretory become restricted to the underlying zona spongiosa, where the stroma cells remain undifferentiated.

It is generally assumed that if pregnancy intervenes and menstruation is prevented these alterations are simply extended. Certainly in endometrial specimens obtained during the first trimester histological examination reveals equivalent areas, decidualised endometrium containing non-secretory glands, termed decidua compacta, and secretory gland containing areas with non-decidualised stroma, termed decidua spongiosa (Fig. 7.1). However, a number of points are relevant to note. First, part of this response to pregnancy must involve tissue growth, since the amount of endometrial tissue recovered during the first trimester is substantially greater than that obtained during the cycle. Secondly, the decidual differentiation of the stroma observed during the late luteal phase may be restricted to the non-conceptual cycle. If this process occurred at an equivalent period in a conceptual cycle decidualisation would be expected to be observed within 2 to 3 days of implantation and thus be inferred to fulfil an important role in early implantation. However, careful examination of early implantation sites has failed to detect decidualisation of the stroma even at 17 days post-conception (F. Leroy, personal communication; Hertig et al. 1956). It is therefore possible either that the hormonal milieu characterising the late luteal phase (i.e. when steroid hormone levels are falling – which does not happen if conception occurs) permits precocious differentiation of the stroma or that the trophoblast locally inhibits decidualisation. These observations imply that decidualisation and its associated secretory products may not have any role in early implantation and suggest that observations on protein-secretory changes during the late luteal phase must be interpreted with caution. It must also be considered whether the predecidual cells observed during the late luteal phase are biochemically identical to decidual cells of pregnancy endometrium. The suggestion that the glandular epithelium and its products may be primarily important during implantation and early pregnancy, whereas the decidual cell and its products are important at a later stage, may be supported by the observation that the majority of active glands involute towards the end of the first trimester. At term the differentiated endometrium contains few active glands and the decidual cell is the principal endometrial cell type.

Of great relevance to the interpretation of studies on protein secretion and its function during pregnancy are the anatomically defined regions of differentiated endometrium forming part of the developing fetoplacental unit. Initially three endometrial regions are apparent: the decidua basalis beneath the developing placenta, the decidua capsularis overlying the implanted embryo, and the decidua

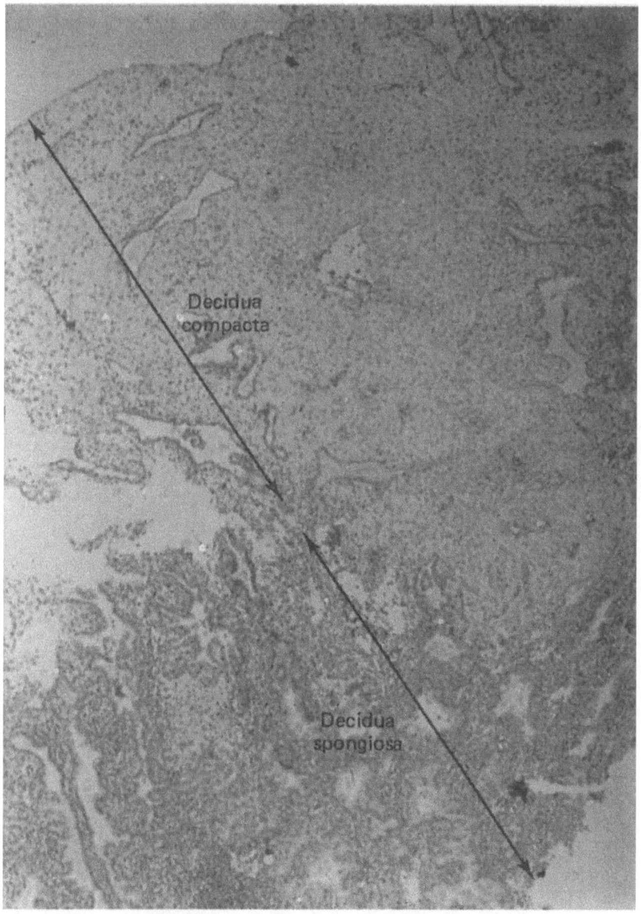

a

Fig. 7.1. a–e Human pregnancy endometrium during first trimester. Sections of a decidua parietalis specimen obtained on week 8–9 post-LMP illustrating the two histologically apparent areas: the decidua compacta and decidua spongiosa (**a**). The decidua compacta containing decidualised stroma with attenuated glands (**b**, **c**) is equivalent to the zona compacta of the menstrual cycle endometrium and is here referred to as decidualised endometrium. The decidua spongiosa containing secretory glands and undifferentiated stroma (**d**, **e**) is equivalent to the zona spongiosa of the menstrual cycle endometrium and is here referred to as secretory endometrium. (Reproduced from Bell 1985, with permission from Academic Press.) (Parts c–e are on pp.98 and 99)

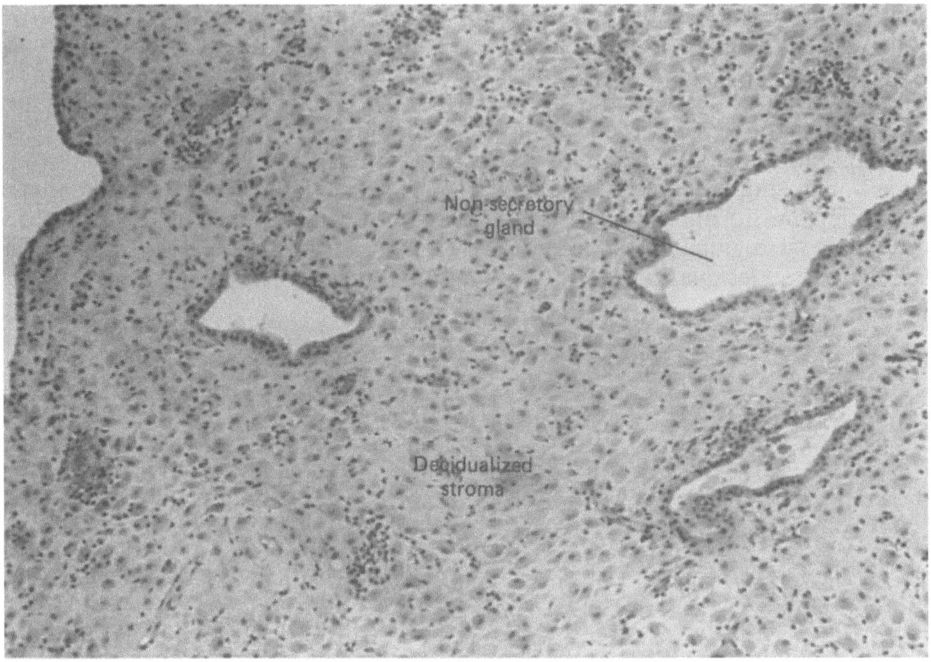

b

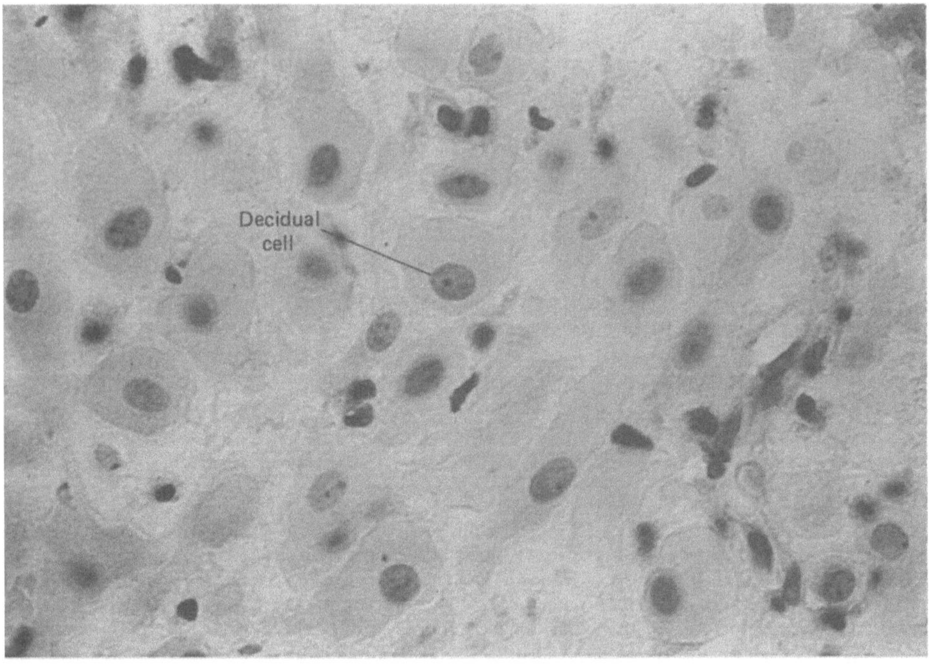

c

Fig. 7.1b and c. See p.97 for legend.

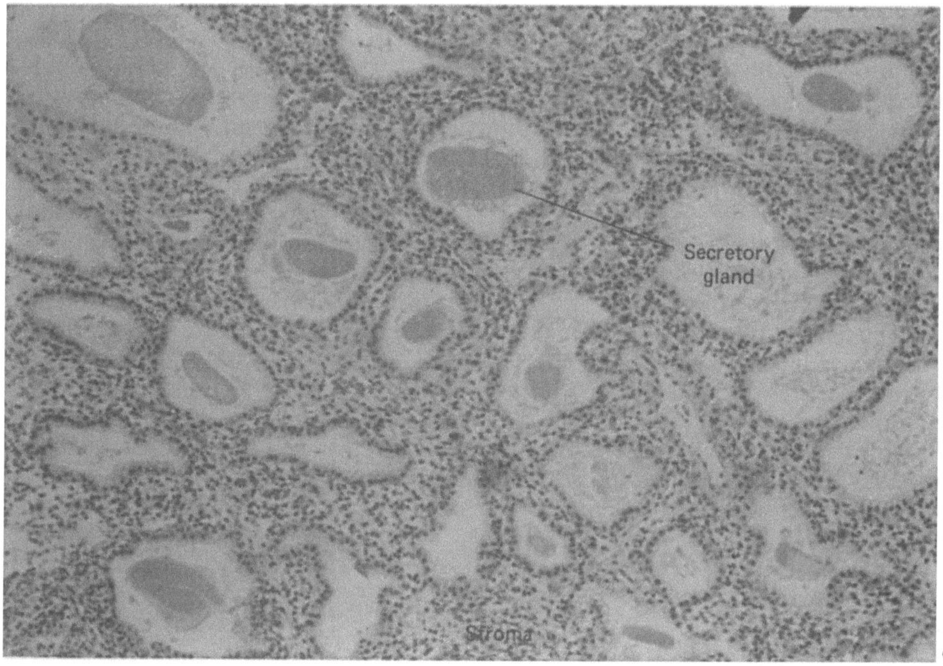

d

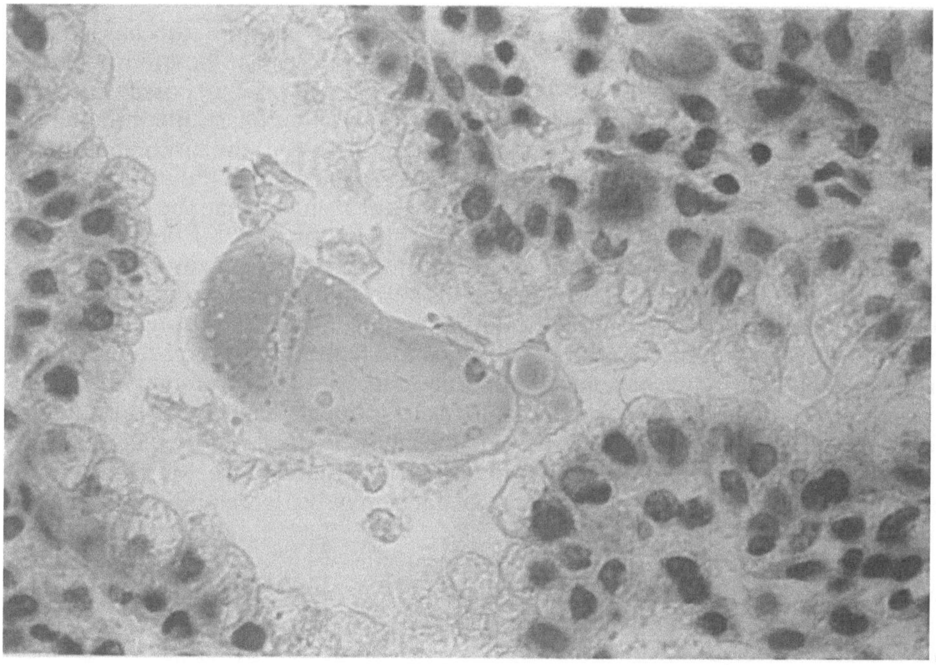

e

Fig. 7.1d and e. See p. 97 for legend.

parietalis in the non-implanted region of the uterus. As early pregnancy proceeds the capsularis region fuses with the opposing decidual parietalis, and the chorion, which has degenerated and contributes to the amniochorion, is brought into intimate contact with the decidua parietalis. Thus, during pregnancy two regions of the differentiated endometrium have developed with potentially different paracrine interactions with the fetoplacental unit – i.e. the decidua basalis associated with the basal plate of the placenta and the decidua parietalis with the amniochorion. Functional differences between these regions could indeed have contributed to differential fates of the chorionic trophoblast, with chorion frondosum developing into the placenta when in contact with the basalis but undergoing atrophy, to form the chorion laeve, when in contact with the superficial capsularis.

These points must be considered with endometrial tissue obtained either during the first trimester or at term (when it is normally represented by the parietalis). Also, endometrial protein synthesis in vivo may be regulated, possibly irreversibly, by the effects of the adjacent trophoblastic tissues. In vitro, however, endometrial tissue is free from the influence of the trophoblast and the pattern of secretion may not reflect that observed in vivo.

Synthesis and Secretion of Proteins During the Menstrual Cycle and Pregnancy: Comprehensive Approach

There are a number of experimental approaches to attempt to characterise proteins secreted by the endometrium. Given the success in identifying uteroglobin in the rabbit (Beier 1982) and the emphasis on the possibility that endometrial secretory proteins may contribute to an alteration in the intrauterine milieu, one approach has been to examine proteins in the uterine luminal flushings and fluid during the menstrual cycle (Schumacher et al. 1979). However, although there are problems with this approach, in that the majority of proteins may not be of endometrial origin, this method is still being pursued (MacLaughlin et al. 1986). A more direct approach is to use acids and methods with high resolving power to detect proteins secreted in vitro by tissue explants during short-term incubation. The main advantages of this approach are that proteins detected represent true products of the tissue, and the proteins would be defined by their physicochemical properties and not by their immunoreactivity. Although one disadvantage of this approach is that it can detect only major products of the tissue (whereas some functions may be mediated by quantitatively minor products such as prolactin), this may be an advantage for the identification of clinical markers, since major products may be detectable in peripheral serum.

This approach has been used to study the profile of endometrial synthesis and secretion in the menstrual cycle (Strinden and Shapiro 1983; Bell et al. 1986a; Heffner et al. 1986) and during pregnancy (Bell et al. 1985a). In two of these studies the effect of short-term exposure to progesterone in vitro was also examined (Strinden and Shapiro 1983; Bell et al. 1986a). It may be concluded from these studies that the synthesis and secretion of no one protein species is

Table 7.1. Major endometrial secretory proteins synthesised and secreted in vitro. (Reproduced from Bell and Smith 1987, with permission from Butterworth Scientific Ltd)

Group	Name of protein and possible equivalents	Molecular weight (kd)	Effect of progesterone in vitro (short-term)	Association with histological type
I	EP6	117		
	EP12	38		
II	EP13	33		Proliferative
	EP14	32(36)		Decidualised
	EP15	28(56)		Secretory
	EP10[a]	50		Decidualised
III	S1	130	+	
	E9, S2	54–58	–	
	EP11, S3	45–50	+	
	S4	35	+	
	S5	28	–	

S1-S5 described by Strinden and Shapiro (1983), using [^{3}H]-leucine as radiolabel.
[a] EP10 apparent as a major secretory protein when [^{3}H]-leucine is used as a radiolabel.

uniquely associated with any phase of the cycle or pregnancy but that the rates of synthesis and secretion change dramatically in certain species. On the basis of these observations a classification of these protein species has been attempted which may indicate a differential value in scientific and clinical applications (Bell et al 1986a) (Table 7.1). The majority of proteins, group I endometrial proteins, showed no dramatic alterations in observed rates of synthesis and secretion in vitro. The group II proteins were defined by the observed enhanced rates of synthesis and secretion associated with endometrial tissue obtained at different physiological stages. Of particular interest were the alterations in endometrial proteins 14 and 15, since at particular stages they represented the major secretory product of the endometrium (Bell et al 1985a, 1986a). Endometrial protein 15 (EP15) was the major protein product synthesised and secreted by the endometrium during the latter half of the mid-luteal and late luteal phase of the menstrual cycle and early first trimester of pregnancy. This protein was not observed by Strinden and Shapiro (1983), whose studies ended, however, on day 18 after the last menstrual period (LMP), before substantial EP15 synthesis has begun. The major secretory protein of the endometrium during the mid to late luteal phase, reported by Heffner et al. (1986), showed similar electrophoretic properties to EP15 in one and two dimensional polyacrylamide gel systems, but these authors suggested that the protein represented glycosylated prolactin (G-PRL). It is likely, however, that the properties of EP15 and G-PRL are similar in the gel system used and that G-PRL is actually a minor product. The pattern of synthesis of EP15 suggests that it is the major secretory protein of the zona spongiosa of the menstrual cycle and the decidua spongiosa (the latter being termed secretory endometrium) (Bell and Drife 1986) (Fig. 7.1). Endometrial protein 14 was never a major secretory protein during the menstrual cycle, although a significant increase in its rate of synthesis and secretion was noted in the final 2 days of the cycle when decidualisation of the zona compacta is most pronounced. The most dramatic alterations were observed during early pregnancy, when EP15 synthesis and secretion fell and EP14 synthesis and secretion increased to the extent that by week 15 post-LMP EP14 was the major secretory protein (Bell et al. 1985a). This profile, together with the report that rates of secretion of this protein by pregnancy endometrial specimens were positively correlated with the proportion of tissue represented by the decidua compacta and the finding that this region tended to be absent in earliest pregnancy endometrial specimens obtained (i.e., 6 weeks LMP) suggested that this protein was the major secretory protein of the decidual-cell-containing decidua compacta, termed decidualised endometrium (Bell and Drife 1986) (Fig. 7.1). This switch from EP15 to EP14 as the major secretory protein of the pregnancy endometrium may support previous discussion in as far as the decidualised endometrium may not make a major contribution to endometrial function during implantation in early pregnancy. The synthesis and secretion of group II proteins was not regulated by short-term in-vitro exposure to progesterone and may primarily be of value in monitoring the function of these histologically defined regions.

In contrast, the synthesis and secretion of group III proteins were modified both negatively and positively by short-term progesterone exposure (Table 7.1). Although no extensive studies on these proteins have been published, they merit attention because their scientific and clinical value will be different from that of group II endometrial proteins.

It is most relevant in this section to discuss the secretory proteins EP15 and 14 since they are quantitatively the most important proteins. Several studies have shown these proteins to be identical to a number of other proteins, including those that have been previously categorised as placental proteins. In some of those studies, however, the proteins have been isolated from different sources and used to raise antisera to develop assays. Since in many studies the proteins have been defined by such antisera and neither these nor their respective antigens have been standardised, the original terminology will be retained here.

Endometrial Protein 15

The major secretory protein during the latter half of the menstrual cycle and early first trimester of pregnancy was identified in vitro, using polyacrylamide gel systems, as a 28 kd glycoprotein exhibiting multiple Mr and pI forms (Bell et al. 1985a, 1986a). However, under native conditions it exists as its dimer of Mr 56 kd and average pI of 4.6 and has been referred to as pregnancy-associated α_2-globulin (α_2-PEG) (Bell et al. 1985b; Bell 1986b). Subsequent studies have implied identity of this protein with a number of others, including progestagen-dependent endometrial protein (PEP), an endometrial cytosolic protein, and placental protein 14 (PP14), although the original properties described for the latter protein were discrepant with those of α_2-PEG and PEP (Table 7.2). Reports have confirmed that PP14 is a secretory protein of the endometrium during the luteal phase of the cycle (Julkunen et al. 1986a) and at term (Julkunen 1986).

The original description that the synthesis and secretion of EP15 was dramatically increased during the mid-luteal phase, more specifically from days 19–21, and thereafter represented the major secretory protein product has been substantiated by other observations. A number of reports confirm that the endo-

Table 7.2. Suggested synonyms of pregnancy-associated endometrial α_2-globulin (α_2-PEG). (Reproduced from Bell 1986a with permission from IRL Press)

Name	Molecular weight	Tissue source	Immunochemical identity (reference)	Other references
Chorionic α_2-microglobulin (CAG-2) or placental specific α_2-microglobulin (PAMG-2)	25 kd	First trimester placental extract	1	5, 6, 7
Placental protein 14 (PP14)	42–43 kd	Term placental extract	2	8
Progestagen associated or dependent endometrial protein (PEP)	48 kd, subunit 27 kd	Pregnancy endometrium synthesis and extract		9
Alpha uterine protein (AUP)	50 kd, subunit 25 kd	Amniotic fluid	3	10
Pregnancy associated endometrial α_2-globulin (α_2-PEG) or endometrial protein 15 (EP15)	56 kd, subunit 28 kd	Pregnancy endometrium synthesis and secretion extract and culture medium	4	11, 12

References: (1) Bohn et al. 1982; (2) Julkunen et al. 1986c; (3) Sutcliffe et al. 1982; (4) Bell and Bohn 1986; (5) Petrunin et al. 1978; (6) Tatarinov et al. 1980; (7) Tatarinov 1982; (8) Bohn et al. 1982; (9) Joshi et al. 1980a; (10) Sutcliffe et al. 1980; (11) Bell et al. 1985a; (12) Bell et al. 1985b.

metrial content of this protein increases dramatically from the early to mid luteal phase (PEP [Joshi et al. 1981]) and throughout the luteal phase (PP14 [Julkunen et al. 1986a]). α_2-PEG content shows a similar increase, but the most dramatic rise in content occurs from the mid to late luteal phase during the final 2–3 days of the cycle (Bell et al., unpublished observations). PEP (Joshi et al. 1982) and PP14 (Julkunen et al. 1986b) have also been detected in peripheral serum during the menstrual cycle, albeit at lower levels than those recovered from intrauterine luminal flushings (Bell and Dore-Green 1987), the levels of which increased during the late luteal phase. We have examined a number of aspects of α_2-PEG synthesis and compartmental distribution in individual patients and have confirmed that endometrial content, luminal flushing content and peripheral serum levels are maximum during the late luteal phase (Bell, Kirwan, Patel and Jackson, unpublished observations). It is considered that the principal route of secretion of α_2-PEG is into the uterine lumen. It is improbable that the protein could have an important influence on the preimplantation blastocyst, but it could be involved in the early implantation process, unless the intra-uterine blastocyst itself could accelerate the onset of synthesis and secretion of the protein.

By means of polyclonal antisera (AUP [Horne et al 1982], PP 14 [Julkunen et al. 1986a]) the protein has been localised in the cycle to the glandular epithelium. We have also examined α_2-PEG localisation during the menstrual cycle with both polyclonal and monoclonal reagents (Waites et al. 1988) (Fig. 7.2). In the functional zone of the endometrium the profile of in-vitro synthesis and secretion and endometrial content was reflected by the intensity of staining in the glandular and luminal epithelium, the most intense staining being observed in the last 2–3 days of the cycle. These cell populations are considered to produce the phase-related fluctuations of synthesis and secretion. It is therefore apparent that α_2-PEG must represent quantitatively the major secretory protein product of the glandular epithelium during this period.

However, α_2-PEG positive glands were detected in the basal layer of the endometrium even in the follicular phase. It is suggested that this glandular population, which synthesizes α_2-PEG, may regulate the basal rates of synthesis and secretion observed during the cycle and the basal levels in serum. No direct evidence is available for the mechanism of regulation of production of this protein by the glandular epithelium in the cycle. However, the cycle-related dramatic increase of synthesis initiated at days 19–21 corresponds to the maximum steroid hormone levels, and the maximum levels of α_2-PEG are found during the period of falling steroid hormone levels at the end of the luteal phase. Synthesis of this protein must presumably be ultimately related to the progesterone-dependent differentiation of the glandular epithelium in the functional zone, since the increased serum levels of PP14 during the luteal phase are not observed during anovulatory cycles (Julkunen et al. 1986b).

During pregnancy α_2-PEG, as EP15, was detected as the major secretory protein synthesised in earliest specimens obtained (i.e., 6 to 8 weeks post LMP), after which in-vitro synthesis and secretion declined dramatically (Bell et al. 1985a). The rates of in-vitro synthesis and secretion of EP15 during the early first trimester were not dramatically different from that observed during the late luteal phase (Bell et al. 1986a). This was reflected by only a two-fold increase in content of this protein between these tissues (Bell et al. 1987). Joshi et al. (1981) reported a 24-fold increase in endometrial content of PEP from mid-secretory to

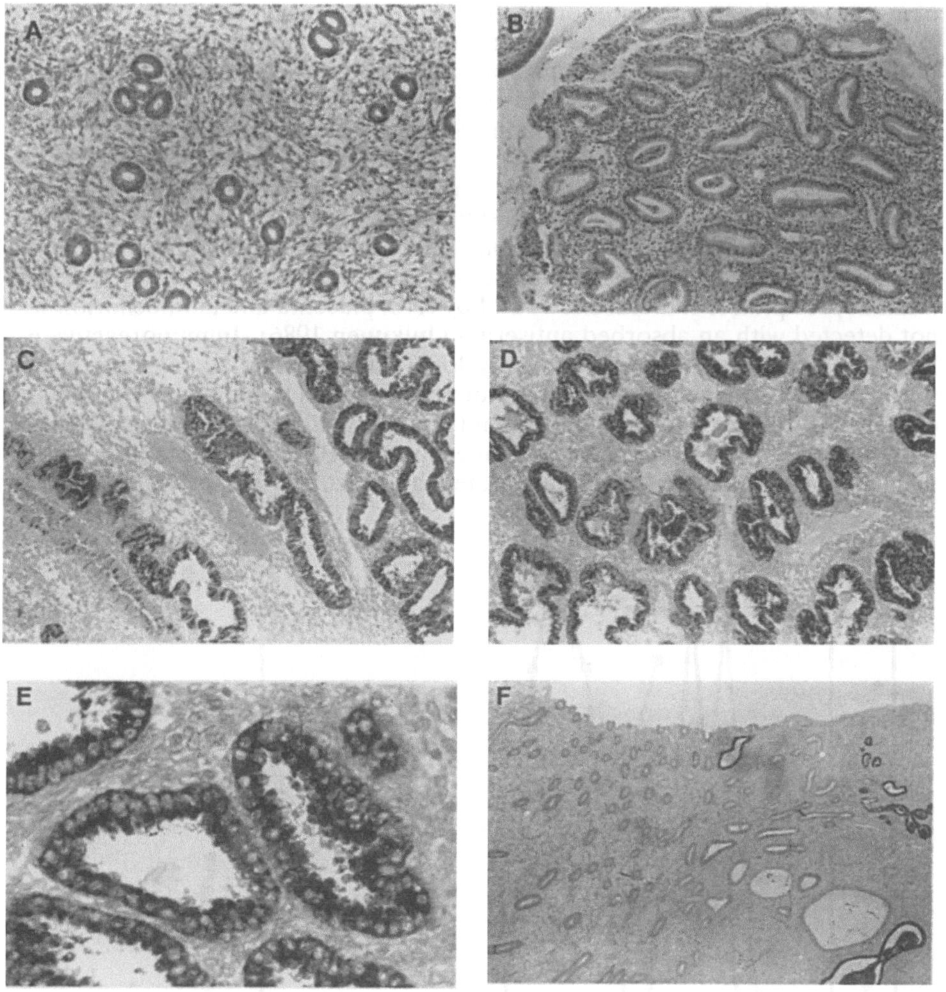

Fig. 7.2. Immunohistological localisation of α_2-PEG in the endometrium during the menstrual cycle employing monoclonal antibodies. A, mid-proliferative endometrium, day 8, × 82; B, early secretory endometrium, day 16, × 82; C, late secretory endometrium, day 24, × 82; D, late secretory endometrium, day 26, × 82; E, as for D except × 328; F, late proliferative endometrium from day 12 hysterectomy, × 26. (Reproduced from Waites et al. [1988] with permission from J.R.F. Ltd.)

first-trimester pregnancy endometrium; but, as we have described, the dramatic increase in endometrial content from the mid to late luteal phase accounts for the greatest proportion of this increase. The profile of levels of immunoreactive protein detected in serum during pregnancy reflects the pattern of in-vitro synthesis and secretion with peak levels detected at weeks 7–9 (PEP [Joshi et al. 1981, 1982], PP14 [Julkunen et al. 1985], α_2-PEG [Bell et al. 1987]). The substantially higher serum levels observed during the first trimester of pregnancy compared with those during the late luteal phase of the cycle, when similar rates of in-vitro synthesis and secretion are found, may be accounted for in part by the greater endometrial tissue mass present in the former situation. That the serum profiles of this protein during pregnancy reflect the pattern of endometrial synthesis (Fig. 7.3) suggests that this tissue represents the major, if not in fact the sole, tissue origin of the protein. Synthesis and secretion of α_2-PEG or immunoreactive α_2-PEG has not been detected by, or in, first-trimester trophoblastic tissue, term placental tissue or amniochorion (Bell, unpublished observations). In contrast, PP14 has been extracted from these tissues, albeit at low levels compared with the endometrium (Julkunen et al. 1985), but this may be due to the non-specificity of the antiserum used (Julkunen 1986). Secretion of immunoprecipitable radiolabelled PP14 by term placenta and amniochorion was not detected with an absorbed antiserum (Julkunen 1986). Immunoreactive α_2-PEG has not been detected with polyclonal and monoclonal reagents in any normal body tissue apart from the reproductive tract, a finding that supports the endometrial origin of the serum protein (Waites, Walker and Bell, unpublished observations). In the pregnancy endometrium, α_2-PEG is principally localised to glandular epithelium, as demonstrated by Mazurkiewicz et al (1981) with

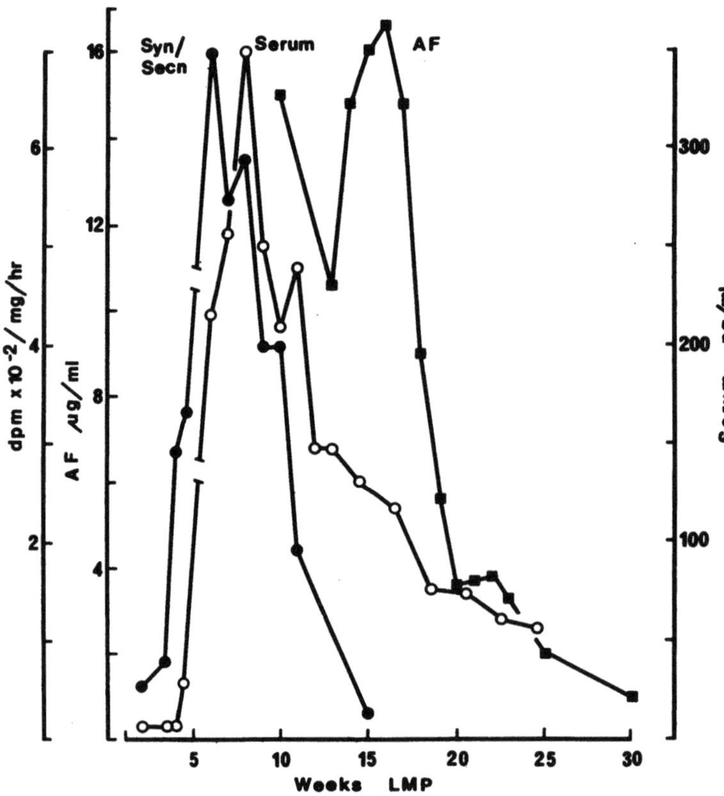

Fig. 7.3. Profiles of rates of in-vitro synthesis and secretion (●), levels in peripheral serum (○) and levels in amniotic fluid (■) of EP15/α_2-PEG. (Reproduced from Bell and Smith 1987, with permission from Butterworth Scientific Ltd.)

polyclonal antisera to PEP. However, we have observed that the highest staining intensity is localised to glandular epithelium in secretory glands of the decidua spongiosa regions, with less intensity in the glands of the decidua compacta and luminal epithelium. The serum profile and pattern of in-vitro synthesis and secretion therefore reflects the development and involution of the secretory gland-containing decidua spongiosa region, which involutes at the end of the first trimester. At term only a small number of α_2-PEG positive glands are observed, reflecting the low rate of synthesis at this time. These observations imply that the glandular epithelium is the principal source of secretory proteins in early first trimester pregnancy and that α_2-PEG is the major secretory protein. During pregnancy, as in the menstrual cycle, the principal route of secretion is towards the uterine lumen, as shown by the high levels of protein detected in amniotic fluid (PAMG-2 [Tatarinov et al. 1980], PP14 [Bohn et al. 1982; Julkunen et al. 1985], AUP [Sutcliffe et al. 1978], PEP [Joshi et al. 1980b]). In a detailed study α_2-PEG was detected consistently at maximum levels during weeks 10–15 and fell rapidly thereafter (Bell et al. 1986b) (Fig. 7.3). However, before this period specimens were either devoid of α_2-PEG or contained maximum levels. If the fetus does not contribute to amniotic fluid levels, α_2-PEG must be transported across the amniochorion, and this transport clearly depends on the fusion of the amnion and chorion.

Thus, EP15 is thought to be the major secretory protein product of the secretory glandular epithelium, the differentiated cell type of the zona spongiosa and decidua spongiosa, and as such provides a marker for this tissue and probably mediates the function of this cell type in implantation and early pregnancy. The failure to find an alternative tissue source of the protein also suggests that measurement of serum levels would provide an indicator of the functioning of this cell type. Since it is also the major secretory protein of the endometrium in early pregnancy, it is also suggested that during this period the principal function of the whole endometrium may be mediated by the secretory glandular epithelium and its major secretory protein. Synthesis and secretion of this protein must also occur in the male reproductive tract, since high levels of immunoreactive protein have been detected in seminal plasma (PAMG.2 [Petrunin et al. 1980], PP14 [Julkunen et al. 1984; Bolton et al. 1986], α_2-PEG [Bell and Patel 1987]). Discussion of the potential function of the protein must take this distribution into account.

Endometrial Protein 14

EP14, a minor secretory product of the endometrium during the menstrual cycle and early first trimester of pregnancy, is the major secretory protein by the end of the first trimester (Bell et al. 1985a, 1986a). Identified in polyacrylamide gel systems as a 32 kd protein of two pI forms, under native conditions it is a 36 kd protein and is termed pregnancy-associated endometrial α_1-globulin (α_1-PEG) (Bell et al. 1985b). Studies have now confirmed that this major secretory protein is identical to placental protein 12 (PP12), although the properties originally described for this protein were different from those of α_1-PEG (Table 7.3). Subsequently PP12 has been shown to be synthesised and secreted by endometrium during the menstrual cycle (Rutanen et al. 1986a), in early pregnancy (Rutanen et al. 1986b) and at term (Rutanen et al. 1985). It is now clear that

Table 7.3. Suggested synonyms of pregnancy associated endometrial α_1-globulin (α_1-PEG). (Reproduced from Bell 1986a, with permission from IRL Press)

Name	Molecular weight	Tissue source	Immunochemical identity (reference)	Other references
Chorionic α_1-microglobulin (CAG-1) or placental specific α_1-microglobulin (PAMG-1)	20kd	First trimester placental extract	1	3, 4, 5
Placental protein 12 (PP12)	25.2kd 51.5kd in SDS/PAGE	Term placental extract		1, 6
Pregnancy associated endometrial α_1-globulin (α_1-PEG) or endometrial protein 14 (EP 14)	36kd 32kd in SDS/PAGE	Pregnancy endometrium synthesis and secretion extract and culture	2	7, 8

References: (1) Bohn et al. 1982; (2) Bell and Bohn 1986; (3) Petrunin et al. 1978; (4) Tatarinov et al 1980; (5) Tatarinov 1982; (6) Bohn and Kraus 1980; (7) Bell et al. 1985a; (8) Bell et al. 1985b.

PP12 (Koistinen et al. 1986) and endometrial cytosolic and secreted α_1-PEG (Bell and Patel, unpublished observations) are identical to the amniotic-fluid somatomedin-binding protein (SmBP) (Chochinov et al. 1977; Nissley and Rechler, 1984). SmBP synthesis and secretion have been said to be a property of all cells and to occur concomitantly with somatomedin synthesis and secretion: binding with the binding protein neutralises activity and ensures that somatomedins act in an autocrine and/or paracrine manner to control growth (Hintz 1984; Nissley and Rechler 1984).

During the menstrual cycle the small elevation in synthesis and secretion during the late luteal phase (Bell et al. 1986a) may be associated with decidualisation of the zona compacta. This profile is supported by the report that endometrial content of PP12 is higher in late secretory than early secretory endometrium, although it is undetectable in proliferative endometrium (Rutanen et al. 1984a). Also, higher rates of release of PP12 by explants in vitro were observed in secretory than in proliferative endometrium, and progesterone stimulated secretion in both (Rutanen et al. 1986a). In the cycle, immunohistological studies have localised immunoreactive PP12 to the glandular epithelial cells of the secretory endometrium (Wahlström et al. 1982; Rutanen et al. 1984b), and Wahlström & Seppälä (1984) reported its appearance from day 4 postovulation. These observations may be discrepant with the suggested primary origin from decidualised endometrium (Bell and Drife 1986; Bell 1986a), and in studies with monoclonal antibodies to α_1-PEG (Bell, Patel and James, unpublished observations) the highest intensity of immunostaining has been observed in stromal cell populations during the final days of the late luteal phase (Waites, Walker and Bell, unpublished observations), corresponding to the period of elevated in-vitro synthesis and secretion (Bell et al. 1986a). Both immunoreactive PP12 (Rutanen et al. 1984a) and α_1-PEG are detectable in peripheral sera during the cycle, although no variation which corresponded to the fluctuations in secretion and content were observed. However, Rutanen et al. (1984a) found higher levels in the luteal phase. This discrepancy probably reflects the minor impact of the increased endometrial synthesis and secretion on peripheral serum levels of the low-molecular-weight form of SmBP found in serum of non-pregnant women and men, and with which it is immunochemically if not biochemically identical.

The temporal switch from EP15 (α_2-PEG) to EP14 (α_1-PEG) as the major secretory protein of the endometrium during weeks 6–12 suggested that α_1-PEG may be principally associated with the decidualised decidua compacta, the predominant form of the endometrium from this time until term (Bell and Drife 1986). That this protein is the major protein synthesised and secreted during early pregnancy is supported by the 77-fold increase in endometrial content of PP12 at this time (Rutanen et al. 1984b). PP12 could, however, be extracted from early placental tissue and from placenta, amnion and chorion at term (Rutanen et al. 1984c). These observations appeared to be supported by the immunohistological localisation of PP12 and PAMG-1 to the syncytiotrophoblast of the placenta (Inaba et al 1980; Wahlström et al. 1982; Tatarinov et al. 1980) and to the amnion and chorion (Inaba et al. 1980). However, studies with α_1-PEG suggest that these tissues are not major sites of synthesis, and in immunohistological studies with monoclonal antibodies no staining has been detected in these tissues. Principal staining was localised to decidual cells of the decidua compacta in endometrial specimens obtained during early pregnancy and at term, supporting this cellular origin of the protein within the endometrium (Waites, Walker and Bell, unpublished observations).

Immunoreactive PP12 (Rutanen et al. 1982; Than et al. 1983) and α_1-PEG (Bell and Patel, unpublished observations) have been detected in pregnancy serum, which at peak levels, achieved during weeks 16–23, are 5–9-fold higher than the non-pregnancy values. Initial increases over non-pregnancy values are observed at weeks 7–8, and after peak values are attained levels generally fall and again rise towards term. Much higher levels of this protein are detected in amniotic fluid and are highest during the second trimester (PAMG-1 [Tatarinov et al. 1980], PP12 [Bohn et al. 1982; Rutanen et al. 1982], α_1-PEG [Bell et al. 1986b]). (Fig. 7.4). In a study on α_1-PEG, highest levels were observed on weeks 20–24 and thereafter fell dramatically until term (Bell et al. 1986b). This detection of α_1-PEG in amniotic fluid is consistent with the identity of PP12/α_1-PEG as SmBP, since SmBP has been isolated from this source (Chochinov et al. 1977; Drop et al. 1984a, b; Povoa et al. 1984). How much the protein synthesised and secreted by the decidualised endometrium contributes to the levels in serum and amniotic fluid during pregnancy is uncertain. Even excluding the possibility

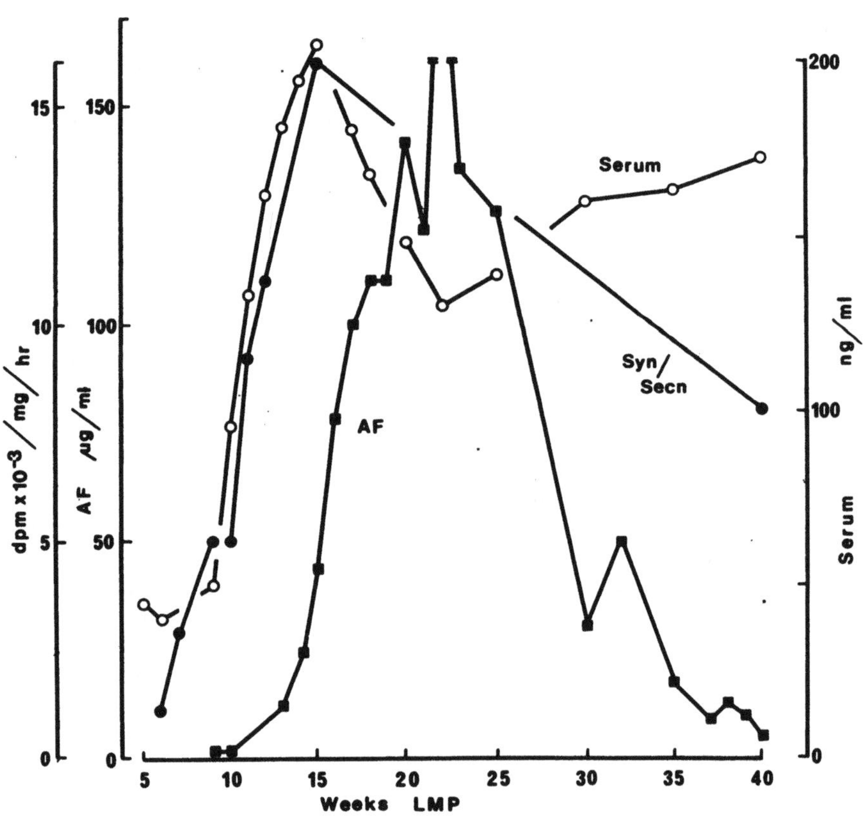

Fig. 7.4. Profiles of rates of in-vitro synthesis and secretion (●), levels in peripheral serum (○) and levels in amniotic fluid (■) of EP14/α_1-PEG. (Reproduced from Bell and Smith 1987, with permission from Butterworth Scientific Ltd.)

of altering production of SmBP from extrauterine sources during pregnancy, a contribution from the fetus must be considered, since the low-molecular-weight SmBP is the major SmBP circulating in the fetus (Nissley and Rechler 1984), and PP12 has been localised immunohistologically to fetal liver cells (Rutanen et al. 1984d). However, certain anti-PP12 sera have been demonstrated to contain unrelated reactivities (Fay et al. 1988). The comparison of the increase in serum and amniotic fluid levels during the end of the first trimester and peak levels during the second trimester with the pattern of in-vitro synthesis and secretion by the endometrium supports the concept that the endometrium is the primary source during this period (Fig. 7.4). Whether the levels of the protein in serum during the third trimester and at term are derived from the same source is uncertain.

In terms of subsequent discussion of function and potential clinical applications, it is suggested that at least during the late first and second trimester the decidual cell is the major source of low-molecular-weight SmBP (PP12/α_1-PEG) in serum, and measurement of levels would indicate decidual cell function. The profile also supports the hypothesis that in pregnancy decidualisation would not occur as early as implied by the appearance of decidualisation in the non-conceptual cycle. In terms of the reported ubiquitous nature of synthesis and secretion of this protein by cells (Nissley and Rechler 1984), then the unique feature of the decidual cell, implied by studies in vitro (Bell et al. 1985a), is that SmBP represents its major secretory protein.

Synthesis and Secretion of Proteins During the Menstrual Cycle and Pregnancy: Other Products

The identification of other secretory products of the endometrium (Table 7.4) has not arisen from a comprehensive analysis and therefore these products may not represent quantitively major products. However, this does not imply that they are not important in mediating functions of the tissue. It must also be considered that in the absence of either identification of specific mRNA in the

Table 7.4. Other proteins suggested or demonstrated to represent secretory products of the endometrium (see text for references)

Prolactin
Relaxin
Diamine/polyamine oxidase
Alkaline phosphatase
PAPP-A
PP 15
PP 10
Basement membrane components
CA 125

tissue or incorporation of radiolabelled aminoacids into physicochemically identifiable protein their identification as secretory products synthesised de novo must remain unproven.

One of the best-characterised secretory products of the differentiated endometrium is prolactin. Its synthesis and secretion by the endometrium has been demonstrated during the menstrual cycle (Maslar and Riddick 1979), in early pregnancy (Maslar et al. 1980; Horvath et al. 1983) and at term (Riddick and Kusmik 1977). The decidual cells of the decidualised endometrium are considered to represent the cellular origin of prolactin (Markoff et al. 1983a; Braverman et al. 1984). Much evidence is available to support the concept that amniotic fluid levels of prolactin originate from the decidualised endometrium, but whether binding sites for prolactin identified on the chorion laeve (Herington et al. 1980; McWey et al. 1982) are involved in prolactin action on this tissue is unknown. The mechanism of regulation of synthesis and secretion of prolactin by the endometrium is different from that involved in the anterior pituitary and has been shown to involve an autocrine inhibitory 38-45 kd protein (Markoff et al. 1983b) and a placentally derived protein (Handwerger et al. 1983). Aspects of the endocrinology of endometrially derived prolactin have been the subject of a number of reviews (Bell 1986a; Healy 1984; Healy and Hodgen 1983; Riddick 1985; Riddick and Daly 1982). Recently, however, decidualised endometrium at term has been shown to synthesise and secrete glycosylated prolactin (G-PRL) as well as prolactin (Lee and Markoff 1986). This form of prolactin (Lewis et al. 1985) has also been detected in amniotic fluid, and during early pregnancy a greater proportion of prolactin exists in this form (cited in Lee and Markoff 1986). Related to this observation may be the report that during the menstrual cycle only the glycosylated form was detected by western blotting in endometrial incubation medium (Heffner et al. 1986). The full implications of these reports have yet to be assessed, since the function of G-PRL compared to prolactin is unknown. However, since available antisera do not distinguish between these forms, previous estimations of prolactin production probably measured both forms.

A second polypeptide hormone reported to represent a secretory product of the endometrium is relaxin (Bigazzi et al. 1980). Although immunohistological studies have implicated a number of cell types, including decidual cells, in its production (Yki-Jarvinen et al. 1985; Koay et al. 1985), at present synthesis has not been demonstrated unequivocally. However, positive staining of the decidualised endometrium for the connecting peptide, prorelaxin, is suggestive of de-novo synthesis (cited by Koay et al. 1986).

Diamine oxidase (also referred to as polyamine oxidase or histaminase), an enzyme whose activity in serum is elevated during pregnancy, is immunohistologically localised to cytoplasm of decidual cells at term (Weisburger et al. 1978). This cellular origin has been supported by the demonstration of secretion of enzyme by first trimester endometrial tissue in vitro (Holinka and Gurpide 1984). Diamine oxidase synthesised and secreted by the decidualised endometrium is considered to produce the serum levels during pregnancy, which achieve peak levels at 20–22 weeks, and also the activity detected in amniotic fluid (for reviews see Morgan 1983; Bell 1986a). Another enzyme secreted in vitro by the decidualised endometrium obtained during pregnancy is alkaline phosphatase. However, for both enzymes definitive biochemical evidence for de-novo synthesis and secretion is lacking.

A number of other placental proteins have been reassessed in terms of their tissue origin. Many reports have been concerned with the evidence supporting an endometrial origin of pregnancy associated plasma protein A (PAPP-A), where decidual cells have been implicated (Bischof et al. 1982, 1984; Schindler et al. 1984). However, definitive evidence (i.e., either isolation of specific mRNA or incorporation of radiolabelled aminoacid and subsequent physico-chemical characterisation of immunoprecipitable protein) is not available. Indeed, the specificity of certain antisera to PAPP-A has recently been questioned (Chemnitz et al. 1986), which is of critical importance if evidence of tissue origin is based principally on antibody dependent assays. Evidence has been presented that two previously designated placental proteins, placental proteins 5 and 10 (PP5, PP10), may represent proteins synthesised and secreted by the endometrium (Butzow et al. 1986; Inaba et al. 1980; Wahlström et al. 1982).

Finally, it is relevant to note reports concerning the endometrial origin of a range of high-molecular-weight proteins which may be involved in local processes within the endometrium and in general may not be detected in the peripheral circulation. Basement membrane components have been detected immunohistologically surrounding decidual cells in the endometrium and have been suggested to be derived from these cells (Wewer et al. 1985). A most interesting development has been the evidence supporting an endometrial origin for the CA125 antigen (Bischof et al. 1986). Raised serum levels of CA125 are observed during the first trimester (Seki et al. 1986), a finding which, in analogy with the serum profile of EP15, may reflect production from the secretory glands.

Discussion

It is apparent that a range of proteins is synthesised and secreted by the endometrium, and although none appears uniquely associated with any stage of endometrium proliferation and differentiation, the degree of alteration in their rates of synthesis and secretion suggests that they could mediate stage-related function. Only a few of the secretory protein products of this tissue have been characterised, and the evidence for de-novo synthesis is inconclusive. This is reflected by the lack of knowledge of in-vivo function and limited clinical research application. However, the synthesis and secretion of a number of these proteins, including the quantitively major products, probably indicates and reflects the function of certain cell types within the endometrium.

In terms of two of the major cells types within the endometrium, the secretory glandular epithelium and the decidualised stromal cell, two proteins – EP 15/α_2-PEG/PEP/PP14 and EP14/α_1-PEG/PP12/SmPB – are proposed to represent the major secretory proteins of these types, respectively. Given the conjecture concerning the time at which decidualisation occurs during a conceptual cycle and the profiles of synthesis and secretion of these proteins, it is apparent that the secretory glandular epithelium and its associated protein would be the principal mediator of endometrial function during implantation and the early first trimester. During this period the human endometrium would be analogous to the endometrium in species exhibiting epitheliochorial placentation, where endometrial differentiation is dominated by secretory glands and decidualisation does not occur. Since the glandular epithelium is thought to be the major or sole tissue source of this protein, serum measurement would reflect activity and functioning of this cell type. The decidual cell, associated with decidualisation, still represents an enigmatic cell, although it is suggested that EP14/α_1-PEG/PP12/SmBP is its major secretory protein. The profiles of synthesis and secretion suggest that decidualisation observed in the menstrual cycle is probably not associated with an equivalent period in a conceptual cycle and may be a later phenomenon. The problem with this protein, as with using prolactin as a decidual cell marker, is that alternative sources of synthesis and secretion exist during pregnancy, although it is proposed that the increased serum levels during the first and second trimesters do reflect function of this cell type. However, given the range of secretory proteins now reported to be derived from the histologically defined decidual cell – i.e. SmBP, prolactin, diamine oxidase – the possibility of heterogeneity in this cell population must be considered. Even if a single cell type produces this range of proteins, the finding of normal levels of a single secretory product must not be taken to imply that other functions of this cell type, mediated by other secretory products, are normal. It must be also considered whether histologically identical cells in the different anatomically defined regions of the endometrium in pregnancy actually synthesise and secrete an identical range of protein products.

Major fundamental questions need to be addressed concerning the synthesis and secretion of proteins by the endometrium, since basic knowledge of regulation up to the level of gene expression and differential tissue localisation will affect the interpretation of in-vivo function and the results of clinical studies. It is apparent from the results of recent investigations that the endometrium, rather than being a passive tissue during pregnancy, through its range of secretory protein products probably has a more active role than has hitherto been suspected.

Acknowledgements. The studies carried out at the Department of Obstetrics and Gynaecology, University of Leicester, were supported by grants from the Trent Regional Health Authority and the Medical Research Council.

References

Beato M, Beier R (1975) Binding of progesterone to the proteins of the uterine luminal fluid: identification of uteroglobin as the binding protein. Biochim Biophys Acta 392: 346–356

Beier HM (1982) In: Beier HM, Karlson P (eds) Protein and steroids in early pregnancy. Springer-Verlag, Berlin, pp 38–71

Bell SC (1983) Decidualization: regional differentiation and associated function. In: Finn, CA (ed) Oxford reviews of reproductive biology, vol 5, Oxford University Press, Oxford, pp 220–271

Bell SC (1985) Comparative aspects of decidualization in rodents and humans: cell types, secreted products and associated function. In: Edwards RG, Purdy J, Steptoe PC (eds) Implantation of the human embryo. Academic Press, London, pp 71–122

Bell SC (1986a) Secretory endometrial and decidual proteins: studies and clinical significance of a maternally derived group of pregnancy-associated serum proteins. Hum Reprod 1: 129–143

Bell SC (1986b) Purification of human secretory "pregnancy-associated endometrial α_2-globulin" from cytosol of first trimester pregnancy endometrium. Hum Reprod 1: 313–318

Bell SC, Bohn H (1986) Immunochemical and biochemical relationship between human pregnancy-associated secreted endometrial α_1- and α_2-globulins (α_1- and α_2-PEG) and the soluble placental proteins 12 and 14 (PP12 and PP14). Placenta 7: 283–294

Bell SC, Dore-Green F (1987) Detection and characterization of human secretory "pregnancy-associated" endometrial α_2-globulin (α_2-PEG) in uterine luminal fluid. J Reprod Immunol 11: 13–29

Bell SC, Drife JO (1986); Secretory proteins of the human uterine endometrium and decidua during menstrual cycle and pregnancy: characterization of pregnancy-associated endometrial α_1- and α-globulins (α_1- and α_2-PEG). In: Hau J (ed) Pregnancy proteins in animals. Walter de Gruyter, Berlin, pp 143–164

Bell SC, Patel SR (1987) Immunochemical detection, physicochemical characterization and levels of human secretory pregnancy-associated endometrial α_2-globulin (α_2-PEG) on seminal plasma in men. J Reprod Fertil 80: 31–42

Bell SC, Smith S (1987) The endometrium as a paracrine organ. In: Chamberlain GVP (ed) Contemporary obstetrics & gynaecology. Butterworths Scientific Ltd, London, pp 273–298

Bell SC, Hales MW, Patel SR, Kirwan PH, Drife JO (1985a) Protein synthesis and secretion by the human endometrium and decidua during early pregnancy. Br J Obstet Gynaecol 92: 793–803

Bell SC, Patel S, Hales MW, Kirwan PH, Drife JO (1985b) Immunochemical detection and characterization of pregnancy-associated endometrial α_1- and α_2-globulins secreted by the human endometrium. J Reprod Fertil 74: 261–270

Bell SC, Patel SR, Kirwan PH, Drife JO (1986a) Protein synthesis and secretion by the human endometrium during the menstrual cycle and the effect of progesterone in vitro. J Reprod Fertil 77: 221–231

Bell SC, Hales MW, Patel SR, Kirwan PH, Drife JO, Milford-Ward A (1986b) Amniotic fluid levels of secreted pregnancy-associated endometrial α_1- and α_2-globulins (α_1- and α_2-PEG). Br J Obstet Gynaecol 93: 909–915

Bell SC, Jackson J, Dore-Green F, Kirwan PH, Drife JO (1987) Development and validation of a radioimmunoassay for human secretory "pregnancy associated endometrial α_2-globulin" (α_2-PEG) and detection in serum during pregnancy. Hum Reprod 2: 389–397

Bigazzi M, Nardi E, Bruni P, Petrucci F (1980) Relaxin in human decidua. J Clin Endocrinol Metab 51: 939–941

Bischof P, Duberg S, Schindler AM, Obradovic D, Weil A, Faigaux R, Hermann WL, Sizonenko PC (1982) Endometrial and plasma concentrations of pregnancy-associated plasma protein-A (PAPP-A). Br J Obstet Gynaecol 89: 701–703

Bischof P, Duberg S, Sizonenko MT, Schindler AM, Hermann WL, Sizonenko PC (1984) In vitro production of pregnancy-associated plasma protein-A (PAPP-A) by trophoblast and decidua. Am J Obstet Gynecol 148: 13–19

Bischof P, Tserg L, Brioschi PA, Hermann WL (1986) Cancer antigen 125 is produced by human endometrial stromal cells. Hum Reprod 1: 423–426

Bohn H, Kraus W (1980) Isolierung und Charakterisierung eines neuen placentaspezifischen Proteins (PP12). Arch Gynecol 229: 279–291

Bohn H, Kraus W, Winckler W (1982) New soluble placental tissue proteins: their isolation, characterization, localization and quantification. Placenta suppl 4: 67–81

Bolton AE, Pinto-Furtado LG, Andrew CE, Chapman MG (1986) Measurement of the pregnancy-associated proteins, placental protein 14 and pregnancy-associated plasma protein A in human seminal plasma. Clin Reprod Fertil 4: 233–240

Bolton AE, Pockley AG, Clough KJ, Mowles EA, Stoker RJ, Westwood OMR, Chapman MG (1987) Identification of placental protein 14 as an immunosuppressive factor in human reproduction. Lancet i: 593–595

Braverman MB, Bagni A, DeZiegler D, Den T, Gurpide E (1984) Isolation of prolactin-producing cells from first and second trimester decidua. J Clin Endocrinol Metab 58: 521–525

Bulmer JN, Sunderland CA (1984) Immunohistological characterization of lymphoid cell populations in the early human placental bed. Immunology 52: 349–357

Butzow R, Aythan H, Julkunen M, Rutanen E-M, Bohn H, Seppälä M (1986) Human endometrium and menstrual fluid contain placental protein S (PP15). Hum Reprod 1: 287–290

Chemnitz J, Folkersen, J, Teisner B, Sinosich MJ, Tornehave D, Westergaard JG, Bolton AE, Grudzinskas JG (1986) Comparison of different antibody preparations against pregnancy-associated plasma protein-A (PAPP-A) for use in localization and immunoassay studies. Br J Obstet Gynaecol 93: 916–923

Chochinov RH, Mariz IK, Hajek AS, Daughaday WH (1977) Characterization of a protein in mid-term amniotic fluid which reacts in the somatomedin C-radioreceptor assay. J Clin Endocrinol Metab 44: 902–908

Drop SLS, Kortleve DJ, Guyda HJ, Posner BI (1984a) Immunoassay of a somatomedin-binding protein from human amniotic fluid: levels in fetal, neonatal and adult sera. J Clin Endocrinol Metab 59: 908–915

Drop SLS, Kortleve DJ, Guyda HJ (1984b) Isolation of a somatomedin-binding protein from preterm anniotic fluid. Development of a radioimmunoassay. J Clin Endocrinol Metab 59: 899–907

Fay T, Jacobs I, Teisner B et al. (1988) Two fetal antigens (FA-1 and FA-2) and endometrial proteins (PP12 and PP14) isolated from amniotic fluid: characterization and distribution in fetal and maternal tissues. Eur J Obstet Gynaecol Reprod Biol (in press)

Handwerger S, Barry S, Markoff E, Barrett J, Conn PM (1983) Stimulation of the synthesis and release of decidua prolactin by a placental polypeptide. Endocrinology 112: 1370–1374

Healy DL (1984) The clinical significance of endometrial prolactin. Aust NZ J Obstet Gynaecol 24: 111–116

Healy DL, Hodgen GD (1983) The endocrinology of human endometrium. Obstet Gynecol Survey 38: 509–530

Heffner LJ, Iddenden DA, Lyttle CR (1986) Electrophoretic analyses of secreted human endometrial proteins: identification and characterization of luteal phase prolactin. J Clin Endocrinol Metab 62: 1288–1295

Herington AC Graham J, Healy DL (1980) The presence of lactogen receptors in human chorion laeve. J Clin Endocrinol Metab 51: 1466–1468

Hertig AT, Rock J, Adams EC (1956). A description of 34 ova within the first 17 days of development. Am J Anat 98: 435–494

Hintz RL (1984) Plasma forms of somatomedin and the binding protein phenomenon. Clins Endocrinol Metab 13: 31–42

Holinka CF, Gurpide E (1984) Diamine oxidase activity in human decidua and endometrium. Am J Obstet Gynaecol 150: 359–363

Horne CHW, Paterson WF, Sutcliffe RG (1982) Localization of alpha uterine protein in human endometrium. J Reprod Fertil 65: 447–450

Horvath K, Falkay G, Sas M (1983) Prolactin synthesizing capacity of the human decidua during the first trimester of pregnancy. Acta Med Hung 40: 1–6

Inaba N, Renk T, Bohn H (1980) Immunohistochemical localization of placental proteins (PP8, 9, 10, 11, 12) in human term placentae. Arch Gynecol 230: 109–121

Joshi SG, Ebert KM, Swartz DP (1980a) Detection and synthesis of a progestagen-dependent protein in human endometrium. J Reprod Fertil 59: 273–285

Joshi SG, Smith RA, Stokes DK (1980b) A progestagen-dependent endometrial protein in human amniotic fluid. J Reprod Fertil 60: 317–321

Joshi SG, Bank JF, Szarowski DH (1981) Radioimmunoassay for a progestagen-associated protein of the human endometrium. J Clin Endocrinol Metab 52: 1185–1192

Joshi SG, Bank JF, Henriques ES, Makarachi A, Matties G (1982) Serum levels of a progestagen-associated endometrial protein during the menstrual cycle and pregnancy. J Clin Endocrinol Metab 55: 642–648

Julkunen M (1986) Human decidua synthesizes placental protein 14 (PP14) in vitro. Acta Endocrinol 112: 271–277

Julkunen M, Wahlström T, Seppälä M, Koistinen R, Koskimies A, Stenman U–H, Bohn H (1984) Detection and localization of placental protein 14-like protein in human seminal plasma and in the male genital tract. Arch Androl 12 (suppl): 59–67

Julkunen M, Rutanen E–M, Koskimies A, Ranta T, Bohn H, Seppälä M (1985) Distribution of placental protein 14 in tissues and body fluids during pregnancy. Br J Obstet Gynaecol 92: 1145–1151

Julkunen M, Koistinen R, Sjoberg J, Rutanen E–M, Wahlstrom T, Seppälä M (1986a) Secretory endometrium synthesizes placental protein 14. Endocrinology 118: 1782–1786

Julkunen M, Apter D, Seppälä M, Stenman U–H, Bohn H (1986b) Serum levels of placental protein 14 reflect ovulation in nonconceptual menstrual cycles. Fertil Steril 45: 47–50

Julkunen M, Raiker RS, Joshir SG, Bohn H, Seppälä M (1986c) Placental protein 14 and progestagen-dependent endometrial protein are immunologically indistinguishable. Hum Reprod 1: 6–8

Koay ESC, Bagnell CA, Bryant-Greenwood GD, Lord SB, Cruz AC, Larkin LH (1985) Immunocytochemical localization of relaxin in human decidua and placenta. J Clin Endocrinol Metab 60: 859–863

Koay ESC, Bryant-Greenwood GD, Yamamoto SY, Greenwood FC (1986) The human fetal membranes; a target tissue for relaxin. J Clin Endocrinol Metab 62: 513–521

Koistinen R, Kalkkinin N, Huhtala M–L, Seppälä M, Bohn H, Rutanen E–M (1986) Placental protein 12 is a decidual protein that binds somatomedin and has an identical N-terminal amino acid sequence with somatomedin-binding protein from human amniotic fluid. Endocrinology 118: 1375–1378

Lee DW, Markoff E (1986) Synthesis and release of glycosylated prolactin by human decidua in vitro. J Clin Endocrinol Metab 62: 990–994

Lewis UJ, Singh RNP, Sinha YN, Vanderlean WP (1985) Glycosylated human prolactin. Endocrinology 116: 359–363

Maclaughlin DT, Santoro NF, Bauer HH, Lawrence D, Richardson GS (1986) Two-dimensional gel electrophoresis of endometrial protein in human uterine fluids: qualitative and quantitative analysis. Biol Reprod 34: 579–585

McWey LA, Singhas CA, Rogol AD (1982) Prolactin binding sites on human chorion-decidua tissue. Am J Obstet Gynecol 144: 435–438

Markoff E, Howell S, Handwerger S (1983a) Inhibition of decidual prolactin release by a decidual peptide. J Clin Endocrinol Metab 57: 1282–1286

Markoff E, Zeitler P, Peleg S, Handwerger S (1983b) Characterization of the synthesis and release of prolactin by an enriched fraction of human decidual cells. J Clin Endocrinol Metab 56: 962–968

Maslar IA, Riddick DH (1979) Prolactin production by human endometrium during the normal menstrual cycle. J Obstet Gynecol 135: 751–754

Maslar IA, Kaplan BA, Luciano AA, Riddick DH (1980) Prolactin production by the endometrium of early pregnancy. J Clin Endocrinol Metab 51: 78–83

Mazurkiewicz JE, Bank JF, Joshi SG (1981) Immunocytochemical localization of a progestagen-associated endometrial protein in the human decidua. J Clin Endocrinol Metab 52: 1006–1008

Morgan DML (1983) Polyamine oxidation and human pregnancy. Adv Polyamine Res 4: 155–167

Nissley SP, Rechler MM (1984) Insulin-like growth factors: biosynthesis, receptors and carrier proteins. In: Hoa Li C (ed) Hormonal proteins and peptides, vol XII. Academic Press, New York, pp 127–203

Petrunin DD, Gryaznova IM, Petrunina YA, Tatarinov YS (1978) Comparative immunochemical and physiochemical characteristics of human chorionic α_1- and α_2-microglobulins. Bull Exp Biol Med USSR 5: 658–661

Petrunin DD, Kozljaeva GA, Mesrjankina NV, Shevchenko OP (1980) Detection of chorionic α_2-microglobulin in the endometrium in the secretory phase of the menstrual cycle and in the male sperm. Akusherstvoi Gynekologija USSR 3: 22–23

Povoa G, Enberg G, Jornvall H, Hall K (1984) Isolation and characterization of a somatomedin-binding protein from mid-term human amniotic fluid. Eur J Biochem 144: 199–204

Riddick DH (1985) Regulation and physiological relevance of non-pituitary prolactin. In: MacLeod RM, Thomer MO, Scapagrini V (eds) Prolactin basic and clinical correlates. Liviana Press, Padova, pp 463–473

Riddick DH, Daly DC (1982) Decidual prolactin production in human gestation. Semin Perinatol 6: 229–237

Riddick DH, Kusmik WF (1977) Decidua: a possible source of amniotic fluid prolactin. Am J Obstet Gynecol 127: 187–190

Robertson WB (1981) The endometrium. Butterworths, London

Rutanen E-M, Bohn H, Seppälä M (1982) Radioimmunoassay of placental protein 12: levels in amniotic fluid, cord blood, and serum of healthy adults, pregnant women and patients with trophoblastic disease. Am J Obstet Gynaecol 144: 460–463

Rutanen E-M, Koistinen R, Wahlstrom T, Sjoberg J, Stenman U-H, Seppälä M (1984a) Placental protein 12 (PP12) in the human endometrium: tissue concentration in relation to histology and serum levels of PP12, progesterone and oestradiol. Br J Obstet Gyaecol 91: 377–381

Rutanen E-M, Koistinen R, Wahlstrom T, Stenman V-H, Seppälä M (1984b) Placental protein (PP12) in the menstrual fluid. Br J Obstet Gynaecol 91: 1025–1030

Rutanen E-M, Koistinen R, Wahlstrom T, Bohn H, Seppälä M (1984c) The content of placental protein 12 in decidua and fetal membranes is greater than in placenta. Br J Obstet Gynaecol 91: 1240–1244

Rutanen E-M, Wahlstrom T, Koistinen R, Sipponen P, Jalonko H, Seppälä M (1984d) Placental protein 12 (PP12) in primary liver cancer and cirrhosis. Tumour Biol 5: 95

Rutanen E-M, Koistinen R, Wahlstrom T, Bohn H, Ranta T, Seppälä M (1985) Synthesis of placental protein 12 by human decidua. Endocrinology 116: 1304–1309

Rutanen E-M, Koistinen R, Sjoberg J, et al. (1986a) Synthesis of placental protein 12 by human endometrium. Endocrinology 118: 1067–1071

Rutanen E-M, Menabawey M, Isaka K, Bohn J, Chard T, Grudzinskas JG (1986b) Synthesis of placental protein 12 by decidua from early pregnancy. J Clin Endocrinol Metab 63: 675–679

Schindler AM, Bordignon P, Bischof P (1984) Immunohistological localization of pregnancy-associated plasma protein-A in decidua and trophoblast: comparison with HCG and fibrin. Placenta 5: 227–236

Schumacher GFB, Holt JA, Reale F (1979) Approaches to the analysis of human endometrial secretions. In: Beller FK, Schumacher GFB (eds) The biology of the fluids of the female genital tract. Elsevier, Amsterdam, p 115

Seki K, Kikuchi Y, Uesato T, Kato K. (1986) Increased serum CA 125 levels during the first trimester of pregnancy. Acta Obstet Gynaecol Scand 65: 583–585

Strinden ST, Shapiro SS (1983) Progesterone-altered secretory proteins from cultured endometrium. Endocrinology 112: 862–870

Sutcliffe RG, Brock DJH, Nicholson LVB, Dunn E (1978) Fetal- and uterine-specific antigens in human amniotic fluid. J Reprod Fertil 54: 86–90

Sutcliffe RG, Bolton AE, Sharp F, Nicholson LVB, Mackinnon R (1980) Purification of human alpha uterine protein. J Reprod Fertil 58: 435–442

Sutcliffe RG, Joshi SG, Paterson WF, Bank JF (1982) Serological identity between human alpha uterine protein and human progestagen-dependent endometrial protein. J Reprod Fertil 65: 207–209

Tatarinov Yu S (1982) Pregnancy-specific $beta_1$-glycoprotein, placenta-specific $alpha_1$- and $alpha_2$-microglobulins. In: Grudzinskas JG, Teisner B, Seppälä M (eds) Pregnancy proteins: biology, chemistry and clinical application. Academic Press, Sydney, pp 179–194

Tatarinov Yu S, Kozljaeva GA, Petrunin DD, Petrunina Yu A (1980) Two new human placenta-specific α-globulins: identification, purification characteristics cellular localization and clinical investigation. In: Klopper A, Genazzi A, Crosignani PG (eds) The human placenta: proteins and hormones. Academic Press, London, pp 35–46

Than GN, Csaba IF, Szabo DG, Bognar ZJ, Aranz AA, Bohn H (1983) Levels of placenta-specific tissue protein 12 (PP12) in serum during normal pregnancy and in patients with trophoblastic tumour. Arch Gynaecol 234: 39–46

Wahlström T, Seppälä M (1984) Placental protein 12 (PP12) is induced in the endometrium by progesterone. Fert Steril 41: 781–784

Wahlström T, Bohn H, Seppälä M (1982) Immunohistochemical studies on pregnancy proteins. In: Grudzinskas JG, Teisner B, Seppälä M (eds) Pregnancy proteins: biology, chemistry and clinical applications. Academic Press, Sydney, pp 415–422

Waites GT, Wood PL, Walker RA, Bell SC (1988) Immunohistological localization of human endometrial secretory protein "pregnancy-associated endometrial α_2-globulin" (α_2-PEG) during the menstrual cycle. J Reprod Fertil 82: 665–672

Weisburger WR, Mendelsohn G, Eggleston JC, Baylin SB (1978) Immunohistological localization of histaminase (diamine oxidase) in decidual cells of human placenta. Lab Invest 38: 703–706

Wewer VM, Faber M, Liotta LA, Albrechtsen R (1985) Immunochemical and ultrastructured assessment of the nature of the pericellular basement membrane of human decidual cells. Lab Invest 53: 624–633

Yki-Jarvinin H, Wahlström T, Seppälä M. (1985) Human endometrium contains relaxin that is progesterone-dependent. Acta Obstet Gynecol Scand 64: 663–665

8. Characterisation, Biological Action and Clinical Studies of Endometrial Proteins

M.-L. Huhtala, M. Seppälä, M. Julkunen and R. Koistinen

Introduction

Placental proteins 12 and 14 (PP12 and PP14) were originally isolated from the placenta (Bohn and Kraus 1980; Bohn et al. 1982) but were later shown to derive predominantly from decidua and endometrium (Rutanen et al. 1986; Julkunen et al. 1986a–c). Both are abundant in amniotic fluid (Rutanen et al. 1982; Julkunen et al. 1985a, b). We have characterised human placental/decidual PP12 and PP14 and purified these proteins from human mid-trimester amniotic fluid. In this chapter we shall review these studies and our recent research on the biological action and clinical significance of PP12 and PP14.

Placental/Decidual PP12

PP12 was originally described as a glycoprotein with a molecular weight of 26.5–52 kd, containing 4.2% carbohydrate (Bohn and Kraus 1980). Two immunoreactive components were described: the major component was designated PP12A and the minor PP12B. Most reports dealing with PP12 refer to PP12A. We have further characterised these two forms. In sodium dodecyl sulphate polyacrylamide gel electrophoresis (SDS-PAGE), PP12A migrates as two bands. The major band has a molecular weight of 34 kd and the minor band 28 kd. Both are immunoreactive in western blot analysis. Only one N-terminal aminoacid sequence, APWQCAPCSADELAL, was obtained for PP12A. This sequence is identical with that of insulin-like growth-factor binding protein (IGF-bp) from human amniotic fluid (Povoa et al. 1984; Koistinen et al. 1986). Both the 34 kd and 28 kd forms of PP12A bind insulin-like growth factor 1 (IGF-1).

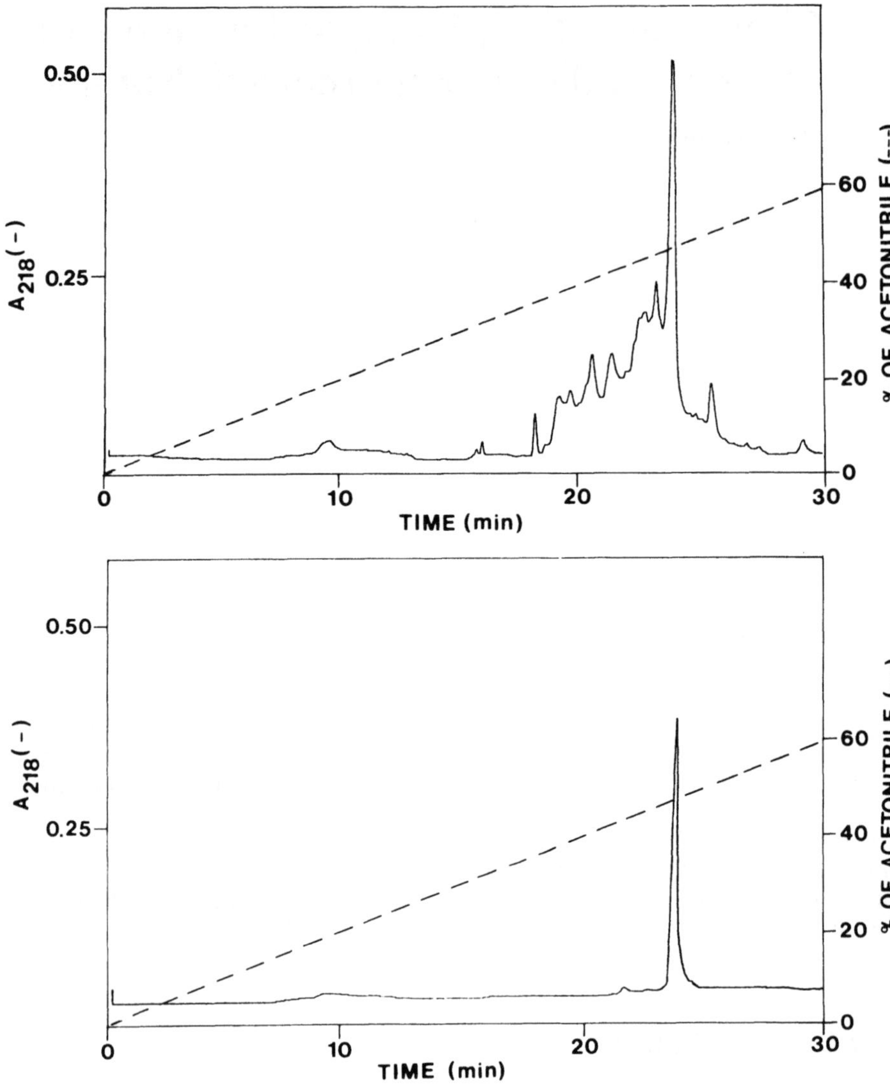

Fig. 8.1. *Upper panel:* HPLC of PP12B. The sample was injected into reversed-phase column and eluted with a gradient of acetonitrile in 0.1% trifluoroacetic acid (*broken line*). *Lower panel:* Rechromatography of the main peak fraction (no. 25). (Published, with permission, from Huhtala et al. 1986.)

PP12B was found to contain several PP12-immunoreactive components. After high-performance liquid chromatography (HPLC) (Fig. 8.1) the major protein peak was analysed in SDS-PAGE (Huhtala et al. 1986). It was homogeneous, with a molecular weight of 21 kd, under both reducing and non-reducing conditions (Fig. 8.2a). In western blot analysis it reacted with antibodies against

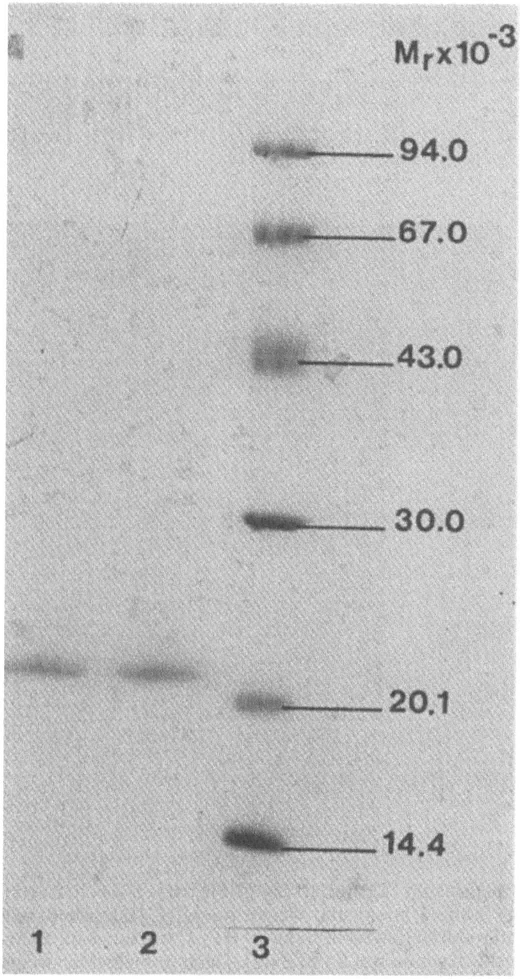

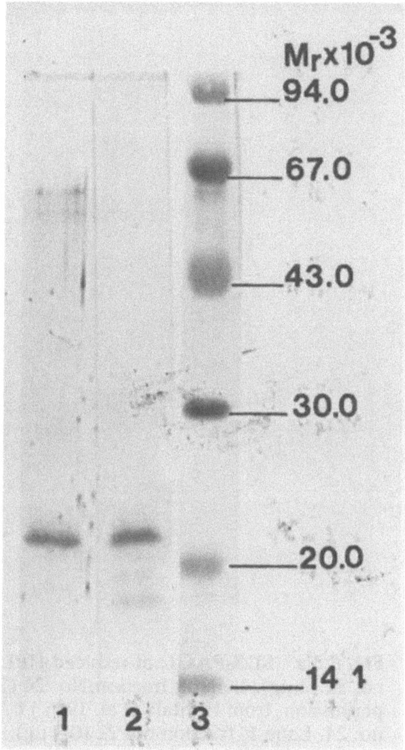

Fig. 8.2.a SDS-PAGE of HPLC fraction no. 25. Lane 1, 3 μg, non-reduced. Lane 2, 3 μg, reduced with 5% 2-mercaptoethanol. Lane 3, molecular weight markers. (Published, with permission, from Huhtala et al. 1986.) **b** Western blot analysis of fraction no. 25. Lane 1, HPLC-purified and reduced fraction no. 25 (0.2 μg) immunoblotted using anti-PP12A antiserum. Lane 2, HPLC-purified and reduced fraction no. 25 (0.2 μg) immunoblotted using anti-PP12B antiserum. Lane 3, molecular weight markers.

PP12A and PP12B, indicating that they share at least one antigenic determinant (Fig. 8.2b). This major HPLC fraction of PP12B binds IGF-1. Minor fractions were also found on HPLC. When analysed in SDS-PAGE they gave molecular weights from 17 to 20 kd (Fig. 8.3a). Again, these fragments were immunoreactive on western blot analysis (Fig. 8.3b). A single N-terminal sequence was obtained for all these various IGF-1-binding components. This sequence is identical to the one described for placental/decidual PP12A. Thus, purification of PP12 from human placental/decidual tissue results in several PP12-immunoreactive IGF-binding components, which share the N-terminal aminoacid

a

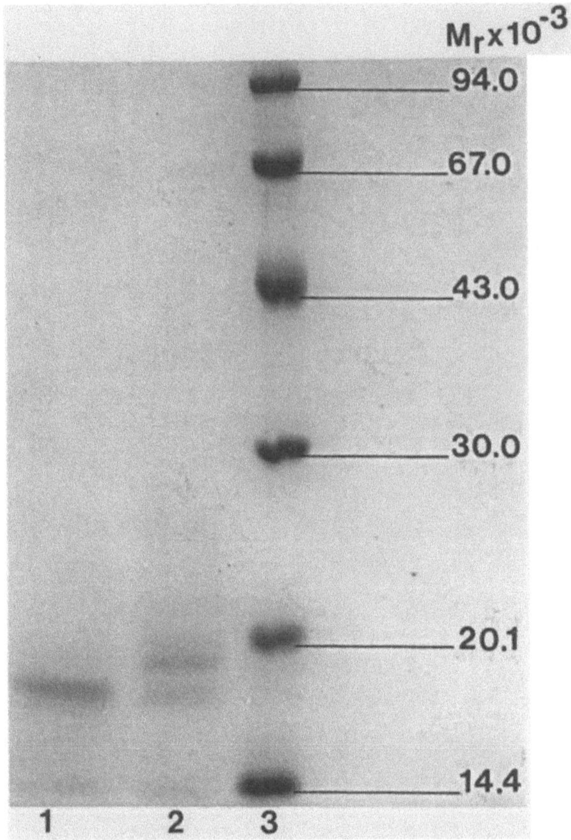

Fig. 8.3.a SDS-PAGE of reduced HPLC-fractions no. 22 and no. 24 of Fig. 8.1. Lane 1, fraction no. 22 (2 μg). Lane 2, fraction No. 24 (2 μg). Lane 3, molecular weight markers. (Published, with permission, from Huhtala et al. 1986.) **b** Western blot analysis of reduced HPLC fractions no. 22 and no. 24. Lane 1, fraction no. 22 (0.3 μg). Lane 2, fraction no. 24 (0.3 μg). Lane 3, molecular weight markers. (Published, with permission, from Huhtala et al. 1986.)

sequence with PP12A but differ from each other in molecular weight. These will be called PP12/IGF-bp here.

Amniotic Fluid PP12

Mid-trimester amniotic fluid was used for purification of PP12/IGF-bp. Purification was in three stages: gel filtration, hydrophobic interaction chromatography and ion-exchange chromatography. The final recovery was 21% (Table 8.1).

Purification was very fast, because concentration steps and buffer exchanges were omitted. After hydrophobic interaction chromatography only one band was observed at 34kd on SDS-PAGE. Purified amniotic fluid PP12 was immuno-

b

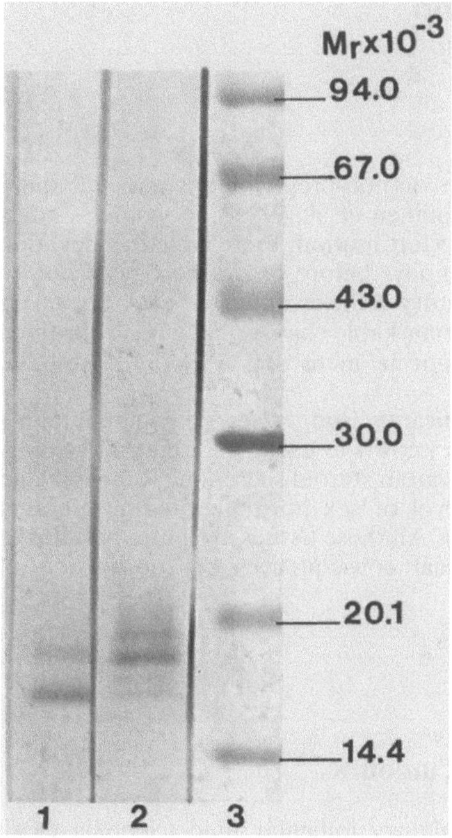

Fig. 8.3b. For legend see opposite.

Table 8.1. Purification of PP12 from human amniotic fluid. Data from Koistinen et al. (1987)

Purification step	Volume (ml)	PP12 (μg)	Recovery (%)
Amniotic fluid	25	1100	100
Sephadex G-100	25	690	63
TSKgel Phenyl-5PW	6	373	34
Mono Q	5	235	21

logically indistinguishable from placental PP12A, and only the 34kd band was obtained on SDS-PAGE. They had similar tryptic peptide maps and identical N-terminal aminoacid sequences (Koistinen et al. 1987). Also the amniotic fluid PP12 was found to bind IGF-1. The K_d for IGF-1-binding was 1 nM (Koistinen et al. 1987). This is of the same order of magnitude as that reported for the IGF-1 receptor (Marshall et al. 1974; Furlanetto and DiCarlo 1984).

Clinical Studies on PP12/IGF-bp

Serum Level of PP12/IGF-bp

PP12/IGF-bp can be detected in serum of non-pregnant women and men. Women have higher levels than men (Rutanen et al. 1982). In women undergoing ovarian hyperstimulation for in-vitro fertilisation, there is a clear elevation of serum PP12 concentration one or two days before or on the day of oocyte retrieval, when there are many preovulatory follicles and the oestrogen levels are high (Seppälä et al. 1987, 1988). No remarkable changes have been observed in serum PP12 concentrations during the normal menstrual cycle (Suikkari et al. 1987).

The levels of PP12/IGF-bp are significantly suppressed in women taking Danazol (Than et al. 1987). This drug is known to cause endometrial atrophy (Lind and Cooke 1976) and to suppress ovarian steroid secretion (Rannevik and Thorell 1984). It also suppresses the level of sex hormone-binding globulin produced by the liver (Nilsson et al. 1982). All these tissues are potential sites of PP12/IGF-bp synthesis. The physiological consequences of suppression of PP12/IGF-bp levels are not yet known.

PP12/IGF-bp in the Ovary and its Tumours

PP12/IGF-bp has been found in preovulatory follicular fluid (Seppälä et al. 1984). The PP12/IGF-bp concentration is higher in those follicles which also contain high oestrogen and progesterone concentrations. Follicular fluid PP12/IGF-bp is physicochemically similar to placental/decidual PP12A. (Koistinen 1984). Within the ovary, PP12/IGF-bp has been detected in luteinised granulosa cells and corpus luteum (Seppälä et al. 1984). Because the PP12/IGF-bp concentration in follicular fluid is about 10-fold higher than that in peripheral blood, it is likely that the protein is formed within the ovary, perhaps in the granulosa cells. Previous studies show that IGF-1 enhances induction of luteinising hormone receptors by follicle stimulating hormone in cultured rat granulosa cells (Adashi et al. 1985). If the same turns out to be true for human granulosa cells, then the role of IGF-bp in this process will be of great interest.

PP12/IGF-bp has been found in ovarian tumours. Benign tumours contain higher levels than malignant (Seppälä and Than 1988). Patients with ovarian cancer may occasionally have elevated serum PP12/IGF-bp levels, particularly those with advanced disease (Iino et al. 1986a–c). After radical surgery the level declines, indicating tumour-relatedness. It is not clear why the levels are higher in benign than malignant ovarian cyst fluids. The same has been observed on CA 125, a marker for ovarian cancer (de Bruijn et al. 1986). It is possible that tumour invasion through the basement membrane causes leakage of proteins into the peripheral circulation.

PP12/IGF-bp in the Endometrium

Studies with immunoperoxidase staining have localised PP12/IGF-bp to the secretory endometrium, where it appears on the 4th postovulatory day (Wahlström and Seppälä 1984). This has been confirmed by direct immunochemical measurement of PP12/IGF-bp in tissue (Rutanen et al. 1984a). The appearance of IGF-bp in the endometrium coincides with the time when proliferation ends and the secretory phase starts. Perhaps the high IGF-binding affinity of PP12/IGF-bp plays a part in this transition by competing for the growth factor with its receptor. Another role of interest is the part PP12/IGF-bp may play in implantation. PP12/IGF-bp appears in the endometrium a few days before implantation, and it is abundant in the endometrium in that phase of the menstrual cycle.

PP12/IGF-bp is rarely detectable in proliferative or atrophic endometrium, but it can be induced by treatment with progestagens (Wahlström and Seppälä 1984). The uterus of postmenopausal women is capable of producing IGF-bp, if sufficiently stimulated by steroids. We have studied the expression of PP12/IGF-bp in endometrial cancer, but the result has been negative in most cases (Seppälä et al. 1983).

PP12/IGF-bp in the Liver and its Tumours

PP12/IGF-bp is detectable with immunoperoxidase staining in fetal but not adult liver. Primary liver cancer cells are PP12/IGF-bp-positive (Rutanen et al. 1984b), and this is compatible with the finding of IGF-bp in a human hepatoma cell line (Povoa et al. 1985). The hyperactive nodules of cirrhosis are also PP12/IGF-bp-positive. Many patients with primary liver cancer have an elevated serum PP12/IGF-bp concentration, and the same has been occasionally observed in patients with other gastrointestinal malignancies. There is no correlation between serum PP12-IGF-bp and alphafetoprotein levels.

PP12/IGF-bp in Pregnancy

In early pregnancy the serum PP12-IGF-bp concentration may rise as early as 40 days after the last menstrual period, but there is great individual variation (Seppälä et al. 1987, 1988). At eight weeks' pregnancy the levels are consistently higher than in non-pregnant women (Rutanen et al. 1982). The levels continue to rise until weeks 22–24 and, after 28 weeks, they decline slightly. In pre-eclampsia serum PP12/IGF-bp concentration is significantly higher than in normal pregnancy (Iino et al. 1986a–c). This may indicate increased decidual contribution, although other sites may also be involved. According to Howell et al. (1985) maternal serum PP12/IGF-bp concentration is inversely correlated with the birth weight of the child, and we have confirmed this observation. Because there is a tendency to a positive correlation between the mothers' IGF-1 levels and the birth weight of their children (Lundin et al. 1985) it is tempting to speculate that increased secretion of IGF-bp decreases the bioavailability of a growth factor, leading to fetal growth retardation.

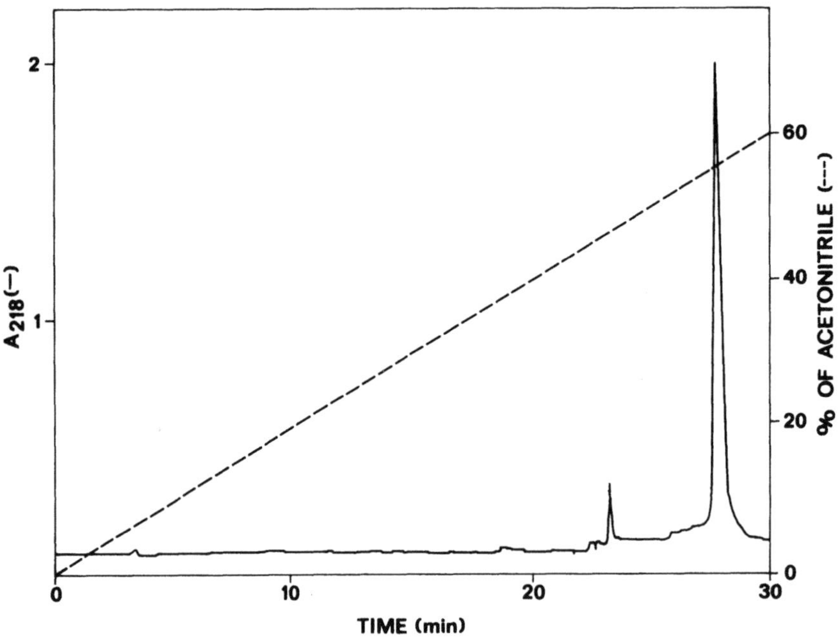

Fig. 8.4. Purification of placental /decidual PP14 on HPLC. The sample (200 μg) was injected onto a reverse-phase column and eluted with a gradient of acetonitrile in 0.1 % trifluoroacetic acid (*broken line*). (Reproduced, with permission, from Huhtala et al 1987.)

Placental Decidual PP14

PP14 was originally described as a glycoprotein with a molecular weight of 42–43kd (Bohn et al. 1982). PP14 contains 17.5% carbohydrate and binds to concanavalin-A (Bolton et al. 1983). Recent studies show that PP14, progestagen dependent endometrial protein (PEP) and α_2-pregnancy-associated endometrial globulin are immunologically related to each other (Julkunen et al. 1986c; Bell and Bohn 1986). We have further characterised placental/decidual PP14 with HPLC, SDS-PAGE and N-terminal sequence analysis (Huhtala et al. 1987). In HPLC the major protein peak consisting of PP14 eluted at 55% acetonitrile (Fig. 8.4). The minor peak eluting at 45% acetonitrile was not PP14-immunoreactive.

When the HPLC-purified PP14 was analysed on SDS-PAGE, a single band was observed with a molecular weight of 28 kd (Fig. 8.5a). This was immunoreactive on western blot analysis (Fig. 8.5b).

The same N-terminal sequence (Fig. 8.6) was obtained for the HPLC-purified and the SDS-PAGE-purified PP14 (Huhtala et al. 1987). This sequence has significant homology with horse, bovine, buffalo, sheep and goat β-lactoglobulins.

Fig. 8.6. Alignment of residues 1–22 of PP14 with residues 1–22 of β-lactoglobulins from horse (LGHO), cow (LGBO), buffalo (LGBUI), sheep (LGSH), and goat (LGGT). Boxes show identities and underlinings conservative mutations between PP14 and β-lactoglobulins.

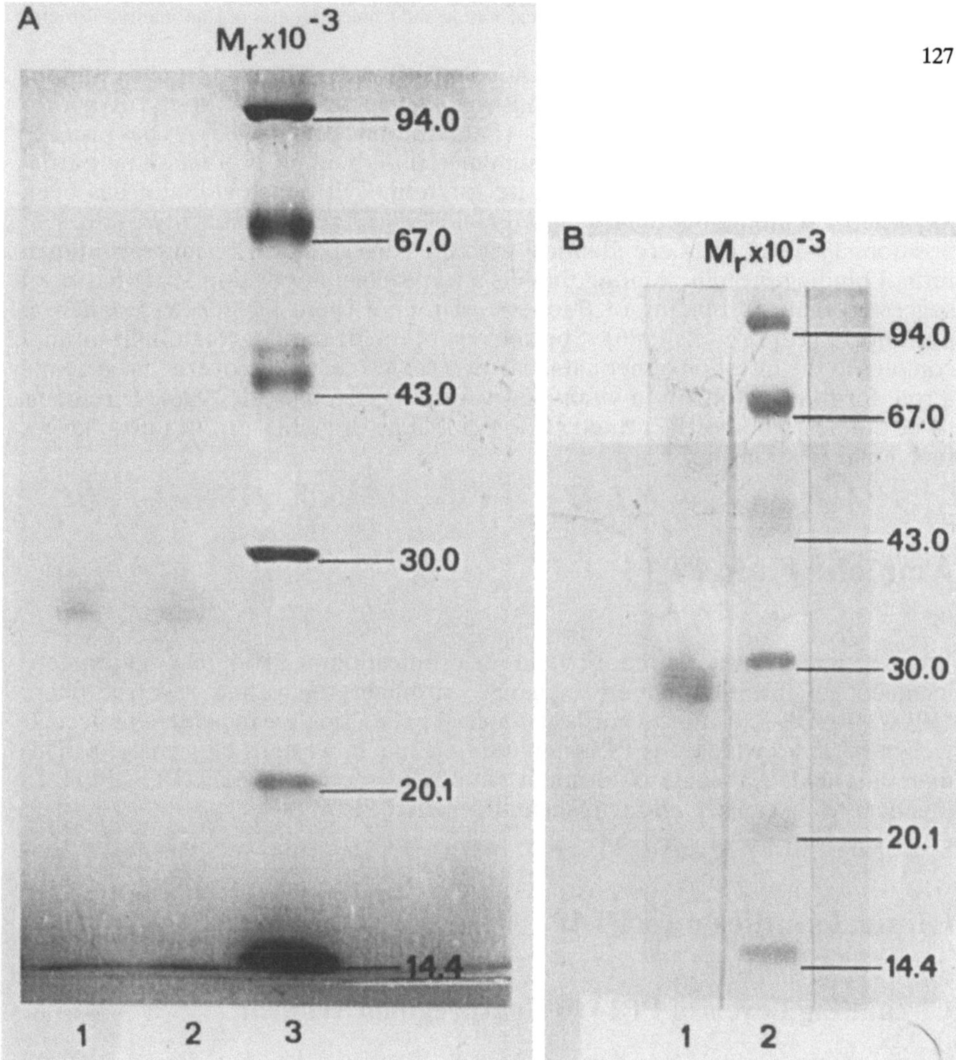

Fig. 8.5.a SDS-PAGE of purified PP14. Lane 1, PP14 (5 μg) non-reduced. Lane 2, PP14 (5 μg) reduced with 2-mercaptoethanol. Lane 3, molecular weight markers. (Reproduced, with permission, from Huhtala et al. 1987.) **b** Western blot analysis of purified PP14. Lane 1, purified and reduced PP14 (0.5 μg) immunoblotted with anti-PP14 antiserum. Lane 2, molecular weight markers. (Reproduced, with permission, from Huhtala et al. 1987.)

	1				5					10					15					20		
PP14	M	D	I	P	Q	T	K	Q	D	L	E	L	P	K	L	A	G	T	W	H	S	M
LGHO	T	N	I	P	Q	T	M	Q	D	L	D	L	Q	E	V	A	G	K	W	H	S	V
LGBO	L	I	V	T	Q	T	M	K	G	L	D	I	Q	K	V	A	G	T	W	Y	S	L
LGBUI	I	I	V	T	Q	T	M	K	G	L	D	I	Q	K	V	A	G	T	W	Y	S	L
LGSH	I	I	V	T	Q	T	M	K	G	L	D	I	Q	K	V	A	G	T	W	Y	S	L
LGGT	I	I	V	T	Q	T	M	K	G	L	D	I	Q	K	V	A	G	T	W	Y	S	L

β-lactoglobulins are the primary component of whey and are found in milk of many animal species. β-lactoglobulins are known to bind a variety of hydrophobic molecules including retinol (Pervaiz and Brew 1985). The primary structures of β-lactoglobulins and human retinol-binding protein show partial homology, and a possible binding site for retinol in β-lactoglobulin has been identified by model-building (Papiz et al. 1986). The aminoacid residues at positions 17–19 of PP14 are identical with the corresponding residues of human retinol-binding protein. Among them is a tryptophan at position 19. It has been suggested that the binding of β-ionone of retinol could include tryptophan at position 19 (Papiz et al. 1986). Specific receptors for the β-lactoglobulin-retinol complex in the intestine of neonatal calves have also been discovered, suggesting a role for β-lactoglobulin in vitamin A transport (Papiz et al. 1986). It remains to be studied whether PP14 and β-lactoglobulins of animal species also have a functional relationship.

Amniotic Fluid PP14

We used mid-trimester amniotic fluid for purification of PP14. The major steps included gel filtration, anion exchange chromatography and reversed-phase HPLC. In SDS-PAGE, the purified material gave a single band with a molecular weight of 28 kd which was PP14-immunoreactive in western blot analysis. This material yielded a single N-terminal aminoacid sequence, MDIPQTKQDL, identical to that observed for placental/decidual PP14.

Clinical studies on PP14

Circulating Levels of PP14 in Non-pregnant Women

PP14 is detectable in serum of non-pregnant women and of men. In ovulatory cycles the levels show a biphasic pattern. Nadir occurs around ovulation, and the levels start rising eight days after ovulation (Julkunen et al. 1986a). High levels are maintained for the first days of the next period (Fig. 8.7). No similar pattern is seen in anovulatory cycles. At the onset of the next period there is hardly any overlap in the PP14 values between previous ovulatory and anovulatory cycles. This indicates that it may be possible, by measurement of serum PP14, to distinguish ovaluatory and anovulatory cycles retrospectively with a blood test (Julkunen et al. 1986a). However, if progesterone secretion is a prerequisite for elevated PP14 levels, then ovulatory cycles with luteal insufficiency should fall between the two extremes. Indeed, recent studies on progestagen-dependent endometrial protein indicate that this is the case (Joshi et al. 1986).

A similar biphasic pattern is seen in women given Clomid and Pergonal for ovarian hyperstimulation before in-vitro fertilisation (Seppälä et al. 1988). The lowest PP14 levels are seen one or two days after the oocyte retrieval, when the progesterone level is already elevated. Because PP14 is synthesised by secretory

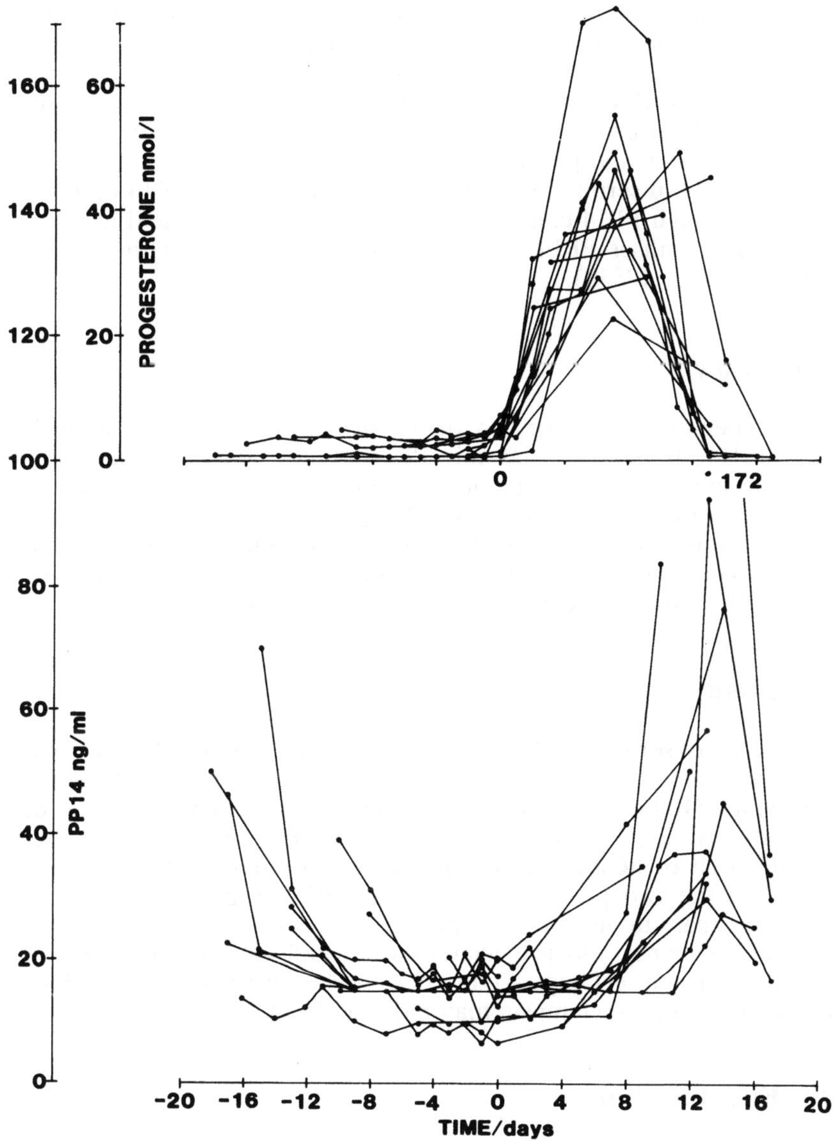

Fig. 8.7. Serum progesterone (*upper panel*) and PP14 concentrations (*lower panel*) throughout ovulatory menstrual cycle. (Reproduced, with permission, from Seppälä et al. 1987.)

endometrium, it is believed that serum PP14 level becomes elevated secondarily to the action of progesterone on the oestrogen-primed endometrium. This finds support from the fact that no similar elevation is seen in anovulatory cycles. Furthermore, women taking micronised natural progesterone have higher serum PP14 levels than comparable women taking placebo. However, it is not known whether endometrium is the only site in the body producing PP14. All the tissues containing progesterone receptors could, in theory, do so. If the endometrium were the only site, then hysterectomy should bring the levels down. While this happens in some cases, the decline is not complete and, in some patients, there is no change at all.

Sustained treatment with Danazol suppresses the circulating PP14 level (Than et al. 1987). This is not surprising, because Danazol causes endometrial atrophy. It may also act at ovarian and pituitary level.

Men also have PP14 in serum. While the origin of circulating PP14 in men is not clear, there is a high PP14 concentration in seminal plasma, and immunoperoxidase staining localises PP14 to the glandular epithelium of the seminal vesicles (Julkunen et al. 1985b).

PP14 in the Ovary

After ovarian hyperstimulation for in-vitro fertilisation preovulatory follicular fluid contains PP14. It is clear that this PP14 cannot come from secretory endometrium, because ovulation has not yet taken place, and the endometrium is supposed to be in the proliferative phase (Seppälä et al. 1985). The levels in follicular fluid are similar to those in serum, and it is possible that follicular fluid PP14 derives from peripheral blood. We have seen elevated serum PP14 levels in some patients with ovarian cancer, but this is not common.

PP14 in Pregnancy

There is a steep rise of serum PP14 concentration starting on the 22nd day of an ovulatory cycle (Julkunen et al. 1986a). This rise continues until about the 10th week of pregnancy (Julkunen et al. 1985a). Thereafter the levels decline but remain elevated until delivery. The pattern of serum PP14 levels during pregnancy resembles that of hCG, but no immunological cross-reaction has been observed between the two proteins. In pre-eclampsia serum PP14 levels are somewhat higher than in normal pregnancy, but considerable overlapping between normal pregnancy and pre-eclampsia precludes any clinical utility of PP14 measurement in pre-eclampsia (Iino et al. 1986a–c.) Like PP12/IGF-bp, elevated PP14 levels in pre-eclampsia may indicate decidual involvement.

PP14 in the Endometrium

Both proliferative and secretory endometria contain PP14 (Julkunen et al. 1986b). In tissue culture, proliferative endometrium releases little if any PP14 into the culture medium, whereas late secretory endometrium releases large amounts (Fig. 8.8). Secretory and decidualised endometria synthesise PP14 (Julkunen et al. 1986b), and PP14 has been suggested to exhibit immunosuppressive activity in the human reproductive system (Bolton et al. 1987).

We studied endometrial biopsy specimens taken from 18 women participating in our in-vitro fertilisation programme, in whom no embryo was obtained for replacement. Biopsies were done at the time of expected embryo replacement. Results of immunoperoxidase staining showed that PP14 appears in the endometrium later than PP12/IGF-bp, usually when endometrial maturation corre-

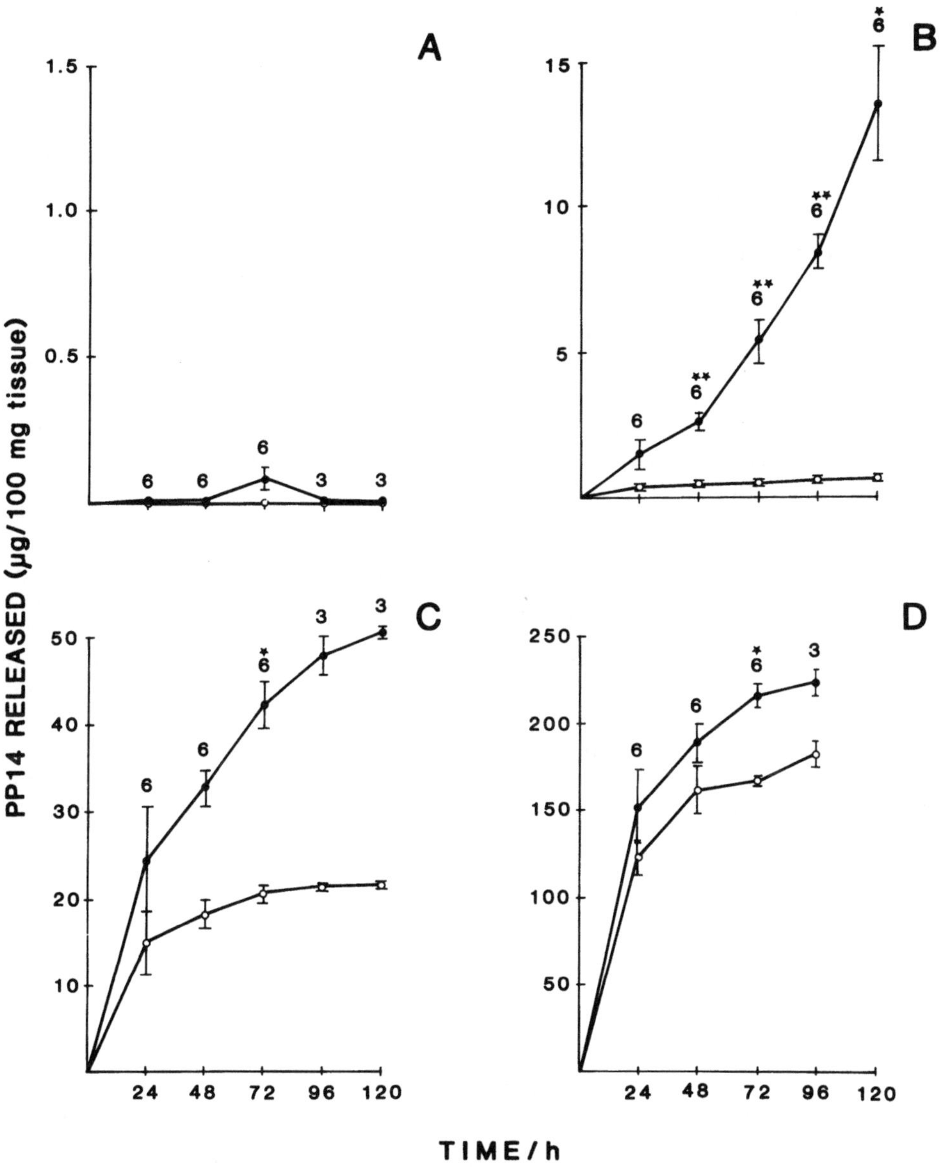

Fig. 8.8. Release of PP14 by endometrial tissue cultures. **a** Proliferative endometrium. **b** Secretory endometrium 4 days after ovulation. **c** Secretory endometrium 7–10 days after ovulation. **d** Secretory endometrium 12 days after ovulation. ●—●, without cycloheximide; ○—○, with cycloheximide. (Reproduced, with permission, from Julkunen et al. 1986a.)

sponds to days 4 or 5 postovulation (Wahlström et al. 1985). An intriguing question is whether these proteins play a part in the opening of the "implantation window". PP14 appears in the endometrium at the time when the serum PP14 level has started to rise.

Summary

The chemical and biological characterisation of endometrial proteins PP12 and PP14 has given new insights into their actions in human reproduction. Analysis of the N-terminal sequence of PP12 disclosed identity to that of insulin-like growth factor-binding protein (IGF-bp) from human amniotic fluid. Purification of PP12 from amniotic fluid confirmed that the immunoreactive amniotic fluid PP12 is identical with the placental/decidual PP12 and is capable of binding IGF-1 with a high affinity. Another form of PP12, designated PP12B, contains several fragments derived from the N-terminal portion of the PP12 molecule. These fragments react with antibodies raised against PP12 and are also capable of binding IGF-1, indicating that the IGF-1-binding domain resides in the N-terminal portion of PP12.

The primary structure of 22 N-terminal aminoacid residues of PP14 shows significant sequence homology with β-lactoglobulins from various animal species, suggesting a common origin of PP14 and β-lactoglobulins. It remains to be studied whether PP14 and β-lactoglobulins also bear other similarities. A role for β-lactoglobulins in vitamin A transport has recently been proposed.

Although the protein parts of PP12 and PP14 have been extensively studied, and some of their biological properties are being disclosed, the physiological role of these proteins in human reproduction is not well understood. The cyclic appearance of PP12 and PP14 in the endometrium, and their abundance in decidualised endometrium and in amniotic fluid, suggest a role in endometrial function and fetal development. Clinically, the circulating levels of PP14 are of great interest, as a clear distinction is observed in the levels between ovulatory and anovulatory menstrual cycles. One of the aims of these and forthcoming studies is to find markers for human endometrial function during the menstrual cycle and pregnancy.

References

Adashi EY, Resnick JCE, Svoboda ME, Van Wyk JJ (1985) Somatomedin-C enhances induction of luteinizing hormone receptors by follicle-stimulating hormone in cultured rat granulosa cells. Endocrinology 116: 2369–2375

Bell SG, Bohn H (1986) Immunochemical and biochemical relationship between human pregnancy-associated secreted endometrial α_1- and α_2-globulins (α_1- and α_2-PEG) and the soluble placental proteins 12 and 14 (PP12 and PP14). Placenta 7: 283–294

Bohn H, Kraus W (1980) Isolierung und Charakterisierung eines neuen plazentaspezifischen Proteins PP12. Arch Gynecol 229: 279–291

Bohn H, Kraus W, Winckler W (1982) New soluble placental tissue proteins: their isolation, characterization, localization and quantification. In: Klopper A (ed) Immunology of human placental proteins. Praeger Publ Placenta, Suppl 4: 67–81

Bolton AE, Chapman MG, Stoker RJ, Andrew CE, Wass D, Bohn H (1983) The radioimmunoassay of placental protein 14 (PP14). Clin Chim Acta 135: 283–291

Bolton AE, Pockley AG, Clough KJ, Mowles EA, Stocker RJ, Westwood OMR, Chapman MG (1987) Identification of placental protein 14 as an immunosuppressive factor in human reproduction. Lancet i: 593–595

de Bruijn HWA, Aalders JG, Kauer F, et al. (1986) The biodistribution of CA 125 in the female gluital tract. Tumour Biol 7: 298
Furlanetto RW, DiCarlo JN (1984) Somatomedin-C receptors and growth effects in human breast cells maintained in long-term tissue culture. Cancer Res 44: 2122–2128
Howell RJS, Perry LA, Choglay NS, Bohn H, Chard T (1985) Placental protein 12 (PP12): a new test for the prediction of the small-for-gestational-age infant. Br J Obstet Gynaecol 92: 1141–1144
Huhtala M–L, Koistinen R, Palomäki P, Partanen P, Bohn H, Seppälä M (1986) Biologically active domain in somatomedin-binding protein. Biochem Biophys Res Commun 141: 263–270
Huhtala M–L, Seppälä M, Närvänen A, Polomäki P, Julkunen M, Bohn H (1987) Amino acid sequence homology between human placental protein 14 and β-lactoglobulins from various species. Endocrinology 120: 2620–2622
Iino K, Seppälä M, Heinonen PK, Sipponen P, Rutanen E–M (1986a) Elevated levels of a somatomedin-binding protein PP12 in patients with ovarian cancer. Cancer 58: 2294–2297
Iino K, Julkunen M, Sjöberg J, Seppälä M (1986b) Is decidua affected in gestational hypertension? Acta Obstet Gynecol Scand 65: 367–370
Iino K, Sjöberg J, Seppälä M (1986c) Elevated circulating levels of a decidual protein, placental protein 12, in pre-eclampsia. Obstet Gynecol 68: 58–60
Joshi SG, Rao R, Henriques E, Raikar R, Gordon M (1986) Luteal phase concentration of a progestagen associated endometrial protein (PEP] in the serum of cycling women with adequate or inadequate endometrium. J Clin Endocrinol Metab 63: 1247–1249
Julkunen M, Rutanen E–M, Koskimies A, Ranta T, Bohn H, Seppälä M (1985a) Distribution of placental protein 14 in tissues and body fluids during pregnancy. Br J Obstet Gynaecol 92: 1145–1151
Julkunen M, Wahlström T, Seppälä M, Koistinen R, Koskimies A, Stenman U–H, Bohn H (1985b) Detection and localization of placental protein 14-like protein in human seminal plasma and in the male genital tract. Arch Androl 12: 59–67
Julkunen M, Apter D, Seppälä M, Stenman U–H, Bohn H (1986a) Serum levels of placental protein 14 reflect ovulation in nonconceptional menstrual cycles. Fertil Steril 45: 47–50
Julkunen M, Koistinen R, Sjöberg J, Rutanen E–M, Wahlström T, Seppälä M (1986b) Secretory endometrium synthesizes placental protein 14. Endocrinology 118; 1782–1786
Julkunen M, Raikar RS, Joshi SG, Bohn H, Seppälä M (1986c) Placental protein 14 and progestagen-dependent endometrial protein are immunologically indistinguishable. Hum Reprod 1: 7–8
Koistinen R (1984) Comparison of three methods for investigating placental protein 12 (PP12)-like immunoreactive substance from preovulatory follicular fluid and purified PP12 from human placenta. Clin Chim Acta 141: 235–240
Koistinen R, Kalkkinen N, Huhtala M–L, Seppälä M, Bohn H, Rutanen E–M (1986) Placental protein 12 is a decidual protein that binds somatomedin and has an identical N-terminal amino acid sequence with somatomedin-binding protein from human amniotic fluid. Endocrinology 118: 1375–1378
Koistinen R, Huhtala M–L, Stenman U–H, Seppälä M (1987) Purification of placental protein PP12 from human amniotic fluid and its comparison with PP12 from placenta by immunological, physicochemical and somatomedin-binding properties. Clin Chim Acta 164: 293–303
Lind T, Cooke DB (1976) How does danazol work? Lancet i: 1401–1402
Lundin G, Povoa G, Hansson U, Luthman M, Persson B, Stangenberg M, Hall K (1985) Somatomedin binding protein during pregnancy in patients with diabetes and pregnancy diabetes. XV Acta Endocrinol Congress, Helsinki, 182
Marshall RN, Underwood LE, Voina SJ, Foushee DB, Van Wyk JJ (1974) Characterization of the insulin and somatomedin-C receptors in human placental cell membranes. J Clin Endocrinol Metab 39: 283–292
Nilsson B, Södergar R, Damber MG, Schoultz B (1982) Danazol and gestagen displacement of testosterone and influence on sex hormone binding globulin capacity. Fertil Steril 38: 48–53
Papiz M, Sawyer L, Elipoulos E, et al. (1986) The structure of β-lactoglobulin and its similarity to plasma retinol-binding protein. Nature 324: 383–385
Pervaiz S, Brew K (1985) Homology of β-lactoglobulin, serum retinol-binding protein, and protein HC. Science 228: 335–337
Povoa G, Enberg G, Jörnvall H, Hall K (1984) Isolation and characterization of a somatomedin-binding protein. Eur J Biochem 144: 199–204
Povoa G, Isaksson M, Jörnvall H, Hall K (1985) The somatomedin-binding protein isolated from a human hepatoma cell line is identical to the human amniotic fluid somatomedin-binding protein. Biochem Biophys Res Comm 128: 1071–1078

Rannevik G, Thorell J (1984) The influence of Danazol on pituitary function and on the ovarian follicular hormone secretion in premenopausal women. Acta Obstet Gynecol Scand, Suppl 123: 89–94

Rutanen E–M, Bohn H, Seppälä M (1982) Radioimmunoassay of placental protein 12: levels in amniotic fluid, cord blood and serum of healthy adults, pregnant women and patients with trophoblastic disease. Am J Obstet Gynecol 144: 460–463

Rutanen E–M, Koistinen R, Wahlström T, Sjöberg J, Stenman U–H, Seppälä M (1984a) Placental protein in the human endometrium: tissue concentration in relation to histology and serum levels of PP12, progesterone and oestradiol. Br J Obstet Gynaecol 91: 377–381

Rutanen E–M, Wahlström T, Koistinen R, Sipponen P, Jalanko H, Seppälä M (1984b) Placental protein 12 in primary liver cancer and cirrhosis. Tumour Biol 5: 95–102

Rutanen E–M, Koistinen R, Sjöberg J, Julkunen M, Wahlström T, Bohn H, Seppälä M (1986) Synthesis of placental protein 12 by human endometrium. Endocrinology 118: 1067–1071

Seppälä M, Than G (1988) Insulin-like growth factor-binding protein in ovarian cyst fluid. Arch Gynecol (in press)

Seppälä M, Rutanen E–M, Siiteri JE et al (1983) Immunological and biological properties and clinical significance of placental proteins PP5 and PP12. Ann NY Acad Sci 417: 368–382

Seppälä M, Wahlström T, Koskimies AI, Tenhunen A, Rutanen E–M, Koistinen R (1984) Human preovulatory follicular fluid, luteinized cells of hyperstimulated preovulatory follicles, and corpus luteum contain placental protein 12. J Clin Endocrinol Metab 52: 505–510

Seppälä M, Koskimies AI, Tenhunen A, et al. (1985) Pregnancy proteins in seminal plasma, seminal vesicles, preovulatory follicular fluid, and ovary. Ann NY Acad Sci 442: 212–226

Seppälä M, Julkunen M, Koistinen R, Huhtala M–L, Rutanen EM (1987) Proteins in the human endometrium. In: Naftolin F, DeCherney AH (eds) The control of follicle development, ovulation and luteal function: lessons from in vitro fertilization. Serono Symposia, Raven Press, New York, 35: 101–109

Seppälä M, Julkunen M, Koskimies A, Laatikainen T, Stenman U–H, Huhtala M (1988) Proteins of the human endometrium: basic and clinical studies toward a blood test for endometrial function. Ann NY Acad Sci (in press)

Suikkari A–M, Rutanen E–M, Seppälä M (1987) Circulating levels of immunoreactive insulin-like growth factor-binding protein in non-pregnant women. Hum Reprod 2: 197–300

Than G, Seppälä M, Bodis J, Szilagyi A, Csaba I (1987) The effect of Danazol on the circulating level of 34K insulin-like growth factor-binding protein (PP12) and secretory endometrial protein (PP14). Hum Reprod 2: 549–551

Wahlström T, Seppälä M (1984) Placental protein 12 (PP12) is induced in the endometrium by progesterone. Fertil Steril 41: 781–784

Wahlström T, Koskimies AI, Tenhunen A, Rutanen E–M, Yki-Järvinen H, Julkunen M, Sjöberg J, Seppälä M (1985) Pregnancy proteins in the endometrium after follicle aspiration for in vitro fertilization. Ann NY Acad Sci 442: 402–407

9. Biological Activity of Placental Protein 14

A.E. Bolton, A.G. Pockley, E.A. Mowles, R.J. Stoker, O.M.R. Westwood and M.G. Chapman

Introduction

During human pregnancy a variety of proteins are present in the circulation which are either exclusive to the pregnant state or at elevated concentrations compared with levels in non-pregnant women. Many of those proteins are of placental origin, some being secreted by the deciduum. The physiological roles of most are not understood, although there is an increasing body of knowledge accumulating in this area.

Human decidual tissue synthesises and secretes two major proteins, which appear to be similar or identical to those isolated from term placenta and are designated PP12 and PP14 (Bohn et al. 1982). PP12 is a somatomedin-binding protein (see Chapter 8), whereas the biological activity of PP14 remains controversial. PP14 shows some aminoacid sequence homology with equine beta-lactoglobin (see Chapter 8), and the structural similarities of beta-lactoglobulins from various species to plasma retinol binding proteins (Papiz et al. 1986) suggest that this may be the function of PP14 (Bolton et al. 1986a). However, an alternative hypothesis – that PP14 plays an immunomodulating role in pregnancy (Bolton et al. 1987) – has been proposed on the basis of the observation that PP14 suppresses the proliferation of lymphocytes in the mixed lymphocyte reaction (Bolton et al. 1986a).

A physiological requirement for an immunosuppressive role can be hypothesised at various stages of the human reproductive process. Spermatozoa are immunogenic (Anderson and Tarter 1982), and the female reproductive tract is not an immunologically privileged site – the presence of antigens in the vagina can result in antibody formation (Bell and Wolf 1967). That sperm do not generally elicit an immune response after repeated insemination may be due to the presence of immunosuppressive factors in seminal plasma (Lord et al. 1977).

Such seminal fluid factors, though widely reported, have not been clearly defined (James and Hargreaves 1984), and indeed this activity may be largely artefactual, caused by the oxidation of polyamines present in seminal plasma by polyamine oxidases in the fetal bovine serum generally used in cell culture tests for immunosuppressive activity (Vallely and Rees 1986).

Because of the nature of the fetus as a semi-allograft expressing paternal antigens, some mechanism is required to explain its immunological non-rejection. The fetus is recognised by the maternal immune system (Rocklin et al. 1973) – indeed, immune recognition, by the mother, of paternal antigens on the fetus appears to be important for fetal survival (Rocklin et al. 1976), perhaps through immune "blocking" by antibodies present in the maternal circulation. Thus the immunosuppression required to prevent rejection of the fetus appears to be at the effector level of the cell-mediated immune reaction. Activation of T lymphocytes is thought to require the interaction of a cell-surface receptor on the T cell with foreign antigen in association with major histocompatibility complex (MHC) glycoprotein molecules (Reinhertz 1987). MHC antigen expression is reduced or absent in the syncytiotrophoblast of the developing placenta (Sunderland et al. 1981), although cells of the blastocyst do express MHC antigens (Rocklin et al. 1979). The presence of molecules suppressing the activation of T lymphocytes, particularly around the time of implantation, could be of physiological importance.

During implantation and pregnancy any immunosuppressive material would be required to exert a localised response within the uterine environment. If it is proposed that such a factor is important at implantation and during the early development and extension of the trophoblast, it is tempting to postulate a decidual origin for this activity. Indeed, immunosuppressive activity has been demonstrated in cultured human decidual cells (Nakayama et al. 1985) as well as in maternal sera (Nicholas et al. 1984; Domingo et al. 1985; Yamashita et al. 1986).

The apparent requirement for a localised immunosuppressive activity, and the demonstration of such activity in deciduum and maternal blood, has led to proposals of an immunosuppressive role for many of the human pregnancy associated proteins. For example, pregnancy associated alpha$_2$-glycoprotein (Stimson 1976), pregnancy associated plasma protein-A (PAPP-A) (Bischof et al. 1982), chorionic gonadotrophin (Han 1974) and alphafetoprotein (Littman et al. 1977) have all been proposed to have immunosuppressive activity, because they suppress lymphocyte proliferation. However, these activities have been disputed, largely on the grounds of the use of unphysiologically high concentrations of material and the possibility that unidentified impurities are the source of the inhibitory activity.

Immunosuppressive action has also been attributed to placental protein 14 (PP14) (Bolton et al. 1986a, 1987). The decidual origin of PP14 (Julkunen et al. 1985) would satisfy one criterion of a potential immunosuppressor of pregnancy – localisation in a relevant site. Furthermore, PP14 is present in relatively high concentrations in seminal plasma (Bolton et al. 1986b).

The experimental base for studies on immunosuppressive factors in pregnancy depends on the use of lymphoproliferation assays and the inhibition of lymphoproliferation. To be able to assess such data, it is important to understand the basis of this methodology, which is unfamiliar to many reproductive biologists and obstetricians.

Lymphoproliferation Assays

Peripheral blood lymphocytes, isolated by density gradient centrifugation, are stimulated either with the mitogen phytohaemagglutinin (PHA) or by mixing lymphocyte preparations from two unrelated donors (the two-way mixed lymphocyte reaction). The resulting lymphoproliferation is assessed by measuring the uptake of tritiated thymidine into the DNA of the dividing cells. In transplantation biology, inhibitors of this proliferation of stimulated lymphocytes (as measured by tritiated thymidine uptake) are considered to have immunosuppressive activity.

All bioassays have a high degree of variability, and lymphoproliferation assays are no exception. Two factors have been identified as contributing particularly to variation in the latter assays, resulting in a high between-sample coefficient of variation. These are differences in frequency of responding units in the cell populations used and large differences in gradients of the cell dose-response curves (Ross et al. 1987). The experimental data from lymphoproliferation assays presented below, generally derived from assays carried out on different days and using blood derived from different individuals from among a pool of donors, reflects this high variability.

In the two-way MLR (Fig. 9.1) T-cell activation requires, from both donors, the presence of accessory cells of the macrophage/monocyte series, as well as T lymphocytes. Interleukin 1 (IL-1) derived from these accessory cells is thought to act on T lymphocytes in two ways – by inducing the production of interleukin 2 (IL-2) (Larson et al. 1980) and by increasing the IL-2 receptor activity on the T cells, possibly by an alteration in receptor affinity rather than by an increase in receptor expression (Oppenheim et al. 1986). Stimulated T cells release IL-2, which results in the expression of further IL-2 receptors and a proliferation of the cells. Although mitogen stimulation of T cells is generally represented as a more simple event (Fig. 9.2), it is by no means clear whether truly isolated T lymphocytes free from contaminating cells will respond to PHA. Thus, although it appears that any differences a putative immunomodulator may show in these two systems could depend upon IL-1 activity (compare Fig. 9.1 with Fig. 9.2), it is now becoming evident that IL-1, either from added or contaminating accessory cells or as an exogenous addition, is required for mitogen stimulation of T lymphocytes (Oppenheim et al. 1986).

From this it is apparent that the lymphoproliferation assays are as yet ill understood at a molecular level, and the designation of a material as an immunosuppressor or immunomodulator solely by its effect on the incorporation of tritiated thymidine during such an assay is simplistic. It is this lack of a full understanding of these assay systems that has probably resulted in the confusion about the "immunosuppressive" activities of the pregnancy-associated proteins.

The aim of the present work was to investigate the action of the decidual protein PP14 on lymphoproliferation assays and to attempt to explore more comprehensively the mode of action of this protein. The greater the detail with which the mode of action of such a material can be identified, the more likely it is that the action is specifically rather than non-specifically immunomodulatory.

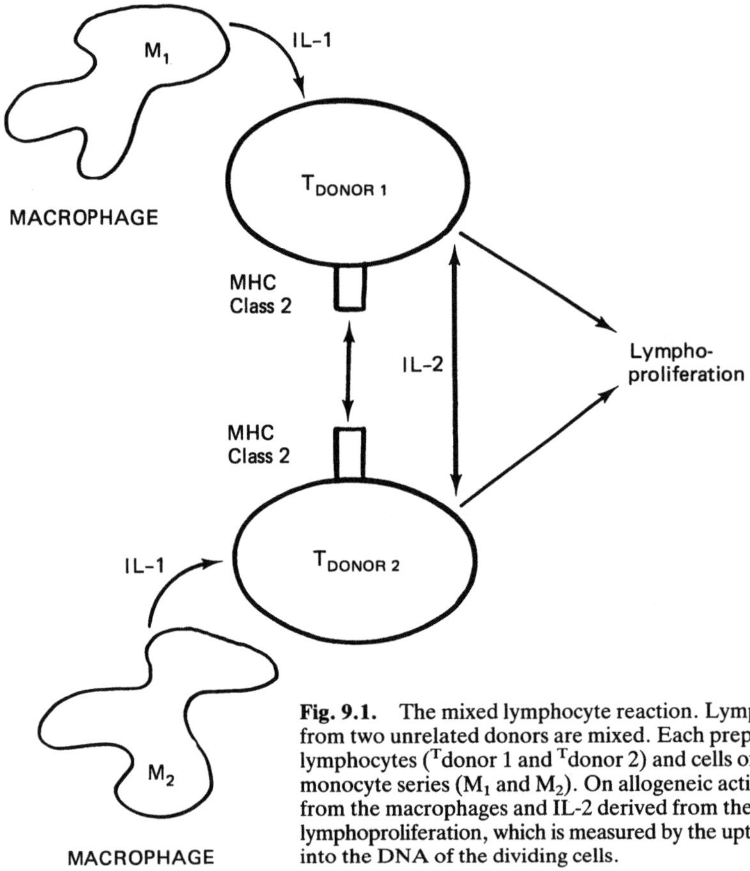

Fig. 9.1. The mixed lymphocyte reaction. Lymphocyte preparations from two unrelated donors are mixed. Each preparation contains T lymphocytes (^{T}donor 1 and ^{T}donor 2) and cells of the macrophage/monocyte series (M_1 and M_2). On allogeneic activation, IL-1 derived from the macrophages and IL-2 derived from the T cells stimulate lymphoproliferation, which is measured by the uptake of ^{3}H-thymidine into the DNA of the dividing cells.

Effect of PP14 on Lymphoproliferation Assays

As much previous work on the effect of pregnancy-associated proteins on the lymphoproliferation response has been suspect because the purity of the protein

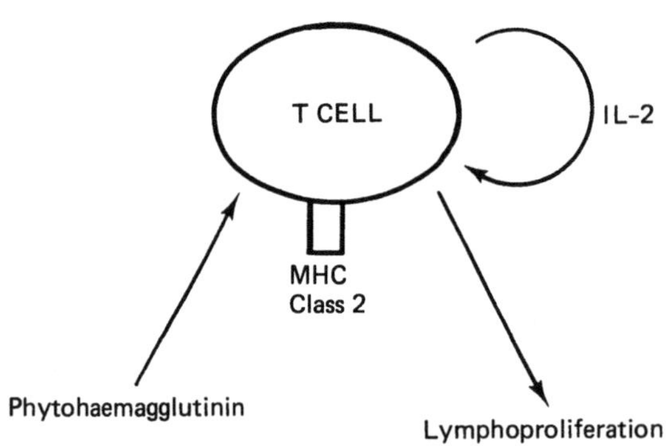

Fig. 9.2. PHA stimulation of T lymphocytes. T cells from a single donor are activated by the lectin phytohaemagglutinin (PHA) to produce IL-2, which stimulates lymphoproliferation.

preparations used has been in doubt, the present work employed an approach involving the specific removal of PP14 from crude tissue extracts by immunoadsorption with a monoclonal anti-PP14 antibody.

Methods

Monoclonal anti-PP14 antibody was derived from spleen cells from mice immunised with crude decidual extract, hybridised with the myeloma cell line P3/NS1/1-Ag4-1 using a polyethylene glycol fusion technique. Hybridoma cells secreting anti-PP14 antibody were identified with a radioimmunoassay procedure (Bolton et al, 1983), and positive cells were cloned by limiting dilution three times. The resulting stable hybridoma cell line was used to induce ascitic tumours in BALB/c mice, and immunoglobulin G (IgG) was isolated from the ascitic fluid by ion exchange chromatography on DEAE-52 and high-performance liquid chromatography on TSK 5000 PW.

After coupling this monoclonal IgG to a solid support (Sephacryl S-300) by the method of Wright and Hunter (1982), the binding specificity of the monoclonal antibody was tested by examination of its binding to purified radioiodinated proteins and its specificity as the labelled antibody reagent in a two-site immunoradiometric assay. No cross-reactivity could be detected with any of the human placental or decidual proteins investigated (Table 9.1). Furthermore, after immunoadsorption of crude decidual extract with this adsorbent, the only protein peak showing reduction in concentration as assessed with two-dimensional immunoelectrophoresis was immunologically indistinguishable from purified PP14.

Table 9.1. Specificity of the monoclonal anti-PP14 antibody[a]

Less than 0.1% cross-reaction with the following:	
Placental and decidual proteins	Enzymes extracted from human placenta
Human chorionic gonadotrophin (hCG)	Alkaline phosphatase
Human placental lactogen (hPL)	Malic dehydrogenase
Schwangerschaftsprotein 1 (SP1)	Sphyngomyelinase
Placental protein 5 (PP5)	Arylamidase
Placental protein 12 (PP12)	Choline acetyltransferase
Pregnancy associated plasma protein-A (PAPP-A)	
Prolactin	

[a] Specificity was investigated in a two-site immunoradiometric assay (all the above) and also by testing the binding of labelled proteins to the monoclonal anti-PP14 immunoadsorbent (hCG, PAPP-A, prolactin).

Mixed lymphocyte cultures and PHA stimulation experiments were carried out as described previously (Bolton et al. 1987; Pockley et al. unpublished observations), using serum-free medium (MB103, Dupont NEN Research Products) to exclude artefacts caused by serum components. Decidual tissue, obtained after legal first-trimester pregnancy termination and identified by histological examination, was homogenised with phosphate buffered saline containing PMSF as a protease inhibitor (100 mg wet weight of tissue/ml). Following removal of cell debris by centrifugation, the extracts were dialysed extensively. Portions (0.5 ml) of each extract were treated with the monoclonal anti-PP14 immunoadsorbent (0.5 ml settled gel volume) or the same volume of unsubstituted Sephacryl S-300. After adsorption the PP14 content was determined in the treated and untreated extracts with the radioimmunossay described elsewhere (Bolton et al. 1983). Extracts, or PP14 preparations purified as described previously (Bolton et al. 1987), were sterilised by filtration through 0.4 μm Millipore membranes and diluted in sterile growth medium before addition to the lymphoproliferation assays.

To investigate the non-specific effect of decidual extracts on cell division, an actively dividing human leukaemic cell line, K562, was cultured in the presence of immunoadsorbed or unadsorbed decidual extract.

Expression of IL-2 receptors by PHA-stimulated T lymphocytes was assessed with flow cytometry using monoclonal anti-Tac antibody (Dako, Clone ACT1) and a fluorescently labelled anti-mouse IgG second antibody.

Results

Decidual extracts inhibited the uptake of tritiated thymidine by both PHA-stimulated lymphocytes (Fig. 9.3) and by mixed lymphocyte cultures (Fig. 9.4). This inhibition appeared to be related to the PP14 content of the extract, and treatment of the extracts with a monoclonal anti-PP14 antibody immunoadsorbent reduced both the measurable PP14 content and the inhibitory effects of the extracts. The suppression of tritiated thymidine uptake by decidual extracts was not apparently caused by non-specific toxicity to the cultured lymphocytes, since cell viability (assessed by trypan-blue exclusion) was not different in control and

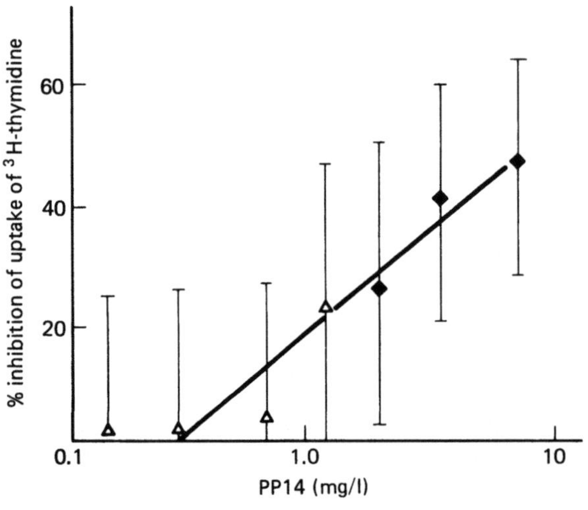

Fig. 9.3. Effect of unadsorbed decidual extract (◆) and decidual extract immunoadsorbed with a monoclonal anti-PP14 antibody adsorbent (△) on the uptake of ^{3}H-thymidine by PHA-stimulated human peripheral-blood lymphocytes. Results are mean ±SD, $n = 12$.

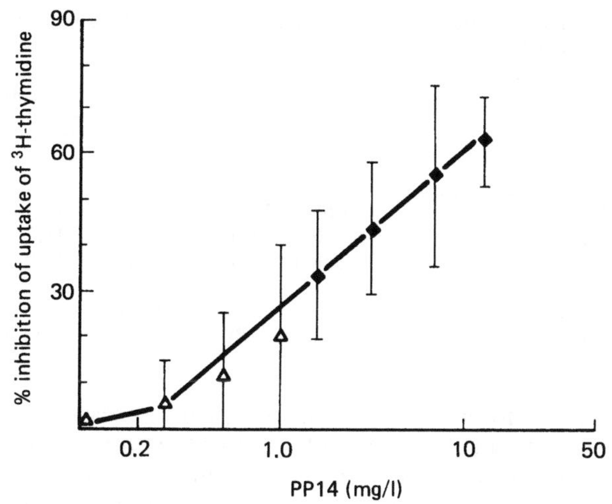

Fig. 9.4. Effect of unadsorbed decidual extract (◆) and decidual extract immunoadsorbed with a monoclonal anti-PP14 antibody adsorbent (△) on the uptake of ^{3}H-thymidine by mixed lymphocyte cultures. Results are mean ± SD, $n = 10$.

Table 9.2. Effect of decidual extract on ^{3}H-thymidine uptake by PHA-stimulated peripheral-blood lymphocytes and the human leukaemic cell line K562

	% uptake of ^{3}H-thymidine	
	K 562	PHA-stimulated lymphocytes
Control	100	100
Decidual extract	94.3 ± 3.6	49.2 ± 2.8

Results are expressed as mean ± standard deviation. The experiment was carried out in 5 replicate using a single culture of cells or a single sample of peripheral-blood lymphocytes

decidual extract-treated cultures. Furthermore, decidual extracts did not inhibit tritiated thymidine uptake by the human leukaemic cell line K562 (Table 9.2), indicating absence of a non-specific effect on cell division.

The activation of T lymphocytes involves cytokine mediators, primarily IL-1 and IL-2. If the synthesis or secretion or the target cell receptor of such an activator is affected by an inhibitor, the lymphoproliferative response will be inhibited, and such inhibition should be reversible by the addition of the activator. One convenient source of a mixture of cytokines is the supernatant collected from normal mixed lymphocyte cultures (conditioned medium). This will contain all the lymphokines required for T lymphocyte activation. The addition of conditioned medium to both mixed lymphocyte cultures and PHA-stimulated T lymphocytes resulted in a reversal of the inhibition of tritiated thymidine uptake by the cells caused by purified PP14 or crude decidual extracts (Table 9.3), suggesting interleukin involvement at some stage of the suppression of lymphoproliferation.

Suppression of the lymphoproliferative response at the level of IL-2 action can be studied by investigating the expression of the IL-2 receptor on T lymphocytes. IL-2 itself leads to the expression of IL-2 receptors, and inhibition of release of IL-2 would thus be reflected in reduced IL-2 receptor expression, as would a

Table 9.3. Effect of conditioned medium on inhibition of ^{3}H-thymidine uptake by decidual extracts and purified PP14

Test system[a]	Inhibitor	% incorporation of ^{3}H-thymidine	
		With conditioned medium	Without conditioned medium
MLR	Purified PP14	70.6 ± 12.4	57.6 ± 11.7
MLR	Decidual extract	90.2 ± 11.7	65.3 ± 9.7
PHA	Decidual extract	84.4 ± 12.0	49.1 ± 4.1

Results are expressed as mean ± standard deviation, $n = 5$
[a] MLR, mixed lymphocyte cultures; PHA, PHA-stimulated lymphocytes.

direct inhibitory effect on receptor expression itself. The presence of decidual extract had no effect on lymphocyte stimulation with PHA or on the expression of IL-2 receptors (measured with monoclonal anti-Tac antibody using fluorescently labelled second antibody and flow cytometry) by PHA-stimulated lymphocytes (Table 9.4). The results suggest that IL-2 is not involved in the inhibiting effect of PP14.

Table 9.4. Effect of decidual extract on expression of IL-2 receptors (Tac antigen) by PHA-stimulated lymphocytes

	% of cells expressing Tac antigen
Control	79.8
Decidual extract	79.3

Results are means of duplicate determinations.

Discussion

Previous investigations of the effect of other pregnancy associated proteins on lymphocyte activation and proliferation, have been criticised, particularly with regard to the specificity of the reactions and the possibility of impurities in the preparations used. We have attempted to overcome this possibility by using crude tissue extracts adsorbed with a monoclonal anti-PP14 antibody as a control for the lymphoproliferation assay systems as well as purified PP14 in these systems. Our data suggest a specific effect of PP14 on tritiated thymidine uptake by activated lymphocytes.

We have also started to investigate the mode of action of PP14 on the inhibition of uptake of tritiated thymidine by stimulated T lymphocytes. We have excluded a non-specific action of decidual extracts on cell division by demonstrating a lack of effect on an actively dividing human leukaemic cell line. The reversal of the inhibitory effect of PP14 and decidual extract on lymphoproliferation by conditioned medium, which contains a variety of cytokines, suggests inhibition at the level of lymphokine synthesis/release or lymphokine receptor expression/binding. As the expression of the Tac antigen of the IL-2 receptor was not inhibited by decidual extract, it may be argued that inhibition is not occurring at the level of IL-2 release or IL-2 receptor expression. However, Tac antigen represents only a low-affinity IL-2 binding site, and high-affinity IL-2 receptors are thought to consist of a 75 kd converter protein in association with Tac (Han 1974). Specific probing of the high-affinity IL-2 receptor will be required before this point can be settled.

As indicated above, the distribution of PP14 in the decidualising endometrium and seminal plasma is compatible with an immunoregulatory role. At the concentrations present in seminal plasma (reference range 15–500 mg/l [Bolton et al, 1986b]), human uterine luminal fluid (about 4 mg/l, S.C. Bell, personal communication) and deciduum (about 9 mg/g wet weight tissue [Bolton et al. 1986a]), PP14 would be expected to exert an effect on T lymphocyte proliferation (Figs. 9.3 and 9.4). Although the highest levels found in the peripheral circulation (up to 2.2 mg/l, median about 1 mg/l [Julkunen et al. 1985]) may be expected to exhibit some inhibition, such high concentrations are present only for a period of about 4 weeks in the first trimester of pregnancy, and even at these concentrations only a minimal effect on T lymphocyte proliferation would be expected. This suggests that any immunoregulatory response would be largely restricted to the lumen of the female tract and the decidual tissue by localisation of the concentration of PP14 present.

Evidence presented here continues to suggest that PP14 is involved in inhibiting T lymphocyte differentiation and proliferation and may therefore be an immunoregulatory molecule. Whether such an activity is expressed in vivo and, if so, whether it is of physiological importance remains to be resolved. If PP14 does prove to be an important immunomodulatory factor associated with the reproductive process, it will have exciting implications in the study of implantation and fertility. As such activity is likely to be highly localised, it is perhaps less likely that the circulating levels of this protein will prove to be diagnostically useful.

Acknowledgements. We thank Dr. H. Bohn, Behringwerke AG, for providing the reagents for the PP14 radioimmunoassay and Dr. R.C. Rees, Department of Virology, University of Sheffield, for providing the K562 cell line and for assistance with flow cytometry.

References

Anderson DJ, Tarter TH (1982) Immunosuppressive effects of mouse seminal plasma components in vivo and in vitro. J Immunol 128: 535–39

Bell EB, Wolf B (1967) Antibody synthesis in vitro by the rabbit vagina against diphtheria toxoid. Nature 214: 423–24

Bischof P, Lauber K, Wurstemburger B, Girard JP (1982) Inhibition of lymphocyte transformation by pregnancy-associated plasma protein-A (PAPP-A). J Clin Lab Immunol 7: 61–65

Bohn H, Kraus W, Winckler W (1982) New soluble placental tissue proteins: their isolation, characterisation, localisation and quantification. Placenta, Suppl 4: 67–81

Bolton AE, Chapman MG, Stoker RJ, Andrew CE, Wass D, Bohn H (1983) The radioimmunoassay of human placental protein 14 (PP14). Clin Chim Acta, 135: 283–291

Bolton AE, Clough KJ, Bohn H, Chapman MG, (1986a) Placental protein 14 in the human reproductive tract. In: Hau J (ed) Pregnancy proteins in animals. Walter de Gruyter, Berlin, pp 165–71

Bolton AE, Pinto-Furtado LG, Andrew CE, Chapman MG (1986b) Measurement of the pregnancy-associated proteins, placental protein 14 and pregnancy-associated plasma protein A in human seminal plasma. Clin Reprod Fertil 4: 233–240

Bolton AE, Pockley AG, Clough KO, et al. (1987), Identification of placental protein 14 as an immunosuppressive factor in human reproduction. Lancet i: 593–595

Domingo CC, Demenech N, Aparicio P, Palomino P (1985) Human pregnancy serum inhibits proliferation of T8-depleted cells and their interleukin 2 synthesis in mixed lymphocyte cultures. J Reprod Immunol 8: 97–110

Han T (1974) Inhibitory effect of human chorionic gonadotrophin on lymphocyte blastogenic responses to mitogen, antigen and allogeneic cells. Clin Exp Immunol 18: 529–535

James K, Hargreaves TB (1984) Immunosuppression by seminal plasma and its possible clinical significance. Immunol Today 5: 357–363

Julkunen M, Rutanen L-M, Koskimies A, Ranta T, Bohn H, Seppälä M (1985) Distribution of placental protein 14 in tissue and body fluids during pregnancy. Br J Obstet Gynaecol 92: 1145–1151

Larson EL, Iscove NN, Coutinho A (1980) Two distinct factors are required for T-cell growth. Nature 283: 664–666

Littman BH, Alpert E, Rocklin RE (1977) The effect of purified alpha-foetoprotein on in vitro assays of cell mediated immunity. Cell Immunol 30: 35–42

Lord EM, Sensabaugh GF, Stites DP (1977) Immunosuppressive activity of human seminal plasma. I. Inhibition of in vitro lymphocyte activation. J Immunol 118: 1704–1711

Nakayama E, Asano S, Kodo H, Miwa S (1985) Suppression of mixed lymphocyte reaction by cells of human first trimester pregnancy endometrium. J Reprod Immunol 8: 25–31

Nicholas NS, Panayi GS, Nouri AME (1984) Human pregnancy serum inhibits interleukin 2 production. Clin Exp Immunol 58: 587–595

Oppenheim JJ, Kovacs EJ, Matsushima K, Durum SK (1986) There is more than one interleukin 1. Immunol Today 7: 45–56

Papiz MZ, Sawyer L, Eliopoulos EE, et al. (1986) The structure of beta-lactoglobulin and its similarity to plasma retinol binding protein. Nature 324: 383–385

Reinhertz EL (1987) T-cell receptors. Nature 325: 660–663

Rocklin RE, Zuckerman JE, Alpert E, David JR (1973) Effect of multiparity on human maternal hypersensitivity to foetal antigens. Nature 241: 130–131

Rocklin RE, Kitzmiller JL, Carpenter CB, Garavoy MR, David JR (1976) Absence of an immunological blocking factor from the serum of women with chronic abortion. N Engl J Med 295: 1209–1213

Rocklin RE, Kitzmiller JL, Kaye MD (1979) Immunobiology of the maternal-fetal relationship. Ann Rev Med 30: 375–404

Ross G, Dixon JB, Veevers A (1987) Variation in the lymphocyte transformation assay. Slope analysis of cell dose–response curves. J Immunol Method 98: 189–193

Smith KA (1987) Two-chain structure of high-affinity IL-2 receptors. Immunol Today 8: 11–13

Stimson WH (1976) Studies on the immunosuppressive properties of a pregnancy-associated alpha macroglobulin. Clin Exp Immunol 25: 199–206

Sunderland CA, Naiem M, Mason DY, Redman CWG, Stirrat GM (1981) The expression of major histocompatibility antigens by human chorionic villi. J Reprod Immunol 3: 323–331

Vallely P, Rees RC (1986) Seminal plasma suppression of human lymphocyte responses in vitro requires the presence of bovine serum factors. Clin Exp Immunol 66: 181–187

Wright JF, Hunter WM (1982) A convenient replacement for cyanogen bromide-activated solid phases in immunoradiometric assays. J Immunol Method 48: 311–325

Yamashita K, Hayashi H, Mure K, Ishikawa M, Shimizu T (1986) Serum immunosuppressive substance in patients with gynaecologic malignancy and in pregnant women. Cancer 57: 69–75

Section V
PROSTAGLANDINS IN REPRODUCTION

10. Prostaglandins and the Establishment of Pregnancy

S.K. Smith and R.W. Kelly

Introduction

The establishment of mammalian pregnancy involves a complex series of events in which the uterine environment is prepared to accept and allow implantation of the embryo. Continuation of the pregnancy depends on rapid recognition by the mother, so that the hormonal support of the endometrium, provided by the corpus luteum, can be maintained until the time of the luteoplacental shift (Csapo et al. 1973). There is considerable variation in the mechanisms by which different species achieve this end with respect both to the systemic, hormonal changes and to the local morphological and biochemical adaptations of the endometrium (Finn 1983; McLaren 1985; Hearn 1986) However, underlying this complexity are certain basic similarities between species, including the ovarian steroids, oestradiol and progesterone, and the need for adaptive changes of the endometrium, termed decidualisation.

Study of this important stage of human development is difficult because of the obvious clinical and ethical difficulties of investigating the embryo–endometrial interactions in vivo and of obtaining human embryos for experimentation in vitro. Nevertheless, there is now considerable interest in the process of implantation in man because of the high rates of failure of in-vitro fertilisation, which seems to arise after the transfer of the embryo into the uterine cavity (Edwards and Steptoe 1983; Yovich et al. 1984). Failure of implantation is not just a problem of artificial fertilisation but occurs with a high frequency in domestic animals (Wilmut et al. 1986) and natural human pregnancies (Short 1979).

Prostaglandins (PGs) synthesised by the uterus in both laboratory and domestic animal species have been implicated in the events that occur in the oestrous cycle and especially in the changes that arise at implantation and in early pregnancy (Kennedy 1983; Bazer et al. 1986). In women, prostaglandins are thought to have a role in menstruation, and disturbances of prostaglandin synthesis are associated with menstrual disorders (Baird et al. 1981; Smith 1986). In this review, the evidence that PGs are involved in implantation in animals will be investigated, and the factors that regulate PG synthesis in the endometrium of women will be described, with reference to their role in the establishment of pregnancy.

Decidualisation

Morphological Changes

In most species, the earliest sign at the site of implantation is an increase in the vascular permeability of the capillaries, as demonstrated by the extravasation of macromolecular dyes such as Evans blue or pontamine blue (Psychoyos and Martel 1985). This change precedes the metamorphosis of the endometrium into decidua. The transformation is essential for the establishment of pregnancy in those species with haemochorial implantation and in which some degree of penetration of the endometrium occurs. In rodents, decidualisation will arise only in response to a stimulus given to an endometrium primed by the ovarian steroids. In the mouse and the rat this involves an initial period of oestradiol stimulation followed by progesterone for four days, after which oestradiol is required again. In other species, such as the hamster, guineapig and rabbit, only progesterone is obligatory for implantation, although oestrogens may assist this process (Finn 1977). The stimulus may be the presence of the embryo next to the endometrium, or an artificial stimulus such as trauma or instillation of oil into the uterine cavity. The non-traumatic stimuli will induce decidualisation only if epithelial cells are present (Lejeune et al. 1981), but trauma will induce these changes in the absence of endometrial cells.

There is considerable evidence that PGs provide the signal for both the vascular changes that precede decidualisation and for the process of decidualisation itself.

Evidence for a Role for Prostaglandins in Implantation

Evidence that PGs are obligatory for the changes in vascular permeability and decidualisation of the endometrium in several species has been reviewed by Kennedy (1983, 1985). This evidence is obtained from observations of decidualisation in normal pregnancy and from the formation of deciduomas in response to artificial stimuli.

Inhibitors of PG synthesis, such as indomethacin, delay or inhibit changes of vascular permeability in the mouse (Lundqvist and Nilsson 1980), rat (Kennedy 1977; Phillips and Poyser 1981), rabbit (Hoffman et al. 1978) and hamster (Evans and Kennedy 1978) and have the same effects on implantation in the mouse (Lau et al. 1973; Saksena et al 1976) and rabbit (El-Banna 1980). These effects may be partially overcome by the administration of PGs in the mouse (Saksena et al. 1976), rat (Oettel et al. 1979) and rabbit (El-Banna 1980). Prostaglandin levels are elevated at the sites of implantation in the rat (Kennedy 1977; Kennedy and Zamecnick 1978), the hamster (Evans and Kennedy 1978) and rabbit (Sharma 1979).

The altered vascular permeability and decidualisation that follows artificial stimuli is prevented by inhibitors of PG synthesis in the rat (Castracane et al. 1974; Sananes 1976, 1981; Kennedy and Lukash 1982) and mouse (Rankin et al. 1979). This effect is reversed by the intrauterine administration of PGs in the rat (Kennedy 1979, 1980a, b) and mouse (Rankin et al. 1979). Finally, deciduogenic stimuli raise PG levels in the endometrium (Rankin et al. 1979; Kennedy 1979, 1980a, b).

The actual type of PG responsible for these changes is not clear. $PGF_{2\alpha}$, PGE_2 and 6-oxo-$PGF_{1\alpha}$ are elevated at the site of implantation (Kennedy 1985) although in the hamster only PGE_2 is raised (Evans and Kennedy 1978). In rats, $PGF_{2\alpha}$ and PGE_2 induce vascular permeability changes and decidualisation (Kennedy and Zamecnik 1978). However, the ability of PGs to induce decidualisation may not depend simply on the capacity of the tissue to synthesise PGs. Kennedy et al. (1983) demonstrated receptors for PGE_2 predominantly in the stromal cells of the rat uterus, and similar receptors have now been identified in the endometrium of pigs (Kennedy et al. 1986) and women (Hofman et al. 1985).

A further possibility is that the embryo itself can synthesise and release PGs (Findlay 1983). The evidence for release of PGs by blastocysts is contradictory. The blastocysts of swine (Davis et al. 1983), rabbit (Pakrasi and Dey 1982) and man (Shutt and Lopata 1981) synthesise PGs. However, mouse and rat blastocysts do not synthesise PGs (Kennedy 1985).

The evidence that one or other of the PGs provides the stimulus for decidualisation of suitably primed endometrium in laboratory species thus appears very strong. However, decidualisation in man does not require a stimulus (Bell 1985) but arises during the course of the normal menstrual cycle. In women, changes in vascular permeability and oedema of the stroma and the formation of predecidual cells around the spiral arterioles begin around day 23 of the cycle and are dependent on the sequential actions of oestradiol and progesterone (Wynn 1977). There is no direct evidence in women that PGs are involved in implantation, but many of the features of decidualisation are similar to inflammation, and PGs and leukotrienes are known to be involved in this mechanism. To understand further the possible role of PGs in implantation in man it is essential to determine the effects of the steroids, oestradiol and progesterone on PG synthesis and metabolism.

Prostaglandin Synthesis and Metabolism

Synthesis and Metabolism

Prostaglandins are not stored in tissues but arise from the rapid metabolism of non-esterified arachidonic acid in the cell (Bito 1975). Arachidonic acid is stored in phospholipids situated in the cell membrane. Phosphatidyl ethanolamine (PE) serves as the substrate for the enzyme, phospholipase A_2 (Samuelsson et al. 1978), which directly cleaves arachidonic acid from the C_2 position of the phospholipid. Alternatively, phospholipase C is capable of cleaving an arachidonic acid rich diacylglycerol (DG) from phosphatidyl inositol (PI), which serves

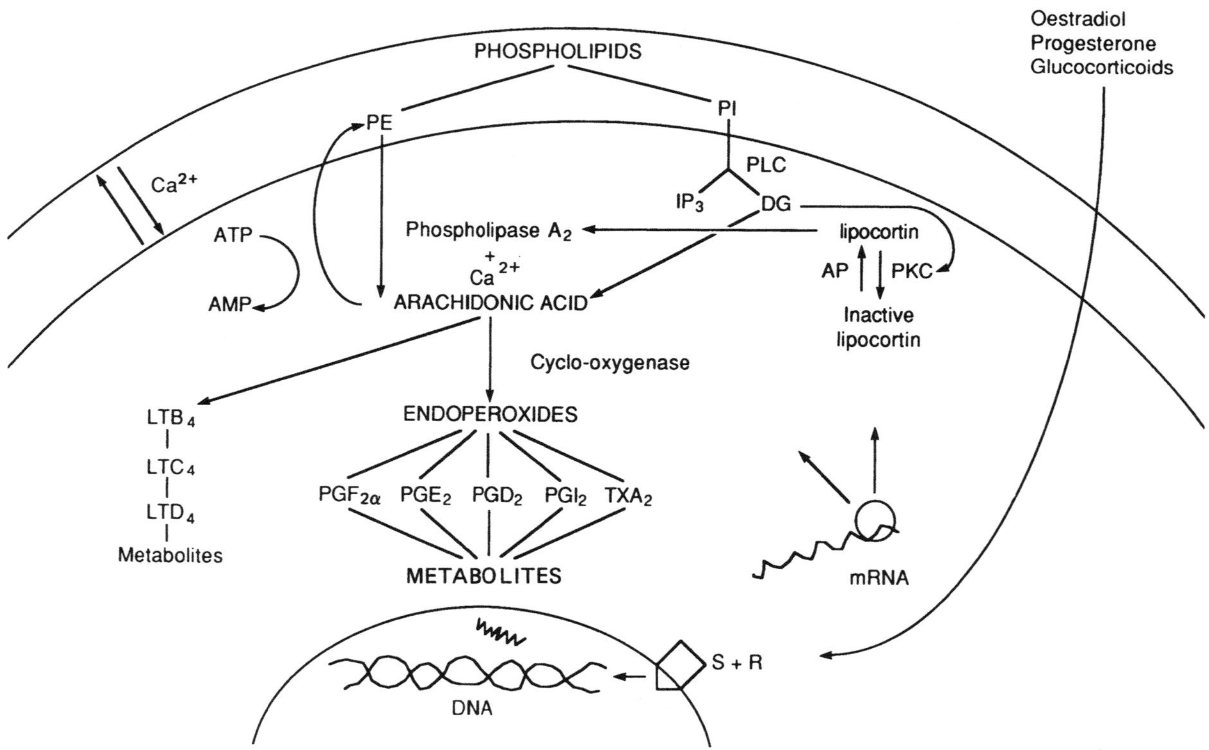

Fig. 10.1. Synthesis and metabolism of prostaglandins and leukotrienes (LTB_4, LTC_4, LTD_4). PGs are derived from phospholipids such as phosphatidyl ethanolamine (PE) and phosphatidyl-inositol (PI) from the cleavage of arachidonic acid by phospholipase A_2 and phospholipase C (PLC). The synthesis is altered by steroids (S) and inhibitory proteins such as lipocortin. Lipocortin is phosphorylated by protein kinase C (PKC) and inactive lipocortin activated by acid phosphatase (AP). IP_3 = inositol tris phosphate. DG = diacyl glycerol.

as a substrate for diacyl and monoacyl glycerol lipases, which form arachidonic acid (Lapetina 1982) (Fig. 10.1).

The release of arachidonic acid from phospholipid stores is assumed to be the rate limiting step in the synthesis of PGs. Phospholipase A_2 activity is catalysed by the intracellular release of Ca^{2+} or the influx of extracellular Ca^{2+}. In most cells, the activity of phospholipase LA_2 is regulated by a group of inhibitory proteins with molecular weights ranging from 15000 to 40000. Glucocorticoids exert their anti-inflammatory action by stimulating the synthesis of the inhibitory proteins. One such protein, lipocortin, has been purified and sequenced (Blackwell et al. 1980; Wallner et al. 1986). Phosphorylation of lipocortin by protein kinase C deactivates lipocortin, and the activity of this enzyme provides another regulatory mechanism for PG release.

Free arachidonic acid may be reacylated into phospholipid stores by the action of diacyl glycerol kinases, but otherwise it serves as a substrate for the PG synthetase enzymes. PG synthetase is a double protein containing fatty acid cyclo-oxygenase and PG hydroperoxidase, although neither enzyme has been identified as a separate protein (Mizumo et al. 1982). Their action is to convert arachidonic acid to PG endoperoxides, from which PGF_{2a}, PGE_2, PGI_2, PGD_2 and thromboxane A_2 (TXA_2) are derived by both enyzmic and non-enzymic mechanisms. These highly active PGs are metabolised, in the case of $PGF_{2\alpha}$ and PGE_2 by 15-PG dehydrogenase to the relatively inactive metabolites 15-keto-13,14-dihydro-$PGF_{2\alpha}$ and 15-keto-13,14-dihydro-PGE_2.

Prostaglandins and Ovarian Steroids

The effects of the ovarian steroids on uterine PG synthesis and metabolism have been extensively studied, yet many of the basic questions concerning the mechanism of action of steroids and the intracellular regulatory systems remain unanswered. Oestradiol stimulates PG synthesis in animals when the endometrium has been primed with progesterone, an action thought to be due to stimulation of PG synthetase activity and perhaps increased phospholipase A_2 activity (Dey et al. 1982). In this section, the results of the actions of the steroids on human endometrium and decidua will be reviewed.

Early studies in vivo measured the amounts of "endogenous" PG in tissue which had been removed for biopsy and rapidly immersed in a fixative such as methanol. This was thought to reflect the capacity of the tissue to respond to the trauma of tissue collection. Cyclical changes in $PGF_{2\alpha}$ content were found with a rise in the luteal phase of the cycle (Downie et al. 1974). PGE_2 levels did not change. This presumed effect of the sequential action of steroids on PG synthesis was mimicked in postmenopausal women given oestradiol alone or with progesterone (Smith et al. 1984). These findings suggested that progesterone promoted the capacity of endometrium to synthesise PGs in response to a traumatic stimulus.

In-vitro studies of the synthesis of PGs by endometrium have yielded conflicting results. Abel and Baird (1980) showed that secretory endometrium released more $PGF_{2\alpha}$ than proliferative endometrium, but Tsang and Ooi (1982) and Schatz et al. (1985) found that proliferative endometrium was the more active tissue. Of particular interest has been the search to identify the source of PG synthesis. Gal et al. (1982) cultured endometrium for more than seven days and

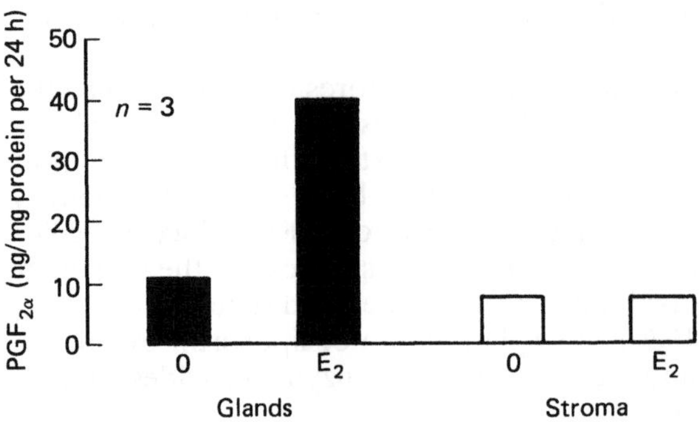

Fig. 10.2. Effect of oestradiol (10^{-8}M) (E_2) on the release of $PGF_{2\alpha}$ from glandular and stromal cells obtained from secretory endometrium.

found that stromal cells were the main source of PG, PGE_2 being the most abundant. However, immunohistochemical techniques have shown that cyclo-oxygenase is situated predominantly in the glandular cells (Rees et al. 1982), and higher levels of $PGF_{2\alpha}$ and PGE_2 are released by glandular cells than stromal cells in culture (J.K. Smith and R.W. Kelly, unpublished).

Oestradiol stimulates $PGF_{2\alpha}$ release from whole tissue obtained from secretory endometrium (Abel and Baird 1980; Schatz et al. 1985) but not proliferative endometrium. However, Schatz et al. (1985) found that oestradiol stimulated $PGF_{2\alpha}$ release from epithelial glands taken from both proliferative and secretory endometrium. Smith and Kelly (unpublished) found that only secretory glandular preparations responded to oestradiol (Fig. 10.2) but confirmed the findings of others using whole tissue (Abel and Baird 1980; Tsang and Ooi 1982) that epithelial glandular cells will not increase PGE_2 synthesis in response to oestradiol. Oestradiol alters the $PGF_{2\alpha}/PGE_2$ ratio by altering the synthesis of $PGF_{2\alpha}$.

Progesterone inhibits histamine stimulated release of PGs from endometrial cellular preparations (Wilson et al. 1986); and suppresses PG release from explants maintained in culture media (Abel and Baird 1980; Kelly and Smith 1987).

Prostaglandins and Implantation

Implantation

Prostaglandin levels are raised in deciduomata of laboratory species and would be expected to be raised in secretory endometrium and decidua of women if it were analogous to the situation in animals. Previous studies have found that secretory tissue releases more PGs in response to trauma than does proliferative tissue, but this is not the case when the separated cells are cultured Schatz et al. (1985) found that epithelial cells from proliferative tissue synthesise more $PGF_{2\alpha}$ than secretory tissue, and these findings have been confirmed for both $PGF_{2\alpha}$ and PGE_2 (S.K. Smith and R.W. Kelly, unpublished). This presumably reflects the inhibition of PG synthesis by progesterone, because although glandular cells from secretory tissue have a greater capacity to metabolise PGs, this does not account for the observed differences between the two phases of the cycle. Preparation of the endometrium in women appears to be associated with diminished rather than enhanced synthesis of PGs. This does not mean that PGs are not important for inducing changes at the site of implantation, but it does mean that the embryo would have to initiate its own local signal.

Human embryos have been shown to release a platelet activating factor (PAF) (O'Neill et al. 1985). Furthermore, the success of embryo transfer is related to the ability of the embryo to release PAF in culture media before transfer (O'Neill and Saunders 1984). PAF does not stimulate the release of $PGF_{2\alpha}$ from either glandular or stromal cells of human secretory endometrium, but it does increase the release of PGE_2 (S.K. Smith and R.W. Kelly, unpublished). PAF also stimulates PGE_2 release from human amnion (Billah et al. 1985). PAF stimulates PG release in neutrophils and platelets by promoting the influx of Ca^{2+} into the cell, which would stimulate phospholipase activity and increase PG endoperoxide levels. It is not clear why this should be converted preferentially to PGE_2, but exogenous AA is converted by epithelial cells primarily to PGE_2. This differential synthesis of PGE_2 could be due to non-enzymatic conversion of endoperoxides or to saturation of $PGF_{2\alpha}$ reductase activity. Changes in the ratio of $PGF_{2\alpha}/PGE_2$ might be important for vascular permeability changes at the site of implantation, especially as the human endometrium has PGE_2 receptors on the stromal cells (Hofman et al. 1985), as is the case with rodents.

Maintenance of Pregnancy

Once implantation has occurred, continuation of pregnancy depends on the secretion of progesterone from the corpus luteum. In species such as the sheep, pig, cow and horse, which have fusion types of implantation, progesterone is required to allow the endometrium to secrete nutritive proteins required by the embryo (Bazer et al. 1986). Withdrawal of progesterone in early pregnancy causes the death of the embryo, and survival depends on its ability to signal its presence to the mother as early as possible (Short 1979).

The observation that hysterectomy in ewes extends the life-span of the corpus luteum indicated that the uterus could influence luteolysis. Furthermore, this effect arises only if the ipsilateral horn of the uterus is removed, indicating a local utero-ovarian mechanism (Moor and Rowson 1966). There is considerable evidence that $PGF_{2\alpha}$ released from the uterus is the luteolysin, since levels are elevated in uterine venous blood at the time of luteolysis in the sheep (McCracken et al. 1981), pig (Bazer et al. 1984), cow (Thatcher et al. 1985) and horse (Sharp et al. 1984). In early pregnancy, the ability of the uterus to release $PGF_{2\alpha}$ is altered, so that reduced amounts reach the ovary. In the pig, this involves redirection of synthesis (Bazer et al. 1986), whereas in the sheep there are changes in the pulsatile release of the PGs from the uterus. It is suggested that this mechanism involves the oestradiol directed synthesis of oxytocin receptors and the action of oxytocin on phospholipase activity (McCracken et al. 1984).

In the sheep and cow, the oestrous cycle is prolonged if extracts of conceptuses are instilled into the uterine cavity (Martal et al. 1979; Ellinwood et al. 1979). In these species, the active agents are a group of immunologically related proteins with molecular weights ranging from 18 to 26 kd (Helmer et al. 1987). Conceptus secretory proteins inhibit $PGF_{2\alpha}$ release in both animals (Fairclough et al. 1984; Thatcher et al. 1985).

The ability of the embryo to alter PG release and to rescue the corpus luteum may involve more subtle effects on uterine PG release. Prostaglandin E_2 may function as a luteotrophin in the ewe (Silvia et al. 1984), and several authors have failed to demonstrate reduced levels of $PGF_{2\alpha}$ in early pregnancy, suggesting that the crucial change is an alteration in the ratio of $PGF_{2\alpha}/PGE_2$ (Rawlings and Hyland 1985). In the sheep, progesterone regulates the ability of the endometrium to synthesise PGE_2 in response to the presence of an embryo (Vincent et al. 1986).

In women, hysterectomy does not result in prolongation of the luteal phase of the menstrual cycle, and uterine secretion of PGs does not induce luteolysis. Suppression of the synthesis of uterine PGs is not required in women to protect the corpus luteum, and in this regard they are quite different from the other species discussed above. However, in women, the demise of the corpus luteum in the non-pregnant cycle is reflected not only in altered steroid levels but also by the shedding of the endometrium at menstruation. This is probably the evolutionary price that women pay for the extensive degree of decidualisation that occurs in the human endometrium during the normal cycle. Bleeding will always arise from the endometrium if progesterone support is withdrawn from an oestradiol-primed endometrium, but the threshold blood levels of progesterone below which bleeding occurs can vary considerably (Smith et al. 1985).

Prostaglandins have long been implicated in the aetiology of menstruation (Baird et al. 1981), since PGs instilled into the uterine cavity induce menstrual bleeding (Wiqvist et al. 1971; Toppozada et al. 1985) and PGs given in early pregnancy induce uterine bleeding and abortion (Karim 1983). The maintenance of pregnancy in women is dependent on the suppression of PG synthesis to prevent menstrual bleeding and early abortion.

Prostaglandins and Early Pregnancy

As expected, the ability of the endometrium to synthesise and metabolise PGs is significantly altered in early pregnancy. After ovulation, presumably as a consequence of the action of progesterone, the amount of PG released from whole tissue extracts in culture is reduced compared with that from proliferative endometrium. Decidualisation of the human endometrium begins in the secretory stage of the cycle, and these changes could simply reflect changes in the ratio of stromal cells to epithelial cells. Experiments in which the glandular epithelial cells are separated from the stromal cells show this not to be the case. In proliferative endometrium the glandular cells produce more $PGF_{2\alpha}$ and PGE_2 than stromal cells, but in secretory endometrium and decidua the ability of the glandular cells to release these PGs is significantly suppressed (S.K. Smith, R. W. Kelly, unpublished) (Fig. 10.3).

Progesterone and PG Synthesis

Progesterone inhibits PG synthesis in the uterus (Abel and Baird 1980; Leaver and Richmond 1984). The study of the action of progesterone on PG synthesis

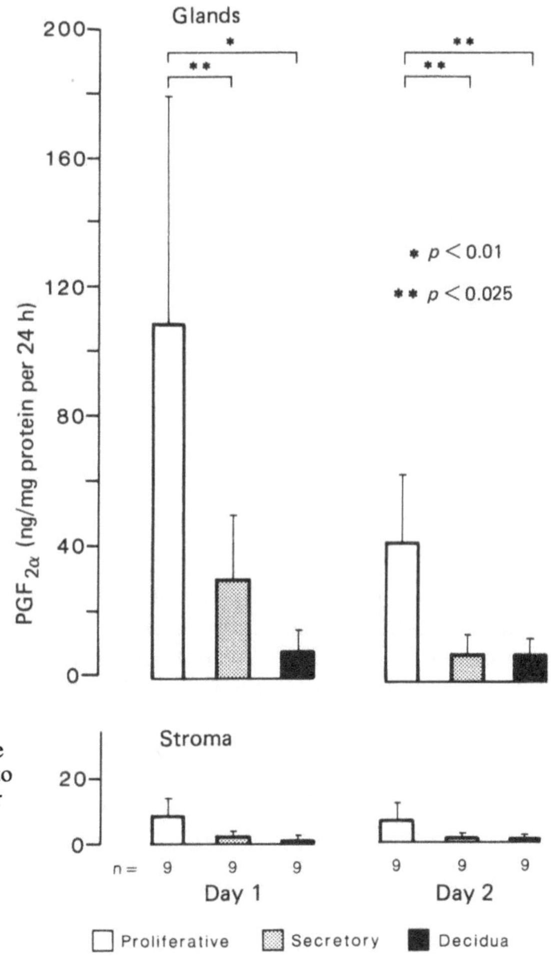

Fig. 10.3. Release of $PGF_{2\alpha}$ from glandular and stromal cells into culture media over 48 h in culture. The proliferative glandular cells have the greatest capacity to release PG, followed by secretory glandular cells. Stromal cells release small amounts of PG, and these amounts are not altered in pregnancy. The suppression of PG release that occurs in pregnancy arises from inhibition of PG synthesis by the glandular cells.

and metabolism has been greatly helped by the introduction of antiprogestins, which compete with progesterone for the nuclear progesterone receptor (for a review see Baulieu 1987). Two such agents, RU 38486 (Roussel-Uclaf) and ZK 98734 (Schering), have been studied by us (Smith and Kelly 1987). Both RU 38486 and ZK 98734 stimulate the synthesis of PGs by the glandular cells of decidua but not the stromal cells. Of particular interest is the finding that both of the antiprogestins stimulate $PGF_{2\alpha}$ and PGE_2 release in the presence of exogenous arachidonic acid, indicating that the effect of progesterone is mediated not only at the level of the release of arachidonic acid, as shown by Wilson et al. (1986), but also at the level of metabolism of arachidonic acid.

These findings are important because inhibition of PG synthesis in most other tissues is mediated by proteins that inhibit phospholipase activity. The morphological changes of implantation are similar to those of inflammation, and glucocorticoids are thought to exert their anti-inflammatory effects by stimulating genomic expression of a group of proteins, called lipocortins, with a molecular weight around 37 kd (Blackwell et al. 1980; DiRosa et al. 1984; Wallner et al. 1986). Although the human uterus can synthesise lipocortins (Gurpide et al. 1986), the inhibition of PG release found in early pregnancy is not caused by glucocorticoids, since they do not share with progesterone an ability to inhibit PG synthesis from endometrium (Kelly and Smith 1987).

These findings suggest that progesterone stimulates the synthesis of an agent which reduces PG release by epithelial cells at the start of the luteal phase. Withdrawal of progesterone support from the endometrium removes this factor and thus allows the release of PG synthesising capacity, resulting in menstrual bleeding. Pregnancy is often associated with a slight fall in progesterone levels at the end of the luteal phase (Lenton et al. 1982) which does not induce uterine bleeding. The ability of the uterus to suppress PG synthesis at this early stage of the cycle could have important consequences for successful implantation after the arrival of the embryo in the uterine cavity.

Conclusions

In laboratory animals, PGs are obligatory for decidualisation induced by non-traumatic stimuli or the presence of the embryo in the uterine cavity. Alterations or redirection of PG synthesis from the endometrium is also required in most species studied, to prevent the demise of the corpus luteum so that the progesterone support of the endometrium can be continued. This occurs in almost all animals studied, but the role of PGs in implantation and the establishment of pregnancy in women remains largely speculative.

It is not clear whether the embryo signals its presence to the endometrium by local messengers such as PAF or whether this affects the eventual implantation. However, mechanisms exist whereby local regulation of vascular permeability, vasodilation and oedema at the site of implantation could be mediated by the embryo itself. The prevention of menstruation by the continued secretion of progesterone from the corpus luteum, associated with a reduction of PG synthesis by endometrium, seems to demonstrate a more obvious change in PG synthesis which would be required for successful pregnancy.

Implantation remains one of the least understood mechanisms in the physiology of reproduction. The evidence obtained from the study of other species suggests that further research into the role of PGs and leukotrienes in implantation should enhance our understanding of the loss in early pregnancy of embryos derived from fertilisation either in vivo or in vitro.

References

Abel MH, Baird DT (1980) The effect of 17β-estradiol and progesterone on prostaglandin production by human endometrium maintained in organ culture. Endocrinology 106: 1599–1606

Baird DT, Abel MH, Kelly RW, Smith SK (1981) Endocrinology of dysfunctional uterine bleeding: The role of endometrial prostaglandins. In: Crosignani PG, Rubin BL (eds) Endocrinology of human fertility: new aspects. Academic Press, London, pp 399–417

Baulieu EE (1987) The antiprogestin steroid RU 486 and human fertility control. Plenum Press, New York

Bazer FW, Marengo SR, Geisert RD, Thatcher WW (1984) Endocrine vs exocrine secretion of $PGF_{2\alpha}$ in the control of pregnancy in swine. In: Edqvist LE, Kindahl H (eds) Prostaglandins in animal reproduction II. Elsevier Science Publishers, Amsterdam, pp 115–132

Bazer FW, Vallet JL, Roberts RM, Sharp DC, Thatcher WW (1986) Role of conceptus secretory products in establishment of pregnancy. J Reprod Fertil 76: 841–850

Bell SC (1985) Comparative aspects of decidualization in rodents and human cell types, secreted products and associated function. In: Edwards RG, Purdy JM, Steptoe PC (eds) Implantation of the human embryo. Academic Press, London, pp 71–121

Billah MM, DiRenzo GC, Ban C, Truong CT, Hoffman DR, Anceschi MM, Bleasdale JE, Johnston JM (1985) Platelet-activating factor metabolism in human amnion and the responses of this tissue to extracellular platelet-activating factor. Prostaglandins 30: 841–851

Bito LZ (1975) Are prostaglandins intracellular, transcellular or extracellular autocoids? Prostaglandins 9: 851–855

Blackwell CG, Carnuccio R, DiRosa M, Flower RJ, Parente L, Persico P (1980) Macrocortin: a polypeptide causing the antiphospholipase effect of glucocorticoids. Nature 287: 147–149

Castracane VD, Saksena SK, Shaikh AA (1974) Effect of IUDs, prostaglandins and indomethacin on decidual cell reaction in the rat. Prostaglandins 6: 243–251

Csapo AI, Pulkinnen MO, Wiest WG (1973) Effects of luteectomy and progesterone therapy in early pregnant patients. Am J Obstet Gynecol 115: 759–765

Davis DL, Pakrasi PL, Dey SK (1983) Prostaglandins in swine blastocysts. Biol Reprod 28: 1114–1118

Dey SK, Hoversland RC, Johnson DC (1982) Phospholipase activity in the rat uterus: modulation by steroid hormones. Prostaglandins 23: 619–630

DiRosa M, Flower RJ, Hirata F, Parente L, Russo-Marie F (1984) Nomenclature announcement, antiphospholipase proteins. Prostaglandins 28: 441

Downie J, Poyser N, Wunderlich M (1974) Levels of prostaglandins in human endometrium during the normal menstrual cycle. J Physiol 236:465–472

Edwards RG, Steptoe PC (1983) Current status of in-vitro fertilisation and implantation of human embryos. Lancet ii: 1265–1269

El-Banna AA (1980) The degenerative effect on rabbit implantation sites by indomethacin. I. Timing of indomethacin action, possible effect of uterine proteins and the effect of replacement doses of $PGF_{2\alpha}$. Prostaglandins 20: 587–599

Ellinwood WE, Net TM, Niswender GD (1979) Maintenance of the corpus luteum of early pregnancy in the ewe. I. Luteotrophic properties of embryonic homogenates. Biol Reprod 21: 281–288

Evans CA, Kennedy TG (1978) The importance of prostaglandin synthesis for the initiation of blastocyst implantation in the hamster. J Reprod Fertil 54: 255–261

Fairclough RJ, Moore LG, Peterson AJ, Watkins WB (1984) Effect of oxytocin on plasma concentrations of 13, 14-dihydro-15-keto prostaglandin F and the oxytocin associated neurophysin during the estrous cycle and early pregnancy in the ewe. Biol Reprod 31: 36–43.

Findlay JK (1983) The endocrinology of the preimplantation period. In: Martini L, James V (eds) Pregnancy and parturition. Current topics in endocrinology, volume 4. Academic Press, New York, pp 35–67

Finn CA (1977) The implantation reaction. In: Wynn RM (ed) Biology of the uterus, 2nd edn. Plenum Press, New York, pp 245–308

Finn CA (1983) Implantation of ova – assessment of the value of laboratory animals for the study of implantation in women. Oxford Rev Reprod Biol 5: 272–289

Gal D, Casey ML, Johnston JM, MacDonald PC (1982) Mesenchyme–epithelial interactions in human endometrium. Prostaglandin synthesis in separated cell types. J Clin Invest 70: 798–805

Gurpide E, Markiewicz L. Schatz F, Hirata F (1986) Lipocortin output by human endometrium in vitro. J Clin Endocrinol Metab 63: 162–166

Hearn JP (1986) The embryo-maternal dialogue during early pregnancy in primates. J Reprod Fertil 76: 809–819

Helmer SD, Hansen PJ, Anthony RV, Thatcher W, Bazer FW, Roberts RM (1987) Identification of bovine trophoblast protein-1, a secretory protein immunologically related to ovine trophoblast protein-1. J Reprod Fertil 79: 83–91

Hoffman LH, DePietro DL, McKenna TJ (1978) Effects of indomethacin on uterine capillary permeability and blastocyst development in rabbits. Prostaglandins 15: 823–828

Hofman GE, Rao CV, De Leon FD, Toledo AA, Sanfilippo JS (1985) Human endometrial

prostaglandin E_2 binding sites and their profiles during the menstrual cycle and in pathological states. Am J Obstet Gynecol 151: 369–375
Karim SMM (1983) Clinical applications of prostaglandins in obstetrics and gynaecology In: Karim SMM (ed) Cervagem, a new prostaglandin in obstetrics and gynaecology. MTP Press, Lancaster, pp 15–34
Kelly RW Smith SK (1987) Glucocorticoids do not share with progesterone the potent inhibitory action on prostaglandin synthesis in human proliferative phase endometrium. Prostaglandins 29:181–196
Kennedy TG (1977) Evidence for a role for prostaglandins in the initiation of blastocyst implantation in the rat. Biol Reprod 16: 268–291
Kennedy TG (1979) Prostaglandins and increased endometrial vascular permeability resulting from the application of an artificial stimulus to the uterus of the rat sensitised for the decidual cell reaction. Biol Reprod 20: 560–566
Kennedy TG (1980a) Timing of uterine sensitivity for the decidual cell reaction: role of prostaglandins. Biol Reprod 22: 519–525
Kennedy TG (1980b) Estrogen and uterine sensitisation for the decidual cell reaction: role of prostaglandins. Biol Reprod 23: 955–962
Kennedy TG (1983) Embryonic signals and the initiation of blastocyst implantation. Aust J Biol Sci 36: 531–543
Kennedy TG (1985) Prostaglandins and blastocyst implantation. Prostaglandin Perspectives 1: 1–3
Kennedy TG, Lukash LA (1982) Induction of decidualization in rats by the intrauterine infusions of prostaglandins. Biol Reprod 27: 253–56
Kennedy TG, Zamecnik J (1978) The concentration of 6-keto prostaglandin $F_{1\alpha}$ is markedly elevated at the site of blastocyst implantation in the rat. Prostaglandins 16: 599–606
Kennedy TG, Martel D, Psychoyos A (1983) Endometrial prostaglandin E2 binding during the estrous cycle and its hormonal control in ovariectomized rats. Biol Reprod 29: 565–571
Kennedy TG, Keys JL, King GJ (1986) Endometrial prostaglandin E_2 binding sites in the pig: characterization and changes during the estrous cycle and early pregnancy. Biol Reprod 35: 624–632
Lapetina EG (1982) Regulation of arachidonic acid production: role of phospholipase C and A_2. Trends Pharm Sci 3: 115–118
Lau IF, Saksena SK, Chang MC (1973) Pregnancy blockade by indomethacin, an inhibitor of prostaglandin synthesis: its reversal by prostaglandins and progesterone in mice. Prostaglandins 4: 795–803
Leaver HA, Richmond DH (1984) The effect of oxytocin, estrogen, calcium ionophore A23187 and hydrocortisone on prostaglandin $F_{2\alpha}$ and 6-oxo-prostaglandin $F_{1\alpha}$ production by cultured human endometrial and myometrial explants. Prost Leuk Med 13: 179–196
Lejeune B, Van Hoeck J, Leroy F (1981) Transmitter role of the luminal uterine epithelium in the induction of decidualization in rats. J Reprod Fertil 61: 235–44
Lenton EA, Sulaiman R, Sobowale O, Cooke ID (1982) The human menstrual cycle: plasma concentrations of prolactin, LH, FSH, oestradiol and progesterone in conceiving and non-conceiving women. J Reprod Fertil 65: 131–139
Lundqvist O, Nilsson BO (1980) Ultrastructural changes of the trophoblast-epithelial complex in mice subject to implantation blocking treatment with indomethacin. Biol Reprod 22: 719–726
McCracken JA, Schramm W, Barcikowski L, Wilson L (1981) The identification of prostaglandin $F_{2\alpha}$ as a uterine luteolytic hormone and the hormonal control of its synthesis. Acta Vet Scand Suppl 77: 71–88
McCracken JA, Schramm W, Okulicz WC (1984) Hormone receptor control of pulsatile secretion of $PGF_{2\alpha}$ from the ovine uterus during luteolysis and its abrogation in early pregnancy. In: Edqvist LE, Kindahl H (eds) Prostaglandins in animal reproduction II. Elsevier Science Publishers, Amsterdam, pp 31–56
McLaren A (1985) The control of implantation. In: Thompson W, Joyce DN, Newton JR (eds) In vitro fertilization and donor insemination. Royal College of Obstetricians and Gynaecologists, London, pp 13–22
Martal J, Lacroix M, Loudes C, Saunier M, Wintenberger-Torres (1979) Trophoblastin an antiluteolytic protein present in early pregnancy in sheep. J Reprod Fertil 56: 63–73
Mizumo K, Yamamoto S, Lands WEM (1982) Effects of non-steroidal anti-inflammatory drugs on fatty acid cyclo-oxygenase and prostaglandin hydroperoxidase activities. Prostaglandins 23: 743–757
Moor RM Rowson LEA (1966) Local uterine mechanisms affecting luteal function in the sheep. J Reprod Fertil 11: 307–310

Oettel M, Koch M, Kurischko A, Schubert K (1979) A direct evidence for the involvement of prostaglandin $F_{2\alpha}$ in the first step of estrone-induced blastocyst implantation in the soayed rat. Steroids 33: 1–8

O'Neill C, Saunders DM (1984) Assessment of embryo quality. Lancet ii: 1035

O'Neill C, Pike I, Porter N, Gidley-Baird AA, Sinosich MJ, Saunders DM (1985) Maternal recognition of pregnancy prior to implantation: methods for monitoring embryonic viability in-vitro and in-vivo. Ann NY Acad Sci 442: 429–439

Pakrasi PL, Dey SK (1982) Blastocyst is the source of prostaglandins in the implantation site in the rabbit. Prostaglandins 24: 73–77

Phillips CA, Poyser NL (1981) Studies on the involvement of prostaglandins in implantation in the rat. J Reprod Fertil 62: 73–81

Psychoyos A, Martel D (1985) Embryo-endometrial interactions at implantation. In: Edwards RG, Purdy JM, Steptoe PC (eds) Implantation of the human embryo. Academic Press, London, pp 197–219

Rankin JC, Ledford BE, Jonsson HT, Baggett B (1979) Prostaglandins, indomethacin and the decidual cell reaction in the mouse uterus. Biol Reprod 249: 399–404

Rawlings NC, Hyland JH (1985) Prostaglandin F and E levels in the conceptus, uterus and plasma during early pregnancy in the ewe. Prostaglandins 29: 933–950

Rees MCP, Parry DM, Anderson ABM, Turnbull (1982) Immunohistochemical localisation of cyclooxygenase in the human uterus. Prostaglandins 23: 207–214

Saksena SK, Lau IF, Chang MC (1976) Relationship between oestrogen, prostaglandin $F_{2\alpha}$ and histamine in delayed implantation in the mouse. Acta Endocrinol 42: 225–232

Samuelsson B, Goldyne M, Granstrom E, Hamberg M, Hammarstrom S, Malmsten C (1978) Prostaglandins and thromboxanes. Annu Rev Biochem 47: 997–1029

Sananes N, Baulieu EE, Le Goascogne C (1976) Prostaglandins as inductive factor of decidualization in the rat uterus. Mol Cell Endocrinol 6: 153–158

Sananes N, Baulieu EE, Le Goascogne C (1981) A role for prostaglandins in decidualization of the rat uterus. J. Endocrinol 89: 25–33

Schatz F, Markiewicz L, Barg P, Gurpide E (1985) In vitro effects of ovarian steroids on prostaglandin $F_{2\alpha}$ output by human endometrium and endometrial epithelial cells. J Clin Endocrinol Metab 61: 361–367

Sharp DC, Zavy MT, Yernon MW, Bazer FW, Thatcher WW, Berglund LA, McDowell KJ (1984) The role of prostaglandins in the maternal recognition of pregnancy in mares. Anim Reprod Sci 7: 269–282

Short RV (1979) When a conception fails to become a pregnancy. In: Maternal recognition of pregnancy. Elsevier Press, Amsterdam, pp 197–219 (Ciba Foundation Symposium 64)

Shutt DA, Lopata A (1981) The secretion of hormones during the culture of human preimplantation embryos with corona cells. Fertil Steril 35: 413–416

Singh EJ, Baccarini IM, Zuspan FP (1975) Levels of prostaglandins $F_{2\alpha}$ and E_2 in human endometrium during the menstrual cycle. Am J Obstet Gynecol 121: 1003–1006

Smith SK (1986) Aetiology of menstrual disorders. Prostaglandin Perspectives 2: 8–9

Smith SK, Kelly RW (1987) The effect of the anti-progestins, RU 486 and ZK 98734 on the synthesis and metabolism of $PGF_{2\alpha}$ and PGE_2 in separated cells from early human decidua. J Clin Endocrinol Metab 62: 527–534

Smith SK Abel MH Baird DT (1984) Effects of 17 beta-estradiol and progesterone on the levels of prostaglandins F_2 alpha and E in human endometrium. Prostaglandins 27: 591–597

Smith SK, Lenton EA, Cooke ID (1985) Plasma gonadotrophin and ovarian steroid concentrations in women with menstrual cycles with a short luteal phase. J Reprod Fertil 75: 363–368

Silvia WJ, Ottobre JS, Inskeep EK (1984) Concentrations of prostaglandins E_2, $F_{2\alpha}$ and 6-keto-prostaglandin $F1_{\alpha}$ in the utero-ovarian venous plasma of non-pregnant ewes. Biol Reprod 30: 936–944

Thatcher WW, Knickerbocker JJ, Bartol FF, Bazer FW, Roberts RM, Drost M (1985) Maternal recognition of pregnancy in relation to the survival of transferred embryos: endocrine aspects. Theriology 23: 129–144

Toppozada M, El-Attar A, El-Ayyat MA, Khamas Y (1985) Management of uterine bleeding by PGs or their synthesis inhibition. Adv Prostaglandin Thromboxane Res 8: 1459–1463

Tsang BK, Ooi TC (1982) Prostaglandin secretion by human endometrium in vitro. Am J Obstet Gynecol 142: 626–633

Vincent DL, Meredith S, Inskeep EK (1986) Advancement of uterine secretion of prostaglandin E_2 by treatment with progesterone and transfer of asynchronous embryos. Endocrinology 119: 527–529

Wallner BP, Mattaliano RJ, Hession C, et al. (1986) Cloning and expression of human lipocortin, a phospholipase A_2 inhibitor with potential anti-inflammatory activity. Nature 320: 77–80
Wilmut I, Sales DI, Ashworth CJ (1986) Maternal and embryonic factors associated with prenatal loss in mammals. J Reprod Fertil 76: 851–864
Wilson T, Liggins GC, Aimer GP, Watkins EJ (1986) The effect of progesterone on the release of arachidonic acid from human endometrial cells stimulated by histamine. Prostaglandins 31: 343–360
Wiqvist N, Bygdeman M, Kirton K (1971) Non-steroidal infertility agents in the female. In: Diczfalusy E, Borell B (eds) Nobel symposium 15, Control of human fertility, p 137
Wynn RM (1977) Histology and ultrastructure of the human endometrium. In: Wynn RM (ed) Biology of the uterus. Plenum Press, New York, pp 341–376
Yovich JL, Stanger JD, Yovich JM, Tuvick AI (1984) Quality of embryos from in-vitro fertilisation. Lancet i: 457

Section VI
REPRODUCTIVE IMMUNOLOGY

11. Current Concepts of Immunoregulation of Implantation

D. A. Clark

Introduction

A successful implantation from the clinical standpoint depends upon a coordinated sequence of events before, at, and immediately after nidation of the conceptus in the wall of the uterus. In man, the conceptus makes contact with the maternal endometrium 5–8 days before the expected time of onset of the next menstrual period. Occult abortion during this time may be viewed as a failure of implantation, just as damage to the conceptus before contact with the endometrium may prevent nidation. It is within this broad time framework that the immunological issues relating to the preimplantation, implantation, and postimplantation stages of mammalian pregnancy will be discussed in this chapter. Current knowledge of the effects of immune responses which occur spontaneously or after deliberate immunisation will also be reviewed. Most of the available information has been obtained from studies of animals, because human peri-implantation uterine contents are seldom accessible for investigation.

Immunological Status of the Conceptus

Cells of the developing conceptus bear antigens rendering it foreign to the mother. Some of these antigens represent major and minor histocompatibility antigens inherited via the paternal genome (Smith et al. 1982; Billington and Bell 1983; Clark et al. 1986b). Other antigens represent organ-specific and embryonal (oncofetal) antigens (Anderson and Alexander 1979; Hamilton 1983; Clark et al; 1986b).

Survival of the conceptus at each stage of its evolution (preimplantation, implantation, and postimplantation) depends upon avoidance of harm from maternal immunity. Avoidance may come about by lack of generation of harmful types of immunity, by lack of accumulation of the effectors of immunity at the site of the embryo or by an insusceptibility of the embryonic tissues to damage by the maternal effector mechanisms causing graft rejection.

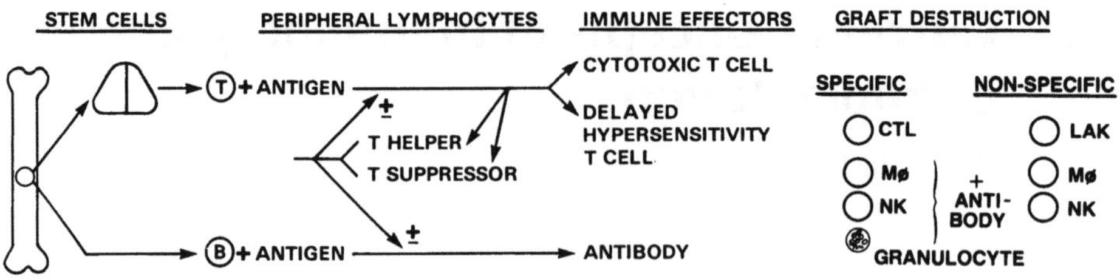

Fig. 11.1. Simple model of the mammalian immune system. The immune system is a cell renewal system originating from stem cells that differentiate into immunocompetent (antigen-reactive) precursors of effector cells. Effector cells playing a major role in destruction of grafts of solid tissue allografts are shown. The T-helper/delayed hypersensitivity cell is particularly important in producing soluble lymphokines, which recruit and activate cytotoxic cell types that directly damage the graft.

Graft rejection as currently understood for adult tissue allografts and tumours is complex. Several cell types are potentially destructive to grafts. They include cytotoxic T cells, lymphotoxin-producing T cells, lymphokine activated killer T cells and NK cells, activated macrophages and their toxins, and antibody in association with complement (rarely) or with cytotoxic cells bearing receptors for the Fc end of antibody (Mason and Morris 1986). Susceptibility or resistance to a particular type of effector mechanism may vary between different types of cells in a graft (Hayry 1984); some cells may be destroyed primarily by non-antigen-specific paraimmune effector mechanisms such as natural killer (NK) cells, natural cytotoxic (NC) cells and macrophages and their toxins. A model of the immune system and the production of effectors of graft rejection is shown in Fig. 11.1.

During pregnancy, the embryo passes through several phases of tissue reorganisation, and each phase is accompanied by alterations in antigen expression and in susceptibility to particular effector mechanisms (Billington and Bell 1983; Heyner 1983; Zuckermann 1986). This evolution is summarised in Fig. 11.2.

Active suppression of immunity may play an important part in protecting the embryo. Some of these events lead to active suppression of the development of maternal immunity. Active suppression may be independently mediated by cells of the female host (Clark et al. 1986a–c; Brierley and Clark 1987), by the embryo (Daya et al. 1986) or as a consequence of a dialogue between the two partners (Slapsys 1987). The pregnant female also undergoes a series of hormonal (Hearn 1986) and immunological (Clark 1985; Clark et al. 1986a–c) changes synchronised with the developments in the embryo.

Preimplantation Phase

Antigen Expression

Major histocompatibility (MHC) antigens class I are expressed at low levels on cleaving eggs and blastocysts, as detected by sensitive antibody methods or by cytotoxic T lymphocyte (CTL) killing (Billington and Bell 1983; Ewoldsen et al.

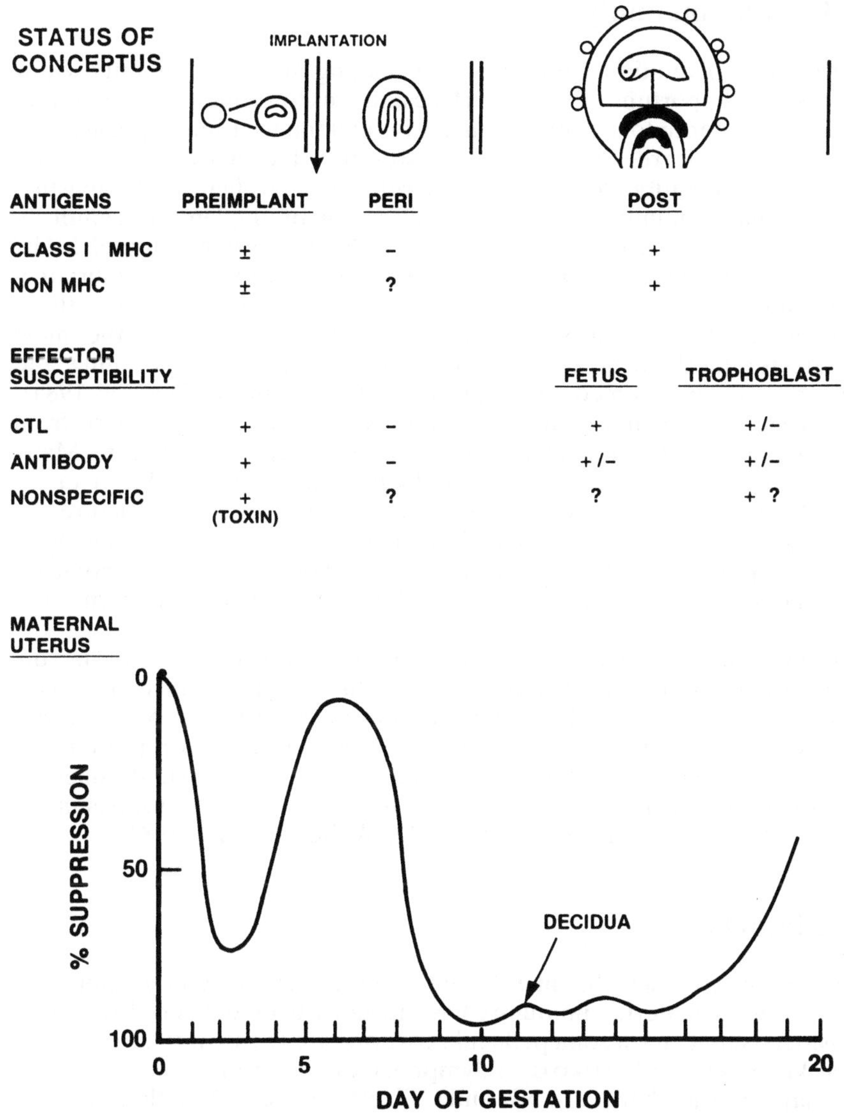

Fig. 11.2. Relationship of embryo development to antigen expression and susceptibility to effector mechanisms (*upper figure*) and suppressor activity in the uterine lining (endometrium/decidua) (*lower figure*). The majority of the data have been obtained from studies in laboratory mice (gestation 20–21 days) and confirmed in human tissue where possible.

1986). Minor (non-MHC) H antigens (and a variety of embryonic antigens such as the F9 determinant) are more readily detected with antibody probes (Heyner 1983). Antigens of the spermatozoal membrane may also be expressed on the membrane of the early cleaving embryo (morula) for the first few cell divisions and react with antibody (Gundersen et al. 1986).

Susceptibility to Immunity

Because the preimplantation embryo is free floating and contains only a few cells, there is little opportunity for cell surface antigens to stimulate the maternal immune system. Spermatozoa are taken up by endometrial macrophages and may produce some degree of maternal sensitisation (Rutherford and Searle 1984), but cytotoxic antibodies and effector cells are not elicited. Pre-existing cytotoxic maternal immunity may, however, affect the preimplantation embryo. If the zona pellucida is removed, cytotoxic T cells recognising paternal MHC antigens can lyse the blastocyst (Ewoldsen et al. 1986). Antibodies to embryonal F9 and possibly sperm associated antigens can also have damaging effects (Hamilton et al. 1979; Allardyce 1984). IgA antibody, which is the main humoral effector mechanism at mucosal surfaces such as the gut, lung and genitourinary tract, is assumed to be primarily responsible (Allardyce 1984). The mechanism by which the female acquires such immunity (apart from deliberate immunisation with sperm or embryonic cells) is unclear. Many embryonic antigen determinants are carbohydrate in structure (Stern et al. 1983); possibly the immunity is stimulated by similar carbohydrate structures on gut bacteria. In individuals who are genetically high responders to these antigens, the precursors of IgA-producing B cells may be activated; these B cells home on mucosal tissue including the uterus, where antibody is released (McDermott et al. 1980).

In addition to antigen specific CTL and antibody, non-specific toxic factors may damage the preimplantation embryo. The soluble toxic products of macrophages, such as interleukin 1 (IL-1), may account for infertility in patients with endometriosis (Morcos et al. 1985). Hill et al. (1986) have also reported that soluble products of T cells may be toxic to the murine blastocyst. Products of polymorphonuclear leucocytes may also exert toxic effects on blastocysts in the uterine lumen (Anderson and Alexander 1979; Waites and Bell 1982).

Defence Mechanisms

Several defence mechanisms that may serve to prevent the initiation or expression of harmful forms of maternal immunity have recently been defined. Preimplantation blastocysts that lack a zona pellucida covering are resistant to direct killing by natural killer cells and cytotoxic T lymphocytes. Before hatching, the zona pellucida may serve as a barrier to NK and CTL effector cells. Whether there is a brief period after hatching when the blastocyst is susceptible to killer cells is unclear.

Our laboratory has been particularly interested in mechanisms blocking the production of CTL by the maternal immune system. During early allogeneic pregnancy in mice, the regional uterine draining lymph nodes (DLN) undergo enlargement; CTL responses are suppressed at the time of implantation, but antibody responses remain normal (Clark et al, 1986b). Study of the uterine lining upstream from these lymph nodes has shown that immediately following mating, suppressor cells appear (Clark 1985; Clark et al. 1987c). These suppressor cells are large, hormonally induced, and act directly without an intervening soluble factor to block CTL generation (Brierley and Clark 1987). A similar cell is present in the endometrium of women during the luteal phase of the menstrual cycle (Daya et al, 1985). In the mouse, these suppressor cells bear T cell markers (Brierley and Clark 1987). If there is no preimplantation embryo within the uterine lumen, suppressor cell activity persists for 7–10 days after initiation of pseudopregnancy. If a blastocyst is present in the uterus, there is an increase in lymphatic flow to the DLN, and suppressor cells disappear from the uterine lining (Clark 1985; Head et al. 1986; Head 1987). It appears that suppressor cells traffic to the DLN, where two components may be identified – a small non-specific suppressor cell and an antigen specific T cell component (Clark et al. 1986b; Thomas and Erickson 1986). It is proposed that suppressor cells in the DLN inhibit the generation of harmful types of immune responses such as CTL (i.e., transplantation immunity) that may be initiated in the DLN. Further, suppression in DLN is selective, for antibody responses in the DLN do not appear to be inhibited (Clark 1985). Searle (1986) has shown that antigenic cells injected into the uterine lumen may elicit cytotoxic cell responses in the spleen but not in DLN. It is possible that antigens that enter maternal blood and go directly to the spleen may not elicit harmful types of responses such as may occur with antigens that go to the DLN.

There is also evidence that the embryo itself may directly suppress immune responses. Blastocysts added to lymphocyte cultures non-specifically block proliferation and generation of cytotoxic T cells (Croy et al, 1983). In vivo, however, the generation of CTL occurs at a considerable distance from the blastocyst, and soluble and diffusible factors would then have to be produced if suppression of CTL generation by embryo products were important for embryo survival. Supernatants from human eggs fertilised in vitro can inhibit maternal T cell reactivity (Daya et al, 1986). This suppression has been associated with two small molecules (5 kd and 1 kd) associated with approximately 40%–60% of fertilised eggs (Daya et al, 1986; Daya and Clark 1988) – a proportion similar to the proportion of conceptions produced by in-vitro fertilisation that implant (Daya et al, 1986; reviewed in Clark et al. 1987a). The IVF eggs are associated with the factor primarily during the first 24 h after fertilisation and to a lesser extent during the second 24 h (Daya et al, 1986). In a collaborative study of more than 200 patients, we have found that pregnancies occurred primarily when cleaving IVF eggs positive for suppression were transferred (Daya et al. 1986 and unpublished). The implantation and pregnancy rates with IVF are lower than with eggs fertilised in vivo. Fertilisation of eggs within fallopian tubes ensures that any factors associated with fertilised eggs stay within the patient, producing higher rates of subsequent pregnancy than IVF (Daya et al, 1986). Therefore, immunosuppressive factors associated with fertilised eggs may play a part in successful human pregnancy. However, the physiological mechanism of factor action need not be immunosuppression – an essential step in implantation is release of vasoactive substances such as histamine (Kennedy 1983). Factors associated with eggs at the time of fertilisation may prepare the uterine lining for subsequent events at implantation. We have not found any evidence of soluble suppressor factors released by the blastocyst itself, perhaps because the pre-implantation blastocyst becomes resistant to direct killing by maternal effector cells and needs no additional protection.

Implantation Phase

Antigen Expression

The developing blastocyst at implantation consists of an inner cell mass that will become the fetus and an outer trophoblast layer that will form the fetomaternal interface including the placenta. In rodents, implantation is triggered by oestrogen (either injected or possibly produced by the blastocyst produced) (Beer and Billingham 1974; Kennedy 1983). Expression of MHC and non-MHC minor H antigens is shut off (Billington and Bell 1983; Leclipteu and Remacle 1983; Marticorena et al. 1983). Shut-off of H antigen expression lasts about 3–4 days in mice. However, certain embryonic antigens are expressed both by the outer trophoblast layer of the implanting embryo and by the adjacent epithelium (and maternal B cells may be sensitised to these antigens [Hamilton et al, 1985]).

Susceptibility to Immunity

There is no information about immunological and paraimmunological factors that may damage the embryo at implantation. Nevertheless, it is very likely that such mechanisms will be defined in future. For example, NK cells arise in the uterine lining at the time of implantation and are then actively down-regulated, (Gambel et al. 1985).

Defence Mechanisms

Several putative defence mechanisms exist at the implantation site that could serve to protect the embryo from "rejection". A zone of decidua develops about the implant, and no macrophages are present in this zone (Head 1987). As already mentioned, suppressor cells disappear from the endometrium at implantation, but the presence of an insulating layer of decidua, plus shut-off of H antigen expression, may afford sufficient protection. However, there may also be direct suppressive effects exerted by the cells of the embryonic trophoblast. Van Vlasselaer and Vandeputte (1984) were able to obtain soluble suppressor factors after trophoblast began to outgrow from blastocysts in vitro (in-vitro implantation) as well as from supernatants from ectoplacental cone trophoblast. Similar factors were not released by fibroblasts. Suppression could be reversed by antiprogesterone antibody. Since murine trophoblast at this stage does not produce progesterone at suppressive levels (and probably never does even in later pregnancy [Sio 1985]), the steroid would appear to be associated with a carrier molecule involved in the suppressive effect. We also show direct suppressive effects of blastocysts and ectoplacental cone (EPC) cells on CTL generation in vitro (Croy et al. 1983) but have not yet been able to obtain convincing evidence of soluble suppressor factors. Indeed, transfers of EPC and embryonic tissue from 7.5 day mouse embryos to the kidney capsule in allo-immune recipients showed that the tissue of the embryo per se was rejected whereas the trophoblast giant cells persisted unharmed (Simmons and Russell 1966). If a soluble suppressor factor were being produced, it should have protected the adjacent embryonic tissue. Taken together, the data suggest that protection of the embryo at implantation results from lack of antigen expression, separation from contact with non-specific effector NK and macrophage cells of the mother form a trophoblast resistant to maternal effector cells, and possibly by a direct suppression of effector cells. Maternally derived suppressive molecules may also include histamine and prostaglandins; no suppressor cells are present, however.

There is some evidence that certain forms of maternal immunity might help rather than hinder implantation. For example, Beer and Billingham (1974) noted that rodent uterine horn that had been sensitised by local injection of allogeneic cells would manifest a marked swelling response to challenge, perhaps through histamine release. Implantation was enhanced in that horn. There is no evidence that antibodies that degranulate mast cells (Rahima and Soderwall 1977) play any part in implantation in normal circumstances.

Postimplantation Phase

Antigen Expression

On day 9.5 of mouse allopregnancy (5 days postimplantation), the placenta is formed, with clear differentiation of trophoblast into a labyrinthine zone, interposed between fetal and maternal blood, and a spongioform layer that contacts decidua between gaps left by degenerating trophoblast giant cells. The spongioform trophoblast expresses class I paternal MHC that can bind anti-H-2 antibody circulating in maternal blood (reviewed in Clark et al. 1986b)

Susceptibility to Immunity

By day 12–13 the cells of the developing fetus that express paternal MHC have become susceptible to CTL (Croy and Rossant 1986). The MHC antigen-bearing spongioform trophoblast may also be susceptible in vitro to killing by CTL, but it is resistant to killing by NK and NC cell (Zuckermann 1986).

In vivo, allogeneic embryos are resistant to maternal immunity to paternal MHC in most circumstances. Intensive alloimmunisation can lead to abortion in some cases, but there is a great deal of variation in susceptibility to this effect between laboratories. Zuckermann (1986) has shown that injection of alloimmune and pregnant mice with rabbit complement can produce acute placental damage with abortion and growth retardation of those embryos surviving abortion. This effect appears to be antibody mediated, is unassociated with any infiltration of the embryo by effector cells, and involves damage to trophoblast, leading to fetal ischemia. Acute ischaemic death, perhaps mediated by vascular spasm or Schwartzman-type reactions, may also be seen with injection of antibody to alphafetoprotein and with a fraction of non-immunoglobulin-containing rabbit serum (Mizejewski and Vonnegut 1984; Waters et al. 1985). Non-MHC histocompatibility antigens, embryonic and trophoblast antigens, may also be expressed at the interface with the mother (Clark et al. 1986b).

Although it is possible that antibodies to these antigens may have effects on the early postimplantation embryo, definitive evidence is lacking. Antibody to H-2 that enter the fetus appears not to be harmful (Bell and Billington 1983). Antibody to antigens of yolk sac may, however, produce damage leading to congenital anomalies (Lerman et al. 1986), and immunisation of mice against F9 antigen (Hamilton 1983) can lead to spontaneous abortion and possibly malformations, as well as preimplantation failure. However, it is quite possible that abortion after implantation (which may also occur with antisperm immunity [Clark 1984; Clark et al. 1987c]) are caused by damage at an earlier stage of pregnancy that retards the rate of embryonic development and deployment of defences against maternal cellular immunity in decidua.

Defence Mechanisms

By the 5th day after implantation in mice, a novel population of suppressor cells appears at the implantation site in the decidua. These cells are small, lymphocytic, contain cytoplasmic granules, lack the surface markers of T and NK cells, and bear receptors for the Fc end of immunoglobulin G (Slapsys et al. 1985; Clark et al. 1986b). Soluble factors of 80–100, 40, and 20kd sizes are released, and these block the response of maternal T and NK cells to the lymphokine interleukin 2 (IL-2 or T cell growth factor [Clark et al. 1986b]) These inhibitory molecules are not prostaglandins (Clark et al. 1986a). IL-2 released by T helper DHT (delayed hypersensitivity T) cells have a key role in the activation of CTL, non-specific LAK (lymphokine activated killer) cells, and NK cells (Fig. 11.1; Clark et al. 1986b). Preliminary studies have also indicated that supernatants from murine pregnancy decidua can also block the action of other lymphokines, such as IL-3, that activate natural cytotoxic (NC) cells and can inhibit the gamma-interferon and LPS-mediated activation of macrophages to cytotoxicity. There may also be a population of large cells in decidua that release suppressor factors (Clark et al. 1986a).

A key point about local suppression in the decidua is that it is dependent upon recruiting signals from trophoblast (Slapsys 1987). Sio (1985) has shown that skin allografts implanted on the decidua of the rat uterus remote from trophoblast were promptly rejected if the animals were immunised; such decidua should contain macrophages and other cells producing prostaglandins and a variety of other putative suppressor molecules, including pregnancy associated plasma protein-A and a factor described by Dearden-Badet (1986). In contrast, grafts placed at the choriodecidual junction of immune rats were protected (Sio 1985). We believe that local recruitment of non-T suppressor cells by trophoblast can account for this protection (Clark et al. 1986a).

In certain circumstances, local suppressor cell activity in decidua is reduced. Xenogeneic trophoblast appears unable in certain circumstances to recruit adequate levels of suppression (Clark 1985; Slapsys 1987) If this happens, resistance to maternal cellular immunity is reduced, and maternal NK cells and CTL can cross the trophoblast and enter the fetus, which then undergoes necrosis and resorbs within its trophoblastic shell (Clark et al 1986c). It is not clear that local suppressor deficiency acts by allowing the development of a maternal immune response that otherwise would have been expressed. Rather, it appears that pre-existing immunity is able to be expressed (Clark et al, 1986c).

Pregnant females may make CTL against virally infected cells distant from the uterus; these CTL recognise virus-modified self class I MHC and may also recognise paternal class I antigens on the fetus. Xenoantigens may be recognised by CTL as if they were alloantigens. In the absence of factors blocking IL-2, any stimulus releasing IL-2 may reactivate T cells, such as CTL, that have previously been sensitised to "modified self". The abortion associated with lymphocytic infiltration of the fetus is seen in the *Mus caroli/Mus musculus* system, in CBA/J mice mated to DBA/2, in B10/J mice mated to B10/A, in horses carrying donkey × donkey conceptuses, and in human IVF embryos failing in the human uterus after implantation (reviewed in Clark et al. 1986b, 1987a). It is not, however, clear that a T-cell infiltrate is essential for abortion, and NK cells and macrophages that kill primitive cells may be involved in attacking embryos unprotected by local suppression (Clark et al. 1986b).

Mechanisms of Fertility Promotion by Maternal Immunity

The emphasis so far in this chapter has been on mechanisms by which maternal immunity may damage the conceptus and mechanisms of protection against this potential damage. The sequence of protective events is summarised in Fig. 11.2. An interesting feature of some of the spontaneous abortion models is that immunisation against paternal antigens may actually prevent the abortion. This result has been obtained in the CBA × DBA/2 spontaneous abortion model, in the B10 × B10/A model, in the horse-donkey system, and in some patients with recurrent unexplained abortion (Clark et al. 1986b, 1987a; Bobe and Kiger 1986; Allen et al. 1986; Chaouat and Clark 1986; Chaouat et al. 1985; Mowbray 1986).

In the CBA × DBA/2 system, protection against abortion is mediated by soluble antigen-specific factors – probably antibodies (Chaouat et al. 1985) – that boost local suppressor activity at the implantation site (Clark et al. 1987b) Immunogenetic studies have shown that non-MHC H antigens are probably important, since third-party male cells (genetically different from the DBA/2 male) are needed for protection (Bobe and Kiger 1986). Indeed, DBA/2 cells given systemically or locally into the uterus appear to boost immunity that leads to "rejection" (Chaouat and Kolb 1986). In the B10 × B10/A system, in contrast, male B10/A spleen cells are effective in immunising against abortion (Chaouat and Clark 1986).

In women with recurrent unexplained first-trimester abortion immunisation with paternal cells has achieved a 70%–80% efficacy in preventing subsequent abortion (Mowbray 1986). The stimulation of circulating antibodies appears to be a correlate of successful immunisation. However, it is unclear if the antibodies detected in vitro are the actual mediators of changes in the uterus that lead to protection. The proportion of abortions occurring spontaneously in human conception is much greater in early pregnancy, particularly during the interval between implantation and the expected time of onset of the next menstrual period (Clark et al. 1987a). Some failures of IVF/embryo transfer are thought to be due to occult abortion, and Mowbray (personal communication) has shown that a pregnancy success rate of approximately 50% can be achieved in such cases by immunising the woman against paternal leucocytes. These data suggest that immunisation may increase the apparent success of implantation by preventing occult abortion immediately after nidation. It seems likely that some cases of unexplained infertility might also be due to occult abortion following unsuccessful nidation and might be considered for immunisation therapy.

How immunisation increases pregnancy success rates is still uncertain. The role of increased local active suppression following implantation has been emphasised (Chaouat et al. 1985; Clark et al. 1987b). Local suppression is, however, dependent on trophoblast cell activity, and Athanassakis and colleagues (Athanassakis and Wegmann 1986; Athanassakis et al. 1987) have proposed that sensitisation of T cells to paternally derived antigens on trophoblast may lead to release of soluble growth factors in the placental bed and stimulate placental trophoblast growth. Stimulatory factors are also present in suppressive supernatants derived from murine postimplantation decidua (Athanassakis and Wegmann 1986). Although the trophic factor(s) in these supernatants may have arisen from T cells in decidua, it can be argued that non-T suppressor cells may also have contributed. For example, elimination of T cells producing placental growth factors in vivo resulted in smaller placentae, but not to the extent that the frequency of resorption was significantly increased (Athanassakis et al. 1987). Decidual suppressor activity persists in T-cell-deficient *Nu/Nu* mice (Clark et al. 1986a, b). Placental trophoblast bears receptors to epidermal growth factor (EGF), and deficiency of EGF leads to abortion (Tsutsumi and Oka 1987). Further, EGF has suppressive activity (Koch et al. 1984). Whether the non-T-cell-derived molecules that mediate the suppression in decidual supernatants also stimulate trophoblastic growth is a subject for further research. Whether there are growth factors acting at the preimplantation (Tzartos and Surani 1979; Papaioannou and Ebert 1986) and implantation stages of pregnancy that can be stimulated by immunisation (Beer and Billingham 1974) is also being investigated.

References

Allardyce RA (1984) Effect of ingested sperm on fecundity in the rats. J Exp Med 159: 1548–1553

Allen WR, Kydd JH, Antczak DF (1986) Successful application of immunotherapy to a model of pregnancy failure in equids. In: Clark DA, Croy BA (eds) Reproductive immunology 1986. Elsevier, Amsterdam, pp 253–260

Anderson DJ, Alexander NJ (1979) Induction of uterine leukocytosis and its effect on pregnancy in rats. Biol Reprod 21: 1143–1152

Athanassakis I, Wegmann TG (1986) The immunotrophic interaction between maternal T cells and fetal trophoblast/macrophages during gestation. In: Clark DA, Croy BA (eds) Reproductive immunology 1986. Elsevier, Amsterdam, pp 91–105

Athanassakis I, Bleachley RC, Paetkau V, Gilbert L, Barr PJ, Wegmann TG (1987) The immunostimulatory effect of T cells and T cell lymphokines in murine fetally derived placental cells. J Immunol 138: 37–44

Beer AE, Billingham RE (1974) Host responses to intrauterine tissue, cellular and fetal allografts. J Reprod Fertil suppl 21: 59–88

Bell SC, Billington WD (1983) Anti-fetal allo-antibody in the pregnant female. Immunol Rev 75: 5–30

Billington WD, Bell SC (1983) Immunobiology of mouse trophoblast. In: Loke YW, Whyte A (eds) Biology of trophoblast, Elsevier, Amsterdam, pp 571–595

Bobe P, Kiger N (1986) Role of minor histocompatibility antigens in fetal tolerance. In: Clark DA,

Croy BA (eds) Reproductive immunology 1986. Elsevier, Amsterdam, pp 246–252
Brierley J, Clark DA (1987) Characterization of hormone-dependent suppressor cells in the uterus of pregnant and pseudopregnant mice. J Reprod Immunol 10: 201–218
Chaouat G, Clark DA (1986) Another model of murine spontaneous abortion and immunological treatment. J Reprod Immunol suppl 1: 135
Chaouat G, Kolb J–P (1986) Immunological concomitants of the CBA/J vaccination against abortion. J Reprod Immunol suppl 1: 134
Chaouat G, Kolb J–P, Kiger N, Stanislawski M, Wegmann TG (1985) Immunological consequences of vaccination against spontaneous abortion in CBA/J mice. J Immunol 134: 1594–1598
Clark DA (1984) Local suppressor cells and the success or failure of the "fetal allograft". Annals d'Immunol (Inst Pasteur) 135D: 321–324
Clark DA (1985) Materno-fetal relations. Immunol Lett 9: 239–247
Clark DA, Damji N, Chaput A, Rosenthal KL, Brierley J (1986a) Decidua-associated suppressor cells and suppressor factors regulating interleukin 2: their role in survival of the "fetal allograft". In: Cinader B, Miller RG (eds) Progress in immunology VI. Academic Press, New York
Clark DA, Slapsys R, Chaput A, et al. (1986b) Immunoregulatory molecules of trophoblast and decidual suppressor cell origin. Am J Reprod Immunol Microbiol 10: 100–104
Clark DA, Croy BA, Rossant J, Chaouat G (1986c) Immune presensitization and intrauterine defences as determinants of success or failure of murine interspecies pregnancies. J Reprod Fertil 77: 633–643
Clark DA, Chaput A, Slapsys R, et al (1987a) Local intrauterine suppressor cells and suppressor factors in survival of the fetal allograft. In: Chacouat G (ed.) Reproductive Immunology: maternal–fetal relationship. INSERM colloque, Paris, 154: 77–88
Clark DA, Kiger N, Guenet JL, Chaouat G (1987b) Local active suppression and vaccination against spontaneous abortion in CBA/J mice. J Reprod Immunol 10: 79–85
Clark DA, Chaput A, Slapsys RM, Brierley J, Daya S, Allardyce R (1987c) Suppressor cells in the uterus. In: Gill TJ, Wegmann TG (eds.) Immunoregulation and fetal survival. Oxford University Press, New York, pp 63–77
Croy BA, Rossant J (1986) Mouse embryonic cells become susceptible to CTL lysis after midgestation. Cellular Immunol 104; 355–365
Croy BA, Rossant J, Clark DA, Wegmann TG (1983) Nonspecific suppression of in vitro generation of cytotoxic lymphocytes by allogeneic and xenogeneic embryonic tissues. Transplantation 35: 627–629
Daya S, Clark DA (1988) Identification of two species of suppressive factor of differing molecular weight released by in-vitro fertilised human oocytes. Fertil Steril 49: 360–363
Daya S, Clark DA, Devlin C, Jarrell J, Chaput A (1985) Preliminary characterization of two types of suppressor cells in the human uterus. Fertil Steril 44: 778–785
Daya S, Lee S, Underwood J, Mowbray J, Craft I, Clark DA (1986) Prediction of outcome following transfer of in vitro fertilized human embryos by measurement of embryo-associated suppressor factor. In: Clark DA, Croy BA (eds) Reproductive immunology 1986. Elsevier, Amsterdam, pp 277–285
Dearden-Badet MT (1986) Comparative study of biological properties of proteins synthesized in vitro by murine decidua and deciduomata. Am J Reprod Immunol Microbiol 10: 20–25
Ewoldsen MA, Ostlie NS, Warner CM (1986) MHC antigens of preimplantation mouse embryos are recognized by cytotoxic T lymphocytes. Fed Proc 45: 990
Gambel P, Croy DA, Moore WD, Hunzicker RD, Wegmann TG, Rossant J (1985) Characterisation of immune effector cells present in early murine decidua. Cell Immunol 93: 303–314
Gunderson GG, Medill L, Shapiro BM (1986) Sperm surface proteins are incorporated into the egg membrane and cytoplasm after fertilization. Dev Biol 113: 207–217
Hamilton MS (1983) Maternal immune responses to oncofetal antigens. J Reprod Immunol 5: 249–264
Hamilton MS, Beer AE, May RD, Vitetta ES (1979) The influence of immunization of female mice with F9 teratocarcinoma cells on their reproductive performance. Transplant Proc 11: 1869–1872
Hamilton MS, Vernon RB, Eddy EM (1985) A monoclonal antibody derived from syngeneically multiparous mouse alters in vitro fertilization and development. J Reprod Immunol 8: 45–49
Hayry P (1984) Intragraft events in allograft destruction. Transplantation 38: 15
Head J (1987) Lymphoid components in the rodent uterus. In: Gill TJ, Wegmann TG (eds) Immunoregulation and fetal survival. Oxford University Press, Oxford pp 46–59
Head J, Miller ST, Kresge CK (1986) Uterine lymphocytic vessels and 1A+ cells: possible role in antigen processing. In Clark DA, Croy BA (eds) Reproductive immunology 1986. Elsevier, Amsterdam, pp 205–210

Hearn JP (1986) The embryo–maternal dialogue during early pregnancy in primates. J Reprod Fertil 76: 809–819
Heyner S (1983) In: Wegmann TG, Gill TJ (eds) Immunology of reproduction. Oxford University Press, New York, pp 77–99
Hill J, Wickersham C, Anderson D (1986) Supernatants from mitogen- and MLC- stimulated lymphocyte cultures inhibit the development of mouse embryos in vitro. J Reprod Immunol suppl 1: 153
Kennedy TG (1983) Embryonic signals and the initiation of blastocyst implantation. Aust J Biol Sci 36: 531–543
Koch JH, Fifis ST, Bender UJ, Moss BA (1984) Molecular species of epidermal growth factor carrying immunosuppressive activity. J Cell Biochem 25: 45–59
Leclipteu T, Remacle J (1983) Disappearance of paternal histocompatibility from hybrid mouse blastocyst at the time of implantation. Febs Lett 157: 277–281
Lerman S, Koszalka TR, Jensen M, Andrew CL, Beckman DA, Brent RL (1986) In vitro studies on the effect of yolk sac antisera on functions of the visceral yolk sac: 1. Pinocytosis and transport of small molecules. Teratology 34: 335–341
McDermott MR, Clark DA, Bienenstock J (1980) Evidence for a common mucosal immune system. II. Influence of the estrous cycle on B immunoblast migration into genital and intestinal tissue. J Immunol 124: 2536–2539
Marticorena P, Hogan B, Dimeo A, Artzt K, Bennett D (1983) Carbohydrate changes in pre- and peri-implantation mouse blastocysts. Cell Different 12: 1–10
Mason DW, Morris PJ (1986) Effector mechanisms in allograft rejection. Ann Rev Immunol 4: 119–145
Mizejewski GJ, Vonnegut M (1984) Mechanisms of fetal demise in mice immunized to murine alphafetoprotein. Am J Reprod Immunol 5: 32–38
Morcos RN, Gibbons WE, Findley WE (1985) Effect of peritoneal fluid in in vitro cleavage of 2-cell mouse embryos: possible role of infertility associated with endometriosis. Fertil Steril 44: 678–683
Mowbray JF (1986) Effect of immunization with paternal cells on recurrent spontaneous abortion. In: Clark DA, Croy BA (eds) Reproductive immunology 1986. Elsevier, Amsterdam, pp 269–276
Papaioannou VE, Ebert KM (1986) Development of fertilized embryos transferred to oviducts of immature mice. J Reprod Fertil 26: 603–608
Rahima A, Soderwall AL (1977) Mast cells in uteri of pregnant young and senescent female golden hamsters. Biol Reprod 17: 523–526
Rutherford PA, Searle RF (1984) Cell-mediated immunity to male-strain histocompatibility alloantigens detected after natural insemination and systemic immunization in the female mouse using the cell-mediated microcytotoxicity test. J Reprod Fertil 72: 543–550
Searle TL (1986) Intrauterine immunization. In: Clark DA, Croy BA (eds) Reproductive immunology 1986. Elsevier, Amsterdam, pp 211–218
Simmons RL, Russell PS (1966) The histocompatibility antigens of fertilized mouse eggs and trophoblast. Ann NY Acad Sci 129: 35–45
Sio J (1985) Allograft reactivity and progesterone involvement at the choriodecidual junction. PhD thesis, University of Texas at Dallas
Slapsys RM (1987) Trophoblast dependent suppressor cell localization. PhD thesis, McMaster University
Slapsys RM, Richards CA, Clark DA (1985) Active suppression of host-versus-graft reaction in pregnant mice. VIII. The uterine decidua-associated suppressor cell is distinct from decidual NK cells. Cell Immunol 99: 140–149
Smith RN, Sternlicht M, Butler GW (1982) The alloantibody response in allogeneically pregnant rats. I. The primary and secondary antibody response and detection of Ir gene control. J Immunol 129: 771–776
Stern PL, Gilbert P, Heath JK, Furth M (1983) A monoclonal antibody which detects a cell surface antigen on murine embryonal carcinoma and early mouse embryo stages may recognize a carbohydrate determinant involving a-linked galactose. J Reprod Immunol 5: 145–160
Thomas IK, Erickson KL (1986) Gestational immunosuppression is mediated by specific Lyt 2+ T cells. Immunology 57: 201–206
Tsutsumi O, Oka T (1987) Epidermal growth factor deficiency causes abortion in mice. Am J Obstet Gynecol 156: 241–244
Tzartos SJ, Surani MAH (1979) Affinity of uterine luminal proteins for rat blastocysts. J Reprod Fertil 56: 579–586

Van Vlasselaer P, Vandeputte M (1984) Immunosuppressive properties of murine trophoblast. Cell Immunol 83: 422–432
Waites GT, Bell SC (1982) Glycogen-induced intrauterine leukocytosis and its effect on mouse blastocyst implantation in vivo and in vitro. J Reprod Fertil 66: 563–569
Waters CA, Singh B, Honore L (1985) An activity derived from rabbit serum causing interruption of pregnancy in mice. Am J Reprod Immunol Microbiol 7: 7–14
Zuckermann F (1986) Studies on the role of placental trophoblast protecting the fetus from immune rejection. PhD thesis, University of Texas at Dallas

12. The Complement System in Normal Pregnancy

B. Teisner, D. Tornehave, J. Hau, J.G. Westergaard
and H.K. Poulsen

Introduction

The complement system is an enzyme cascade including more than 20 plasma proteins. The system is the humoral mediator of immunologically initiated inflammation and is a key factor in defence against infections (Brown and Frank 1981).

Activation of the complement system can be initiated via two pathways (Fig. 12.1): – the classical pathway (C1, C4 and C2); and the alternative pathway (C3b, factors B, D and P). Activation via both of these pathways results in activation of the terminal sequence of the cascade (C5, C6, C7, C8 and C9) with

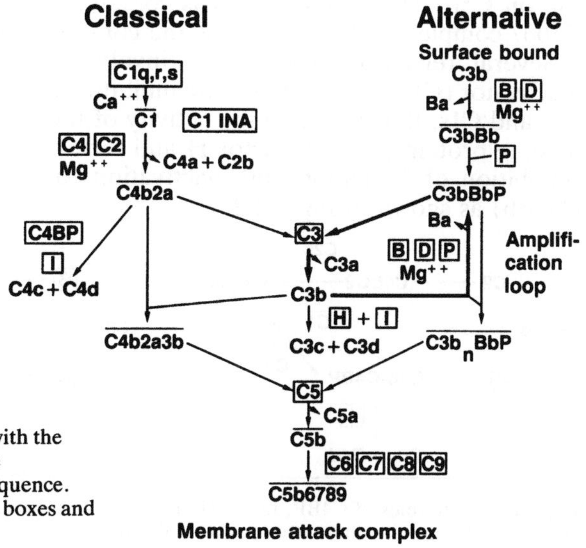

Fig. 12.1. The complement system with the classical and alternative pathways, the amplification loop and the terminal sequence. Native components are indicated with boxes and enzymes with bars.

formation of the membrane attack complex (MAC). The activation of the complement cascade is the result of proteolytic cleavage of components, with formation of enzymatically active protein-protein complexes and biologically active split products. Classical pathway activation is initiated when C1 binds to the Fc part of immunoglobulins in a complex with the corresponding antigen (immune-complexes, IC), and bound C1 is activated, resulting in the formation of the enzyme $\overline{\text{C1s}}$. $\overline{\text{C1s}}$ acts primarily on C4, which is split into C4a and C4b. C4b then binds to a cell membrane or an IC and forms a complex with C2 (C4bC2). Thereafter $\overline{\text{C1s}}$ acts on the C2 part of the C4bC2 complex, and a smaller fragment (C2b) is cleaved off to form a new enzymatically active complex, $\overline{\text{C4b2a}}$. As C3 is the substrate for $\overline{\text{C4b2a}}$, this complex is referred to as the C3-convertase of the classical pathway, responsible for cleaving C3 into C3a and C3b.

Activation of the complement system via the alternative pathway depends on the presence of C3b and can be initiated by a number of substances and surfaces. Activation via the alternative pathway does not involve the components of the classical pathway (C1, C2 and C4) and can take place in the absence of specific antibodies. C3b bound to an "activator" substance forms an Mg^{2+} dependent complex with factor B, C3bB, and this complex is cleaved by factor D into Ba and the enzyme $\overline{\text{C3bBb}}$. $\overline{\text{C3bBb}}$ is a labile C3 convertase which is stabilised by factor P. $\overline{\text{C3bBbP}}$ is referred to as the C3 convertase of the alternative pathway. It is also referred to as the amplification convertase, since C3b produced by the classical as well as the alternative pathway may result in further C3b generation.

The C3 molecule has a key role in the complement cascade. Thus: (i) C3 is situated at the junction point of the two pathways; (ii) C3 is the substrate for convertases produced in both pathways; (iii) C3 is present in high concentrations (1 g/l); and (iv) the cleavage product of C3 (C3b) is an integral part of the amplification convertase. The MAC consists of C5, C6, C7, C8 and C9. The C3 convertases of both the classical ($\overline{\text{C4b2a}}$) and the alternative pathways ($\overline{\text{C3bBb}}$) act in excess of C3b as a C5 convertase, resulting in C5a and C5b formations. C5b binds C6 and C7 and the C5b67 complex binds to membranes via a binding site on C5b. The lytic complex is completed by binding of C8 and C9 to the C5b67 complex, with cell lysis as the consequence.

Several regulator proteins are involved in the complement cascade. C1 inactivator (C1 INA) is an enzyme inhibitor that can form stable complexes with C1r and C1s, thus blocking the activity of these enzymes. Three proteins – (C4 binding protein (C4BP), factor H and factor I – are involved in the regulation (inhibition of formation and inactivation) of the C3 convertases (C4b2a and C3bBb) as shown in Fig. 12.2.

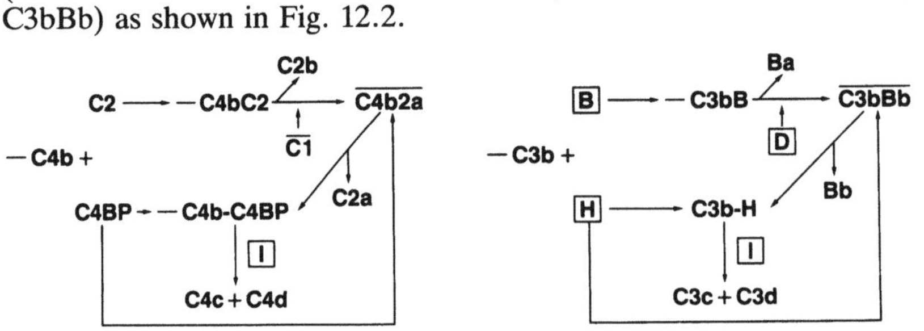

Fig. 12.2. The roles of C4BP, factor H and factor I in the regulation of the C3 convertases of the classical (*left*) and the alternative (*right*) pathways.

The aim of the present study was to examine complement activation and changes in complement factor and regulator proteins during normal pregnancy.

Materials and Methods

Blood Samples

EDTA plasma (10 mmol/l ethylene diamine tetra-acetic acid) was obtained from 20 non-pregnant women, 20 women in each of the first, second and third trimesters of apparently normal pregnancies, and 20 women 12 weeks post partum. The samples were kept at −80 °C until assayed.

Quantitative Electroimmunoassays

C4d and C3d were measured by rocket immunoelectrophoresis (RIE), using an intermediate gel technique (Teisner et al. 1983; Folkersen et al. 1985).

Factor B, C3, C4, C5 and the regulator proteins factor H and I were measured by conventional RIE (Teisner et al. 1984). Results are given in AU/ml, 100 AU/ml being the concentration in a pool of EDTA plasma from non-pregnant women.

Immunohistochemical Studies

Normal placenta was obtained at term vaginal delivery and during legal abortion performed at week 12. The tissues were fixed in 4% (w/v) paraformaldehyde for 4–6 h, washed, dehydrated, cleared in toluene and embedded in Paraplast at 56 °C (Chemnitz et al. 1984; Tornehave et al. 1984)). Staining was carried out with and without trypsin pretreatment (Tornehave et al. 1984).

Results and Comments

The levels of C3 significantly increased during the second and third trimester (Fig. 12.3). However, as the C3d levels (Fig. 12.4) remained unchanged throughout pregnancy, there was no evidence for local complement activation. Measurement of circulating C3d is generally accepted as an indicator of such activation (Teisner et al. 1984). However, when the change in plasma volume is taken into consideration, increased formation of C3d cannot be excluded.

There was no significant change in the circulating levels of C4d, which is an indicator of classical pathway activation. Similarly, the concentrations of C4, factor B and factor I remained unchanged during pregnancy.

By contrast, a significant increase in C5 (a component of MAC) was observed during the second and third trimester. The most striking change among the

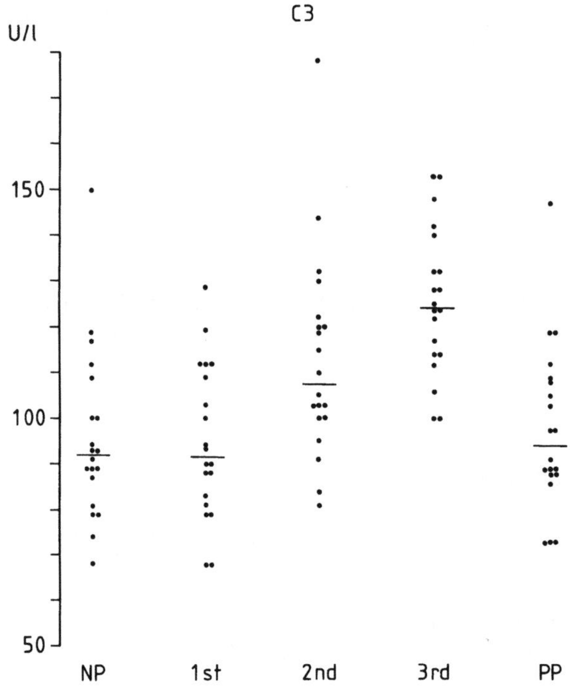

Fig. 12.3. Circulating levels of C3 during normal pregnancy. NP, non-pregnant women; 1st, first trimester; 2nd, second trimester; 3rd, third trimester; PP, 12 weeks post partum.

complement factors was a substantial increase in the regulator protein factor H (Fig. 12.5).

Factor H is involved in the regulation of the amplification loop by acting on C3b as the first step in C3b inactivation (inhibition of C3bB formation). Moreover, factor H can act directly on already generated amplification convertase (inactivation of $\overline{\text{C3bBb}}$) (Fig. 12.6).

There are several reports on circulating antitrophoblast antibodies in maternal serum during normal pregnancy (Guesdon et al. 1970; Davies 1985; Davies and Browne 1985), and the placenta is thus a potential target for complement mediated cytolysis.

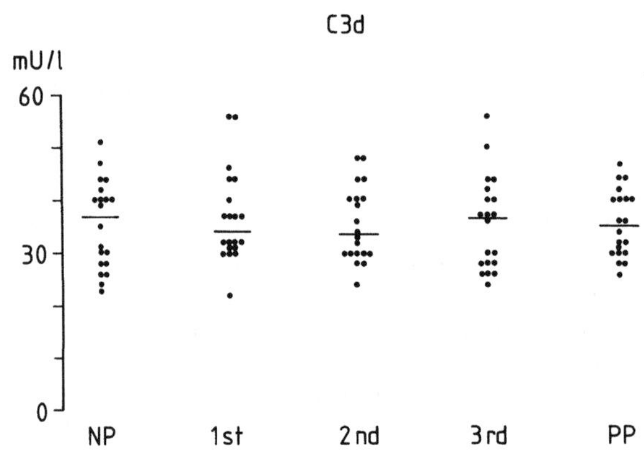

Fig. 12.4. The circulating levels of C3d during normal pregnancy. For explanation, see Fig. 12.3.

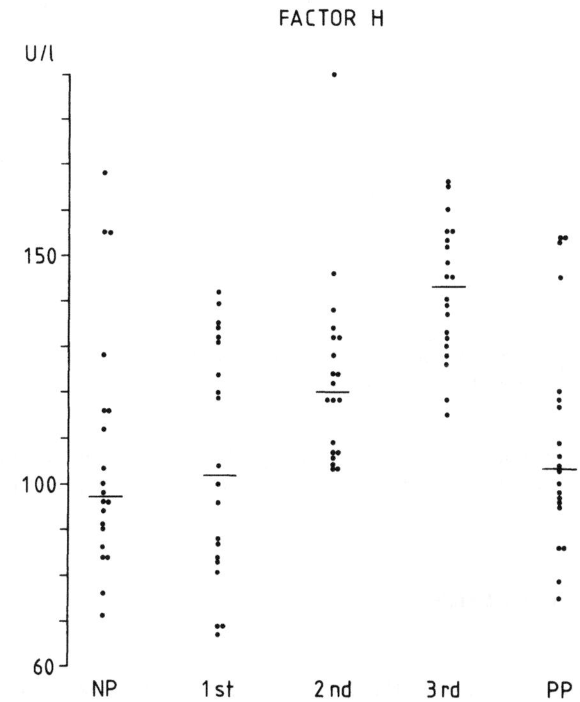

Fig. 12.5. Circulating levels of the regulator protein factor H. For explanation, see Fig. 12.3.

We examined the placenta by the indirect immunoperoxidase technique using rabbit anti-factor-H IgG and found factor H exclusively on the surface of the syncytiotrophoblast. The possible significance of factor H on the placental surface is shown in Fig. 12.7.

The presence of IgG in the placenta is well documented (Johnson et al. 1977). This may result in complement activation which, in the absence of any protective mechanism, will result in formation of the MAC and lysis. However, factor H on

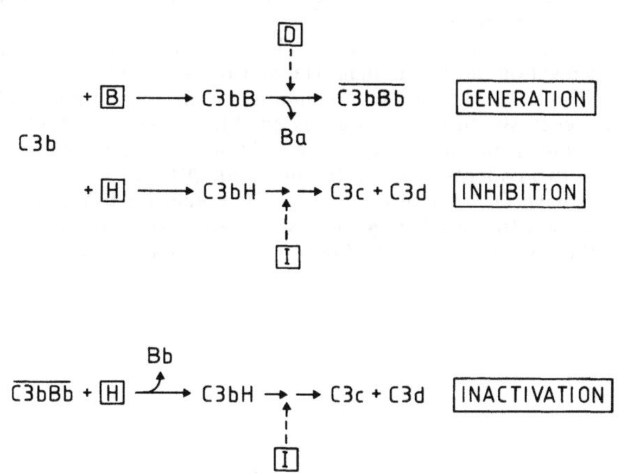

Fig. 12.6. Factor H in the regulation of the C3 convertase of the alternative pathway.

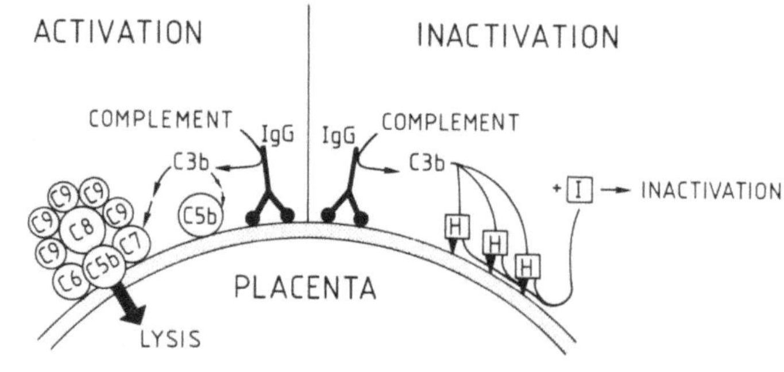

Fig. 12.7. A hypothesis to explain the biological significance of the presence of factor H on the surface of the chorionic villi.

the trophoblast surface may protect by inactivating C3b. In other words, factor H on the placental surface protects against complement mediated lysis.

References

Brown EJ, Frank MM (1981) Complement activation. Immunol Today 3: 129–134

Chemnitz J, Tornehave D, Teisner B, Poulsen HK, Westergaard JG (1984) The localisation of pregnancy proteins (hPL, SP-1 and PAPP-A) in intra- and extrauterine pregnancies. Placenta 5: 489–494

Davies M (1985) An ELISA for the detection of maternal anti-trophoblast antibodies in human pregnancy. J Reprod Immunol 77: 109–118

Davies M, Browne CM (1985) Anti-trophoblast antibody responses during normal pregnancy. J Reprod Immunol 7: 285–297

Folkersen J, Teisner B, Petersen NE, Sim RB, Jessen E (1985) Preparation of antibodies against the fourth complement component (C4) and development of a direct electroimmunoassay for quantification of C4d. J Clin Lab Immunol 16: 163–167

Guesdon JP, Leake NH, Burt RL (1970) Naturally occurring antibody to placental protein in human pregnancy. Am J Obstet Gynecol 108: 1056–1062

Johnson PM, Natvig JB, Vetehede UA, Faulk WP (1977) Immunological studies of human placentae: the distribution and character of immunoglobulins in chorionic villi. Clin Exp Immunol 30: 145–153

Teisner B, Brandslund J, Grunnet N, Hansen LK, Thellesen I, Svehag SE (1983) Acute complement activation during an anaphylactoid reaction to blood transfusion and the disappearance rate of C3c and C3d from the circulation. J Clin Lab Immunol 12: 63–67

Teisner B, Brandslund J, Rasmussen JM, Poulsen LO, Svehag SE (1984) Factor I deficiency and C3 nephritic factor: Immunochemical findings and association with Neisseria meningitidis infections in two patients. Scand J Immunol 20: 291–297

Tornehave D, Chemnitz J, Teisner B, Folkersen J, Westergaard JG (1984) Immunohistochemical demonstration of pregnancy associated plasma protein A (PAPPA) in the syncytiotrophoblast of the normal placenta at different gestation ages. Placenta 5: 427–432

13. Spontaneous and Recurrent Abortion: Epidemiological and Immunological Considerations

L. Regan

Introduction

About one in four women who conceive undergo a miscarriage during their reproductive life (Warburton and Fraser 1964). Thus, establishing the cause and assessing the risk of subsequent pregnancy loss for women who present with a history of early abortion is a continuing problem. Yet despite the recent advances in reproductive medicine, most clinicians working in this area are still unable to answer satisfactorily the two fundamental questions: "Why did the abortion occur?", and "What is the chance of it happening again?". These apparently simple questions have proved difficult to answer. There are inconsistencies in the terminology used and a lack of generally acceptable numerical data. Furthermore, since the aetiology is usually not understood, the implementation of treatment becomes empirical, and the results of such regimens are inconclusive.

The term "abortion" was originally used to describe a pregnancy that terminated before the fetus was viable. The World Health Organisation (1977) defined abortion as "the expulsion or extraction from its mother of a fetus or an embryo weighing 500 g or less", which approximates to 20–22 completed weeks of gestation. As the incidence of abortion reaches its lowest level at 20–27 weeks, most epidemiological data will correspond reasonably well to the WHO definition. Here, spontaneous abortion will be defined as the loss of a pregnancy before 20 weeks' gestation and recurrent abortion as the loss of three or more consecutive pregnancies before 20 weeks.

In the vast majority of spontaneous abortions, the cause remains unknown. Fetal chromosome anomalies are cited as accounting for 50%–60% of first-trimester losses (Boue et al. 1975; Lauritsen 1976; Geisler and Kleinebrecht 1978), but only 10%–20% of products of conception yield successful culture. Hence percentages quoted are a percentage of those karyotyped successfully. Moreover, the risk of a second abortion following the abortion of a fetus with an abnormal karyotype is less than the risk incurred by women who have had karyotypically normal abortuses (Boue et al. 1975; Lauritsen 1976; Geisler and Kleinebrecht 1978).

The finding of a balanced reciprocal translocation in one of the parents affects the prognosis for a subsequent pregnancy, but the incidence is less than 5% (Warburton and Strobino 1987), even in those couples who have experienced repeated abortions. Structural abnormalities of the genital tract are an important and potentially remediable group (McDonald 1980), but their contribution to the overall abortion rate is small (Tho et al. 1979). Similarly, maternal systemic disorders are uncommon, and there is little evidence to support the frequent implication of both diabetes (Crane and Wahl 1981) and thyroid disease (Prout 1975). An infectious agent is frequently sought but rarely found. The possibility of a disturbed endocrine environment is attractive (Horta et al. 1977), but most studies have concluded that the administration of exogenous hormones (human chorionic gonadotrophin and progesterone), in an attempt to increase corpus luteum efficiency, have no beneficial effect (Goldzieher 1964).

It has been suggested that the commonest aetiology in this group of women is an immunological alteration which prevents them from mounting a protective response towards their fetal allograft (Mowbray and Underwood 1985). The nature of this putative maternal immune response has yet to be established, but if it is related causally to recurrent abortions it might be remediable. Thus, immunological therapy is already being explored (Taylor and Faulk 1981; Mowbray et al. 1985), further emphasising the need to establish the incidence of recurrent abortion and to identify those women for whom therapy might be beneficial.

Incidence of Spontaneous Abortion

Some 15% of clinically recognised pregnancies abort spontaneously (Roth 1963; Warburton and Fraser 1964; Petersson 1968; Poland et al. 1977; Stray-Pederson and Stray-Pederson 1984), although the use of highly sensitive hCG assays, detecting implantation before the first missed period (Tamada and Fukwai 1976), has led some to suggest a loss rate as high as 60% (Miller et al. 1980; Edmonds et al. 1982). This figure is probably an overestimate.

Abortion may be a more common outcome of first pregnancies than later pregnancies (Glass and Colbus 1978). The increased risk of abortion after a previous abortion is reported consistently (Warburton and Fraser 1964; Petersson 1968; Lauritsen 1976; Poland et al. 1977) but there are no generally accepted figures for the risk. An hypothesis has been supplied that all abortions can be attributed to random factors and that a large percentage of recurrent abortions will therefore be purely fortuitous (Huisjes 1984). Using an estimate that 15% of clinically recognised pregnancies will abort, he calculated that 2.3% of women will abort two successive pregnancies, and similarly that 0.34% would be expected to abort three consecutive pregnancies. Another theoretical formulation estimated the same figure at 73% (Malpas 1938). However, it is now recognised that not all women have the same propensity to abort, the risks being higher or lower depending on age, parity and variables such as chromosome abnormality of the conceptus (James 1963; Lauritsen 1976; Alberman et al. 1975; Roman et al. 1980). Moreover, abortion tends to recur in the same women and the reproductive characteristics of "repeaters" are well known (Fitzsimmons et al.

1983; Warburton and Strobino 1987). Thus, there are factors which distinguish women with repeated abortion, and "recurrent abortion" is a distinct entity.

Studies of Recurrence Risks

The incidence of abortion with no previous history of an unsuccessful pregnancy is around 12% (Table 13.1), whereas values from 25%–50% are quoted for those women who have aborted on three previous occasions. There are numerous studies on the recurrence risk after the first, second, third and subsequent pregnancy loss, but the results vary with the population sample and the method of data analysis used. For example, Warburton and Fraser (1964) interviewed over 2000 women consulting a medical genetics unit and analysed 10000 of their accumulated pregnancies. Their figures included all women with previous abortions, irrespective of whether or not a successful pregnancy intervened. By contrast, Poland et al. (1977) collated their data from hospital records alone, noting subsequent pregnancy outcome only in 472 patients who had had successive abortions. Retrospective studies can provide no information on first pregnancies, and since they are centred on hospital populations, early abortions will be underrepresented. The prospective studies have not included primigravidae, nor have they enrolled their patients before conception.

Table 13.1. Probability of spontaneous abortion. Percentage of women aborting in relation to number of previous abortions

Type of study	Number of previous abortions			
	0	1	2	3+
Retrospective studies:				
Stevenson et al. (1959)		16.8	19.2	26.2
Warburton and Fraser (1964)	12.3	26.2	32.2	30.2
Leridon (1976)	15.2	22.0	35.3	
Poland et al. (1977)		19.0	35.0	47.0
Naylor and Warburton (1978)	11.0	20.3	29.2	37.0
Cohort studies:				
Shapiro et al. (1970)	10.9	18.0		
Awan (1974)	10.4	22.1	27.4	
Prospective studies:				
Boue et al. (1975)		13.8		24.9
Harger et al. (1983)			17.4	29.2
Fitzsimmons et al. (1983)			31.3	45.7

Despite the shortcomings of such studies, most authors have found the likelihood of a successful pregnancy after three previous abortions to be in the region of 60%; this leads many gynaecologists to question whether treatment is justifiable. However, few women who have aborted repeatedly are prepared to accept that they have been the victims of chance. There is pressure to investigate and treat a patient who may have been statistically unlucky (Vlaanderen 1977).

This point is well illustrated by the enormous public demand that has recently developed for "immunotherapy" and further emphasises the need to establish the basic facts of this situation.

Immunotherapy

Some women who abort recurrently are thought to have an immune deficiency that prevents them from mounting a protective response towards their fetus. Reports of successful pregnancy outcome after immunisation with whole donor blood or partner's lymphocytes are accumulating (Taylor and Faulk 1981; Beer et al. 1985; Mowbray et al. 1985; Taylor et al. 1985; Unander et al. 1985). Success rates for subsequent pregnancy outcome average 75%–80%. Only two studies have included controls, and in these the successful pregnancy outcomes were 64% (Beer et al. 1985) and 34% (Mowbray et al. 1985), compared with 72% and 78%, respectively, for treated patients.

The mechanism whereby immunisation confers its effects is presumed to be secondary to the maternal immunological response to allogeneic (paternal) lymphocytes. Some workers believe that the production of an antipaternal cytotoxic antibody after treatment is a good prognostic factor for the subsequent pregnancy (Mowbray et al. 1985; Mowbray et al. 1987), and most centres offering immunisation will accept only patients who are negative for antipaternal cytotoxic antibody before treatment. The use of this criterion presumes that the absence of the antibody (the presence of which by definition demonstrates that the mother has immunologically recognised the paternal component of her fetus) is important in the aetiology of her recurrent abortions. Mowbray et al. (1987) state that "only women with no antibodies to paternal T and B lymphocytes were offered treatment, on the basis that they had not shown the normal antibody response to repeated pregnancies". They go on to suggest that women who seroconvert after white-cell infusion have an improved outcome and that this immunological protection persists for up to a year after treatment. By contrast, patients who remain negative for cytotoxic antibodies and conceive more than 80 days after treatment are likely to miscarry again and should be offered booster injections. However, if the development of cytotoxic antibody is essential for successful pregnancy outcome, why do some immunised women who go on to have successful pregnancies never produce detectable antibody? There is no documentation of the prevalence and significance of such antibodies in normal pregnancy, so the clinical assumptions described above are unjustifiable. The Cambridge study was begun to resolve some of the questions about the role of cytotoxic antibody and its diagnostic value. We recruited a group of pregnant women, including primigravidae, in order to:

(i) establish the spontaneous abortion rate in the community;
(ii) assess the risks of recurrent spontaneous abortion;
(iii) examine some immunological parameters in normal human pregnancies, with particular reference to the incidence and time of development of antipaternal cytotoxic antibody.

Cambridge Early Pregnancy Loss Study Recruitment

All patients in Cambridge receive their maternity care at one hospital. In order to recruit a representative cross-section of the community, a radio and poster appeal was started in February 1986, asking women who were contemplating a pregnancy in the near future to come forward. In addition, recruits were drawn from the infertility, general gynaecology and postnatal clinics, together with women who had had a recent miscarriage in the maternity hospital.

All volunteers were seen, and the purpose of the study was explained to them. As soon as the pregnancy was diagnosed an early blood sample was taken and arrangements were made for serial blood samples. Ninety-five per cent of the pregnancies were confirmed by ultrasound before the end of the eighth gestational week. Blood samples were also taken at the time of booking (between 10 and 12 weeks' gestation), at 18, 28 and 32 weeks, and at the time of delivery. After delivery, the woman and her partner were seen 4 weeks post-partum, when the final maternal serum sample and fresh paternal blood for cytotoxicity testing were collected. Between February 1986 and August 1987 a total of 456 women were enrolled. Pregnancies were declared by 234 of the women – 117 who came through the radio and poster appeal, 44 from the infertility clinic (half from the in-vitro fertilisation [IVF] programme) and 73 who had had a recent miscarriage.

Pregnancy Outcome

Full information is available on 226 pregnancies (Fig. 13.1). There have been 130 live births, 23 spontaneous abortions, one therapeutic termination for fetal abnormality diagnosed at 20 weeks' gestation and two ectopic gestations. In addition 70 pregnancies beyond 20 weeks' gestation are yet to be completed.

All the abortions in this population occurred before 14 weeks' gestation (Fig. 13.2), and all patients who had completed 20 weeks' gestation have been

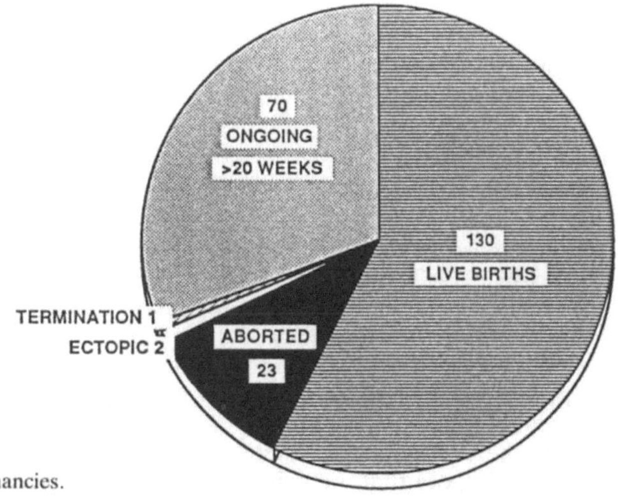

Fig. 13.1. Outcome of 226 pregnancies.

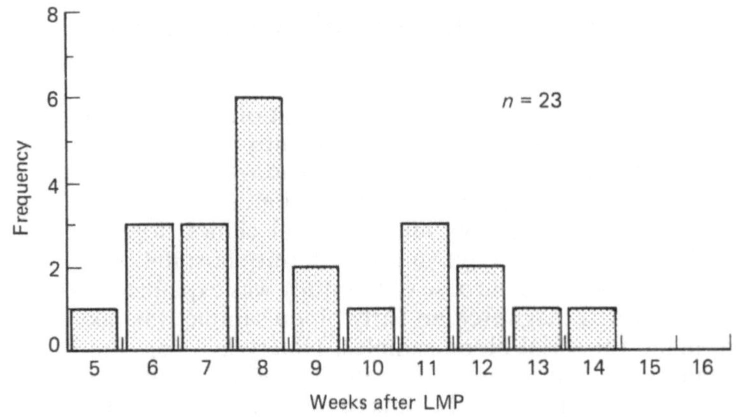

Fig. 13.2. Frequency of spontaneous abortion by gestational age.

included, since the chances of abortion after this stage are minimal. The outcome of 223 patients, of whom 200 had potentially live births (including the 70 ongoing pregnancies) is shown in Table 13.2. (Because the classification for the ectopic pregnancies and the termination is debatable the three patients affected have been excluded from the analysis.) Although the crude spontaneous abortion rate is 10.3% there is a demonstrable increase in the abortion rate when the pregnancies are grouped by gravidity (taking no account of the outcome of any previous pregnancy), an effect which is most pronounced after the second pregnancy.

Table 13.2. Outcome of 223 pregnancies by gravidity

Outcome of indexed pregnancy	1st pregnancy	2nd pregnancy	3rd pregnancy	⩾4th pregnancy	Total
Live	51	65	45	41	200
Abortion	3	4	8	8	23
Total	54	69	53	49	223
Spontaneous abortion rate	5.6%	5.8%	15.1%	16.3%	10.3%

The increase is well documented (Warburton and Fraser 1964; Naylor and Warburton 1979; Roman et al. 1980). Most women in our society are aiming for an average of two children. Once they have achieved this goal, they discontinue further attempts at pregnancy (James 1974). On the other hand, women who have been unsuccessful will embark on more pregnancies to achieve their aim, and pregnancy will therefore occur at progressively older ages (James 1974; Resseguie 1974).

Since there are no published prospective data on first pregnancies it is interesting to note that the primigravid loss rate in our population was consider-

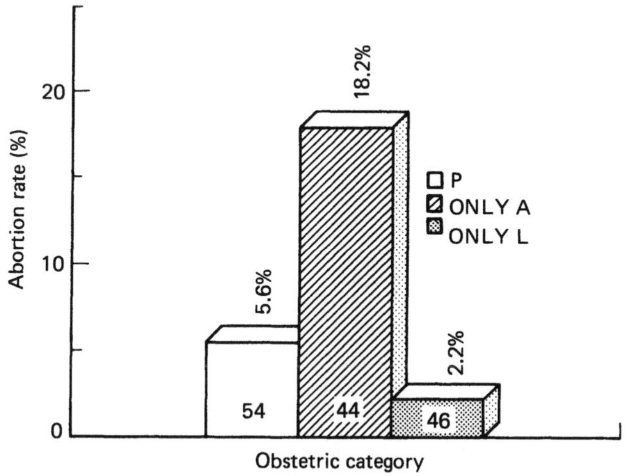

Fig. 13.3. Abortion rate in women who have only aborted (ONLY A) or have had only live births (ONLY L), compared with primigravidae (P).

ably lower than the overall abortion rate. When it is remembered that this group included some of the IVF pregnancies, the finding is even more remarkable, since a 1-in-4 pregnancy wastage is quoted for such patients. Thus, gravidity influences the spontaneous abortion rate but provides little information about past obstetric history. To answer one of the initial questions – what advice should be given to a woman who has experienced pregnancy losses and is planning to embark on further pregnancies – it is necessary to re-examine the data with reference to past productive outcome. Though the numbers are relatively small, some interesting trends emerge.

1. Women with previously unsuccessful histories were at greater risk of aborting the current pregnancy (18.2%) than were primigravidae (5.6%). In contrast, women with previous live births had a lower abortion rate (2.2%) than the primigravidae (Fig. 13.3).

2. In women who had only aborted previously, there was a cumulatively higher chance that the current pregnancy would be aborted when their past history recorded one previous abortion (11.5%), two abortions (29.4%) and three abortions (36.4%) (Fig. 13.4).

3. The outcome of the last pregnancy had a direct influence on the outcome of the current pregnancy (Fig. 13.5). Women whose last pregnancy had been successful had a low abortion rate (3.3%) comparable to that of women whose previous pregnancies had all been successful (2.2%). Both groups aborted less frequently than primigravidae (5.6%). Those women whose last pregnancy had aborted had a higher risk of a further abortion (17.5%) – a rate similar to that of women with a history of abortions only (18.2%). Thus there is some evidence that a successful last pregnancy has a protective effect on the current pregnancy. Conversely, a detrimental effect was observed if the last pregnancy had been unsuccessful. An immunological event provides one plausible explanation of this observation.

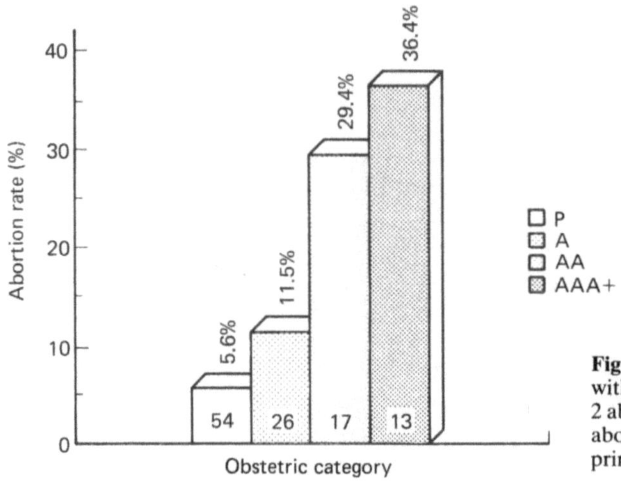

Fig. 13.4. Abortion rate in women with a history of only 1 abortion (A), 2 abortions (AA) and 3 or more abortions (AAA+), compared with primigravidae (P).

Incidence of Antipaternal Cytotoxic Antibody in the Completed Pregnancies

The results presented here include all those women who had completed a pregnancy by the end of August 1987: 120 live births and 21 spontaneous abortions. The test used was the microdroplet lymphocytotoxic assay (Mittal et al. 1969). All serum samples were tested against the partner's freshly prepared peripheral-blood lymphocytes taken four weeks post partum.

All samples were tested in quadruplicate, and the results were compared with those of positive and negative control sera. The results were read by two independent observers, and positive results were recorded with lymphocyto-

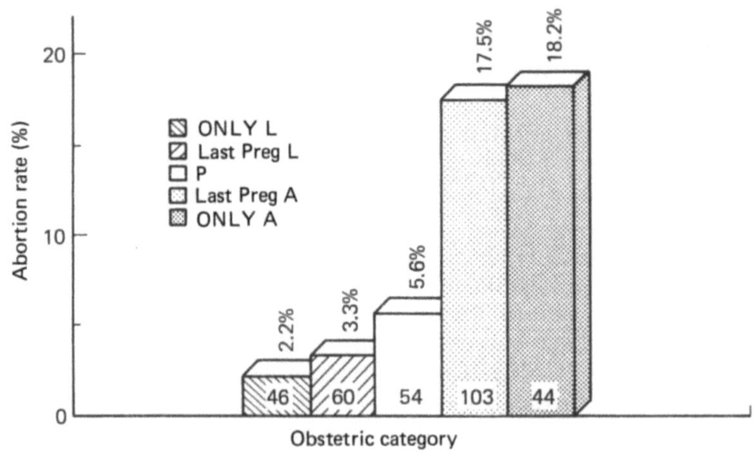

Fig. 13.5. Abortion rate in women with a history of only live births (ONLY L), those whose last pregnancy ended in a live birth (Last Preg L), those whose last pregnancy was aborted (Last Preg A), and those who have only aborted (ONLY A), compared with primigravidae.

toxicity levels in excess of 20% above negative control. Only 38% of the women who had live births produced detectable antibody (Table 13.3), considerably lower than the figure of 64% previously reported (Beard et al. 1983). This argues against an absolute necessity for antipaternal cytotoxic antibody for successful pregnancy. Although another study (Gelabert et al. 1981) found that women with spontaneous abortions had a higher percentage of positive sera than primigravidae, the percentage of positive antibody results in our aborting group was only 9.5%, which might at first be interpreted as supportive evidence implicating causally the lack of antibody in abortion. However, the 21 abortions described here all occurred before 14 weeks' gestation (Fig. 13.2). Hence an equally plausible explanation for the low incidence of antibody in this aborting group would be that the cytotoxic antibody does not appear until later in pregnancy.

Table 13.3. Frequency of antipaternal cytotoxic antibody in 141 completed pregnancies

Outcome of pregnancy	No. of patients	No. antibody positive	% antibody positive
Live	120	46	38.3
Abortion	21	2	9.5
Total	141	48	34.0

Time Course of Development of Antipaternal Cytotoxic Antibody

The possibility that cytotoxic antibody production is a time dependent event was confirmed by examining the gestational intervals at which the 120 women with live births developed antibody for the first time (Fig. 13.6). Only four patients

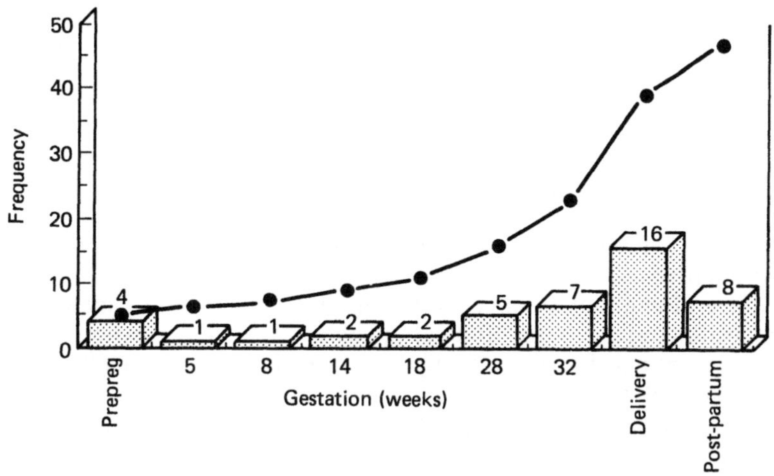

Fig. 13.6. Time of first detection of antipaternal cytotoxic antibody in 120 live births. The number of patients with positive tests at each gestational time interval is denoted above each bar. The cumulative frequency of antipaternal cytotoxic antibody is shown by the *continuous line*.

had antibody present in their pre-pregnant sample. Two of these women had received blood transfusions in the past and one had had ectopic gestation, suggesting that all three had been "sensitised" previously. In the current pregnancy, one patient developed antibody at five weeks, one at eight weeks, and two became positive at 14 and 18 weeks. The suggestion that early antibody production correlates with a past history of abortion (Gelabert et al. 1981) could not be confirmed, since most positive results were found only after the pregnancy had reached the third trimester, the incidence rising from 28 weeks' gestation and peaking at delivery. This contrasts with previous studies, which have shown that antibody is increased in the first trimester and falls towards term (Vives et al. 1976). In eight of the live births antibody was detectable only in the four-week post-partum serum.

Three important points are emphasised by this study. First, only four patients developed antibody before 14 weeks, the time by which all the abortions in this study group had occurred. Secondly, the overall incidence of antibody production in the population of live births involved only one-third of the patients. Thirdly, of those patients who did produce antibody, most became positive after 28 weeks' gestation. In addition, it can be inferred from the finding of only four positive prepregnant serum samples that this antibody is a transient and pregnancy related phenomenon.

Table 13.4 Frequency of antipaternal cytotoxic antibody in 141 completed pregnancies by gravidity

Outcome of indexed pregnancy	1st pregnancy	2nd pregnancy	3rd pregnancy	4th pregnancy	Total
Live	36.7% (11/30)	42.4% (14/33)	40.0% (14/35)	31.8% (7/22)	38.3% (46/120)
Abortion	33.3% (1/3)	0 (0/4)	0 (0/7)	14.3% (1/7)	9.5% (2/21)
Total	36.4%	37.8%	33.3%	27.6%	34.3%

Incidence of Antibody by Gravidity

In contrast to previous evidence that the incidence of antibody-positive patients increases with cumulative successful pregnancies (Van Rood et al. 1958; Ahrons 1973), the percentage of antibody positive women in groups analysed by maternal gravidity demonstrated no such relationship (Table 13.4). Even when it is considered that the third and fourth gravidity groups included women who had aborted in the past and would therefore be less likely to have produced antibody, the percentages show little increase above the 38% incidence of positive results for live births in the current pregnancy. The primigravidae in this population had a similar incidence of antibody production to that found in the multiparous women. Moreover, if infertility patients are considered separately, since not all fall into the primigravid grouping, 55% of the 31 patients so far tested became antibody positive during the current pregnancy.

Conclusions

The overall incidence of spontaneous abortion in this study was lower than in previous studies (10.3% overall; 5.6% for primigravidae), possibly reflecting the fact that the population was recruited prospectively and preconceptually. Alternatively, it may reflect the fact that this was a well-motivated group of women who, as a consequence, have an inherently lower risk of spontaneous abortion. The pregnancy loss rate increased with gravidity, in agreement with previously cited studies, and the outcome of the previous pregnancy was correlated with the outcome of a future pregnancy, suggesting that a previously successful pregnancy may offer some protective effect.

Sixty-two per cent of the patients who had live births never produced detectable cytotoxic antibody during their pregnancy, but the corresponding figure for the women in the aborting group was 90.5%. However, this result should not be interpreted as supportive evidence for a causal relationship between lack of antibody and recurrent abortion because all of these women had pregnancy losses before 14 weeks, whereas the majority of positive antibody results in the live birth group occurred after 28 weeks' gestation. Thus the low incidence of antibodies in the spontaneously aborting group arose simply because they had insufficiently long gestations to produce a response.

Finally, the results indicate that circulating antipaternal cytotoxic antibody is not necessary for the maintenance of pregnancy, nor does this antibody provide a reliable marker for a pregnancy that is destined to be successful (Regan and Braude 1987). It may be concluded that the absence of antipaternal cytotoxic antibody is not causally related to recurrent abortion and the use of this antibody to monitor immunisation treatment and predict future pregnancy outcome should be avoided until its significance is more fully understood.

Acknowledgements. I thank Dr. Peter Braude and Dr. Martin Johnson for their practical help and advice and Dan Hill, Janet Currie, Anne Cant and Soo Pickering for their technical assistance.

This work is being supported by an MRC Programme grant awarded to Dr. Johnson and Dr. Braude.

References

Ahrons S (1973) Leukocyte antibodies: occurrence in primigravidae. Tissue Antigens 1: 178–183

Alberman E, Elliott M, Creasy M, Dhadial R (1975) Previous reproductive history in mothers presenting with spontaneous abortions. Br J Obstet Gynaecol 82: 366–374

Beard RW, Braude PR, Mowbray JF, Underwood JL (1983) Protective antibodies and spontaneous abortion. Lancet ii: 1090

Beer AE, Semprini AE, Ziaoyu Z, Quebberman JF (1985) Pregnancy outcome in human couples with recurrent spontaneous abortions: HLA antigen profiles; HLA antigen sharing; female serum MLR blocking factors; and paternal leukocyte immunisation. Exp Clin Immunogenet 2: 137–153

Beer AE, Quebbeman JF, Semprini AE (1987) Immunopathological factors contributing to recurrent

spontaneous abortion in humans. In: Bennett MJ, Edmonds DK (eds) Spontaneous and recurrent abortion. Blackwell Scientific, Oxford, pp 90–108
Boué J, Boué A, Lazar P (1975) Retrospective and prospective epidemiological studies of 1500 karyotyped spontaneous human abortions. Teratology 12: 11–26
Crane JP, Wahl N (1981) The role of maternal diabetes in repetitive spontaneous abortion. Fertil Steril 36: 477–479
Edmonds DK, Lindsay KS, Miller JF, Williamson E, Wood PJ (1982) Early embryonic mortality in women. Fertil Steril 38: 447–453
Fitzsimmons J, Jackson D, Wapner R, Jackson L (1983) Subsequent reproductive outcome in couples with repeated pregnancy loss. Am J Med Genet 16: 583–591
Geisler M, Kleinebrecht J (1978) Cytogenetic and histologic analyses of spontaneous abortions. Hum Genet 45: 239–251
Gelabert A, Balasch J, Ercilla G, et al. (1981) Abortion may sensitise the mother to HLA antigens. Tissue Antigens 17: 353–356
Glass RH, Colbus MS (1978) Habitual abortion. Fertil Steril 29: 257–268
Goldzieher JW (1964) Double blind trial of a progestin in habitual abortion. J Am Med Assoc 188: 651–654
Horta JLH, Fernandez JG, Soto de Leon B, Cortes-Gallegos V (1977) Direct evidence of luteal insufficiency in women with habitual abortion. Obstet Gynecol 49: 705–708
Huisjes JH (1984) Spontaneous abortion. Churchill Livingstone, Edinburgh
James WH (1963) Notes towards an epidemiology of spontaneous abortion. Am J Hum Genet 15: 223–240
James WH (1974) Spontaneous abortion and birth order. J Biosoc Sci 6: 23–41
Lauritsen JG (1976) Aetiology of spontaneous abortion. A cytogenetic and epidemiological study of 288 abortuses and their parents. Acta Obstet Gynecol Scand 52: 1–29
Malpas PJ (1938) A study of abortion sequences. J Obstet Gynaecol Br Empire 45: 932–949
McDonald IA (1980) Cervical cerclage. Clin Obstet Gynaecol 7: 461–479
Miller JF, Williamson E, Gordon YG, Grudzinskas JG, Sykes A (1980) Fetal loss after implantation. Lancet ii: 554–556
Mittal KM, Mickey MR, Singal PP, Terasaki PI (1969) Serotyping for homotransplantation. XVIII. Refinement of microdroplet lymphocyte cytotoxicity test. Transplantation 6: 904–912
Mowbray JF, Underwood JL (1985) Immunology of abortion. Clin Exp Immunol 60: 1–7
Mowbray JF, Gibbings C, Liddell H, Reginald PW, Underwood JL, Beard RW (1985) Controlled trial of treatment of recurrent spontaneous abortion with paternal cells. Lancet i: 941–943
Mowbray JF, Underwood JL, Michel M, Forbes PB, Beard RW (1987) Immunisation with paternal lymphocytes in women with recurrent miscarriage. Lancet ii: 679–680
Naylor AF, Warburton D (1979) Sequential analysis of spontaneous abortion. II. Collaborative study data show that gravidity determines a very substantial increase in risk. Fertil Steril 31: 282–286
Petersson F (1968) Epidemiology of early pregnancy wastage. Svenska Bokforlaget, Norstedts
Poland BJ, Miller JR, Jones DC, Trimble BK (1977) Reproductive counselling in patients who have had a spontaneous abortion. Am J Obstet Gynecol 127: 685–691
Prout TE (1975) Thyroid disease in pregnancy. Am J Obstet Gynecol 122: 669–676
Regan L, Braude PR (1987) Is antipaternal cytotoxic antibody a valid marker in the management of recurrent abortion? Lancet ii: 1237
Resseguie LJ (1974) Pregnancy wastage and age of mother among the Amish. Hum Biol 46: 633–639
Roman EA, Alberman E, Pharoah POD (1980) Pregnancy order and reproductive loss. Br Med J i: 715
Roth DB (1963) The frequency of spontaneous abortion Int J Fertil 8: 434
Stray-Pederson B, Stray-Pederson S (1984) Etiologic factors and subsequent reproductive performance in 195 couples with a prior history of habitual abortion. Am J Obstet Gynecol 148: 140–146
Tamada T, Fukwai Z (1976) Early detection of pregnancy around the first missed menstrual period by 2 hour haemagglutination test. Acta Obstet Gynecol Jap 23: 23–26
Taylor C, Faulk WP (1981) Prevention of recurrent abortion with leukocyte transfusions. Lancet ii: 68–70
Taylor CG, Faulk WP, McIntyre JA (1985) Prevention of recurrent spontaneous abortions by leukocyte transfusion. J Roy Soc Med 78: 623–625
Tho PT, Byrd JR, McDonough PG (1979) Etiologies and subsequent reproductive performance of 100 couples with recurrent abortion. Fertil Steril 32: 389–395
Unander AM, Lindholm A, Olding LB (1985) Blood transfusion generates/increases previously absent-weak blocking antibody in women with habitual abortion. Fertil Steril 44: 766–771

Van Rood JJ, Eernisse JG, Van Leeuwen A (1958) Leucocyte antibodies in sera from pregnant women. Nature 181: 1735–1736
Vives J, Gelabert A, Castillo R (1976) HLA antibodies and period of gestation: decline in frequency of positive sera during last trimester. Tissue Antigens 7: 209–212
Vlaanderen W (1977) Het begrip "habituele abortus", een fictie? Nederlands Tijdschrift voor Geneeskunde 121: 439–442
Warburton D, Fraser FC (1964) Spontaneous abortion risks in man: data from reproductive histories collected in a medical genetics unit. Hum Genet 16: 1–25
Warburton D, Strobino B (1987) Recurrent spontaneous abortion. In: Bennet MJ, Edmonds DK (eds) Spontaneous and recurrent abortion. Blackwell Scientific, Oxford, pp 193–213
World Health Organization (1977) Recommended definitions, terminology and format for statistical tables related to the perinatal period. Acta Obstet Gynaecol Scand 56: 247–253

14. Contraception: An Immunological Approach

W. R. Jones

Vaccinating for Contraception

The principle of immunological contraception is based on manipulation of the body's natural defences to block or interrupt the reproductive process. The rationale for an immunological approach to human fertility control has been established over many years by the induction of immunological infertility in experimental animals and by the recognition of the role of immunological factors in some infertile men and women (Jones 1982; Ada et al. 1985).

The advantages of a contraceptive vaccine include:

1. The use of a non-pharmacologically active agent.
2. Ease and convenience of administration, making it suitable for distribution by paramedical personnel.
3. Long-lasting (6–12 months) but potentially reversible effect.
4. Acceptability of the "vaccine" principle – of particular importance in developing countries.
5. Reduced patient failure, since protection is not dependent on constant positive action by the recipient.
6. Large-scale synthesis and manufacture of a vaccine at low cost.

An immunological approach to contraception is based on the establishment of an immune response to specific reproductive tract antigens. The selection of such targets for an immunological attack must depend on certain criteria. A target antigen must be specific to the reproductive process. It should be available in pure form from natural sources in adequate amounts or be capable of synthesis. Finally, immunisation against the antigen should prevent or disrupt fertilistion or implantation without undesirable side-effects. In other words, a contraceptive vaccine should interrupt the reproductive process as early as possible and should not result in an immune reaction against other body tissues.

The development of an immunological contraceptive method may involve several approaches:

1. The characterisation of reproductive tract antigens for tissue specificity, chemistry, availability and suitability for synthesis.
2. The possible modification of the natural structure of target antigens to improve immunogenicity.
3. A search for evidence of naturally occurring immunity to potential antigens in patients with otherwise unexplained infertility.
4. The use of naturally occurring antibodies or of monoclonal antibodies in the isolation and purification of target antigens.
5. The synthesis of adequate amounts of antigen(s) by conventional or recombinant DNA techniques to allow preclinical and clinical testing.
6. The development of acceptable carriers and adjuvants and/or recombinant live vectors in the preparation of a vaccine for human use.
7. The establishment of adequate experimental models, including subhuman primates, for the study of efficacy and safety of a vaccine.

Immunological Disruption of Implantation

Although an immunological strategy for contraception operating at or around the time of implantation may not have the ethical appeal of a prefertilisation method, such an approach will still have a major utility in both Western and developing countries. Theoretically, implantation might be prevented or disrupted by both active and passive immunisation against antigens in the conceptus or its environment. From a practical viewpoint, however, active immunisation is a more attractive proposition although, with the advent of human monoclonal antibodies, another option may be added to postcoital contraception and menstrual induction methods.

The contraceptive immunomanipulation of implantation may be implemented by: (i) a direct cytotoxic attack on the peri-implantation embryo; (ii) indirect distortion of the hormonal or immunological environment for successful implantation, or (iii) blocking "signals" which operate within the genital tract to facilitate embryo transport, implantation and early embryonic survival.

Potential target antigens for an anti-implantation vaccine are listed in Table 14.1. The two most feasible targets are human chorionic gonadotrophin (hCG) and the membrane antigens of trophoblast.

The Development of an Anti-hCG Vaccine

Background and Pre-trial Studies

By far the most advanced approach to vaccine development has involved immunisation against the placental hormone hCG. This hormone appears to be

Table 14.1. Anti-implantation immunocontraception: possible target antigens

Trophoblast antigens	Embryonic antigens/signals	Other antigens
Hormones:	Carcinofetal antigens	Tubal proteins
Human placental lactogen	Differentiation antigens	Endometrial proteins
Human chorionic gonadotrophin	Early pregnancy factor, etc.	Decidual proteins
Gonadotrophin releasing hormones		Sperm antigens
Steroids		
Cytoplasmic proteins:		
Schwangerschaftsprotein 1		
Placental protein 5, etc.		
Enzymes		
Cell membrane antigens:		
Defined by monoclonal antibodies		

restricted to pregnancy (and to some neoplasms), and its early appearance and vital, though incompletely defined, role in the maintenance of gestation make it a theoretically ideal target antigen. An enormous amount of money and scientific effort has now been invested in several lines of research which have ultimately led to the establishment of immunity to hCG as a promising approach for investigation in clinical trials (Stevens 1986).

hCG is a glycoprotein with a molecular weight of approximately 38000 daltons and a carbohydrate content of 30%. The hCG molecule consists of two dissimilar, non-covalently linked, subunits designated alpha and beta. These subunits have been dissociated and purified and their primary structures established. The alpha subunit consists of 89-92 amino acids and is similar in structure to that of pituitary glycoprotein hormones, except for its content of β-galactose and sialic acid. There are, however, great chemical differences in the beta subunits of hCG (β-hCG) and those of pituitary hormones, such as follicle stimulating hormone and thyroid stimulating hormone. However, β-hCG and the beta chain of human luteinising hormone (hLH) have 94 (85%) of the first 110 aminoacids in common; but whereas β-hLH has only 115 aminoacids, β-hCG has 145. Except for one position between residues 111 and 115, the 35 aminoacid sequence of the carboxy-terminal (CT) end of β-hCG is not represented in β-hLH. These extra residues confer antigenic properties on hCG differing from hLH and provide the means of producing an immunogen capable of provoking antibodies with maximum specificity for hCG.

There are several possible mechanisms by which a vaccine directed against hCG might exert antifertility effects. The most likely mechanism is the stimulation of antibodies that neutralise the luteotrophic effect of the target hormone. This would result in regression of the corpus luteum and expulsion of the peri-implantation embryo in association with an apparently physiological subsequent menstruation. Another possible action is a direct cytotoxic effect on the hCG-producing cells of the peri-implantation blastocyst. This could be mediated by cytotoxic antibodies or by a cellular reaction, either of which would need to be directed against hCG determinants at or on the trophoblast cell membrane.

Whatever its particular mode(s) of action, data on actively immunised animals (notably marmosets and baboons) indicate that an hCG vaccine is capable of blocking fertility at an early stage of pregnancy without discernible clinical alterations in menstrual cycling. The method is therefore likely to be a highly acceptable birth-control strategy.

The most refined approach to the development of a specific hCG vaccine, and that pursued by the World Health Organisation Task Force on Birth Control Vaccines, is based on the use of synthetic peptides of the unique 35 aminoacid CT chain of β-hCG, which is not present in hLH. Several peptides representing this region of the molecule have been evaluated as potential antigens in an anti-fertility vaccine.

Initial studies were conducted with enzyme-cleaved and purified peptide chains from the β-hCG C-terminus. Antibodies generated to a 37 aminoacid peptide (sequence 109–145 in the beta subunit) were highly cross-reactive with intact hCG in vitro and neutralised its biological activity in vivo. Antibodies to shorter peptides (less than 35 aminoacids) reacted specifically with hCG but had little neutralising ability. These findings provided a rationale for the further assessment of CT peptides of at least 35 aminoacid residues as the basis for a vaccine that would elicit a specific immune response to hCG but not to hLH. In view of the obvious logistic problems in obtaining adequate quantities of natural peptides, further studies utilised molecules generated by solid-phase synthesis with a view to eventual large-scale production at relatively low cost to ensure sufficient vaccine supplies.

Subsequent studies with synthetic peptides confirmed the necessity of using preparations containing 35 aminoacid residues in order to generate the desired immune response (Stevens et al. 1981a). Shorter peptides probably represent only part of a determinant existing on the native hormone and generate antibodies of low affinity and poor or absent neutralising ability.

In order to ensure adequate immunogenicity it has been necessary to conjugate these peptides to carriers of higher molecular weight. Many classical methods of protein conjugation are available (glutaraldehyde, carbodiimide, diisothiocyanate), and these have been used successfully in the preparation of conjugates involving larger molecules such as the intact beta subunit of hCG. Unfortunately, none of these methods proved suitable for the specific coupling of hCG peptides, since they yielded ill-defined preparations of poor immunogenicity and resulted in unpredictable peptide or carrier polymerisation during the conjugation procedures. Newer methods produced conjugates of constant and predictable composition capable of eliciting higher, although still somewhat variable, antibody responses (Lee et al. 1980). Several carriers have been used, including polymerised flagellin, tetanus toxoid, and diphtheria toxoid; the last eventually being chosen for inclusion in the final vaccine preparation.

None of the antisera to CT peptides has shown detectable cross-reactivity with hLH or with any other purified pituitary hormones. In contrast to anti-β-hCG antibodies, an antibody raised against a 37 aminoacid peptide (β-hCG sequence 109–145) showed no evidence of pituitary localisation on immunofluorescence testing.

Antibodies to hCG peptides are not species-specific. They react with baboon and chimpanzee chorionic gonadotrophin (CG) but at lower levels than with hCG. Assays using partially purified CG from these species suggest a weak cross-reactivity with anti-hCG peptides (baboon 2%–10%, chimpanzee 5%–20%). This cross-reactivity has made it possible to assess the efficacy, and certain aspects of the safety, of an hCG peptide vaccine in non-human primates, notably the baboon.

Significant antifertility effects have been observed in baboons actively immunised against various conjugates of CT hCG peptides administered in Freund's complete adjuvant (Stevens et al. 1981b). Further immunogenicity studies in inbred strains of mice, rabbits and baboons have led to the formulation of a vaccine with constituents (particularly the adjuvant and vehicle) that is potentially suitable for use in man (Stevens et al. 1981c). This vaccine contains a 109-145 aminoacid CT peptide conjugated to diphtheria toxoid and dissolved, together with a synthetic muramyl dipeptide analogue as adjuvant, in the aqueous phase of a squalene-mannide monooleate-saline emulsion.

The most recent antifertility study in baboons using this preparation showed two pregnancies in 44 matings in immunised animals, compared with 14 pregnancies in 20 matings in control animals injected with adjuvant and vehicle. These results indicate that an antifertility effect of greater than 90% can be achieved by active immunisation of baboons with CT peptide, providing that an antibody response of appropriate titre, affinity and neutralising ability can be obtained. Given the low (2%–10%) cross-reactivity of anti-hCG antibodies with baboon CG in this heterologous system, these data are encouraging.

This prototype vaccine has been subjected to detailed toxicological and immunosafety testing. The data from these studies were used to request permission for phase I clinical trials from drug regulatory authorities. Toxicological assessment involved routine acute mouse and rat, subacute rat and rabbit muscle irritancy studies. Immunosafety studies were conducted in baboons receiving three intramuscular injections of 0.5 ml of prototype vaccine at 28-day intervals. Clinical, haematological, biochemical and urine parameters were assessed together with in-life testing for hypersensitivity, tissue antiantibodies, immune complexes and hormone antibody cross-reactivity. Histopathology was extensively assessed following sacrifice 90 days after the last injection. Additional studies involved a separate group of baboons hyperimmunised with the vaccine or its individual components by six injections at fortnightly intervals.

Phase I Clinical Trial

Thirty previously sterilised women aged between 30 and 45 years were recruited for a phase I safety and immunogenicity trial at the Flinders Medical Centre, Adelaide. The trial protocol called for a three-month pre-injection interval followed by the primary immunisation consisting of two injections separated by six weeks and a follow-up period of six months after the first injection. The volunteers were allocated to five groups of six subjects based on a progressive escalation of vaccine dose from 50 μg of conjugate in 25 μl of vehicle to 1.0 mg in 0.5 ml. The protocol for the trial, which is now in its concluding phase, is described by Jones (1986).

To date there have been no significant adverse effects, either serological or clinical. The emulsion vehicle has provoked minimal local reaction, and the injection site has invariably been unidentifiable after one week. Putatively contraceptive levels of anti-hCG antibodies have been provoked for periods of up to six months in the lower-dose groups. The initial results confirm the immunogenicity of this synthetic peptide vaccine, the first to be administered systemically in man (Steward and Howard 1987).

Ongoing and Related Studies

As part of the phase I trial the following studies have been initiated, using appropriately timed blood samples from immunised subjects: (i) the development of human monoclonal antibodies to hCG and diphtheria toxoid; (ii) analysis of T/B cell interactions; (iii) lymphoblast transformation in response to hCG and the vaccine components; and (iv) interleukin-2 production by sensitised lymphocytes.

Ongoing studies in the WHO programme include: the delineation of new peptide immunogens from hCG, the development of improved delivery systems (e.g., biodegradable microcapsules) and the establishment of a homologous model in the baboon for pre-phase-2 assessment of teratogenicity.

Trophoblast Membrane Antigens as Vaccine Targets

Apart from hCG, other cytoplasmic glycoproteins in trophoblast, such as human placental lactogen, Schwangerschaftsprotein 1 and placental protein 5, have proved to be unsuitable as target antigens on efficacy testing in sub-human primates (Jones 1982). Interest has therefore centred on trophoblast surface (membrane) antigens. These have the advantage of being expressed at only one anatomical site following fertilisation and of being in intimate contact with maternal blood at an early stage of pregnancy, perhaps within 10 days of ovulation. In addition, there is now clear evidence that human villous trophoblast cells fail to express major-histocompatibility-complex antigens but exhibit surface expression of tissue specific antigens of potential usefulness in vaccine development. Much of this evidence derives from studies utilising mouse monoclonal antibodies (Anderson et al. 1987). It is only recently, however, that putatively specific monoclonal antibodies generated in our laboratory have given promise of accelerated progress towards the isolation and characterisation of relevant trophoblast membrane antigens (Mueller et al. 1986, 1987). At present three such monoclonal antibodies are under study, all of which are expressed on first-trimester human syncytiotrophoblast. Two are also present in term placenta and localise on marmoset and baboon trophoblast, thereby providing the basis of animal models for the assessment of vaccines. Further studies are planned to assess their reactivity with peri-implantation baboon embryos.

Monoclonal antibodies such as these can be used to devise a plan for the development of a contraceptive vaccine based on trophoblast membrane antigens:

1. Generation of specific mouse monoclonal antibodies of the IgG class.
2. Isolation and preliminary analysis of antigenic fractions recognised by monoclonal antibodies.
3. Further preparative isolation and characterisation of antigens using monoclonal antibodies as immunoaffinity reagents.
4. Testing of immunogenicity and efficacy of putative antigen(s) in marmosets and/or baboons.

5. Preparation and immunogenicity assessment of natural fragments of the native polypeptide antigen(s).
6. Sequencing and synthesis of potential antigens.
7. Selection of best candidate and scaling up of antigen production.
8. Alternatively, utilisation of recombinant DNA technology to generate antigen selected by probing cDNA libraries with specific antibodies raised against antigenic fractions derived in (3) above.
9. Selection of delivery system and initiation of preclinical studies.

Acknowledgements. I acknowledge my co-investigators in the hCG Vaccine Trial, Professor J. Bradley and Dr. S. Judd, and wish to thank Ms. E.H. Denholm and Mr. R.M.Y. Ing for their assistance with the trial. I am grateful to Dr. U.W. Mueller and Dr. C.S. Hawes for allowing me to discuss their work on trophoblast membrane antigens and to Mrs. H. Brown for typing the manuscript.

References

Ada GL, Basten A, Jones WR (1985) Prospects for developing vaccines to control fertility. Nature 317: 288–289

Anderson DJ, Johnson PM, Alexander NJ, Jones WR, Griffin PD (1987) Report of WHO-sponsored workshop on monoclonal antibodies to human sperm and trophoblast antigens. J Reprod Immunol 10: 231–257

Jones WR (1982) Immunological fertility regulation. Blackwell Scientific, Oxford

Jones WR (1986) Phase I clinical trial of an anti-HCG contraceptive vaccine. In: Clark DA, Croy BA (eds) Reproductive immunology. Elsevier, Amsterdam pp 184–187

Lee AC, Powell JE, Tregear GW, Niall HD, Stevens VC (1980) A method for preparing β-hCG COOH peptide-carrier conjugates of predictable composition. Mol Immunol 17: 749–756

Mueller UW, Hawes CS, Jones WR (1986) Cell surface antigens of human trophoblast: definition of an apparently unique system with a monoclonal antibody. Immunology 59: 135–138

Mueller UW, Hawes CS, Jones WR (1987) Identification of extravillous trophoblast cells in human decidua using an apparently unique murine monoclonal antibody to trophoblast. Histochem J 19: 288–296

Stevens VC (1986) Current status of antifertility vaccines using gonadotropin immunogens. Immunol Today 7: 369–374

Stevens VC, Cinader B, Powell JE, Lee AC, Koh SW (1981a) Preparation and formulation of a human chorionic gonadotropin antifertility vaccine: selection of a peptide immunogen. Am J Reprod Immunol 1: 307–314

Stevens VC, Powell JE, Lee AC, Griffin PD (1981b) Antifertility effects from immunization of female baboons with C-terminal peptides of human chorionic gonadotropin. Fertil Steril 35: 98–103

Stevens VC, Cinader B, Powell JE, Lee AC, Koh SW (1981c) Preparation and formulation of a human chorionic gonadotropin antifertility vaccine: selection of adjuvant and vehicle. Am J Reprod Immunol 1: 315–321

Steward MW, Howard CR (1987) Synthetic peptides: a next generation of vaccines. Immunol Today 8: 51–58

Section VII
CLINICAL ASPECTS

15. Early Pregnancy and its Failure after Assisted Conception: Diagnosis by Ultrasonic and Biochemical Techniques

A.F. Riddle, I. Stabile, V. Sharma, S. Campbell, B.A. Mason and J.G. Grudzinskas

Introduction

Monitoring in early pregnancy can be performed by both ultrasound and biochemical techniques. The diagnosis of pregnancy has traditionally been made by the detection of human chorionic gonadotrophin (hCG) in blood or urine, although Schwangerschaftsprotein 1 (SP1) can also be used (Grudzinskas et al. 1977), especially when exogenous hCG has been administered in the luteal phase (Grudzinskas et al. 1986). Transabdominal ultrasound cannot detect a pregnancy until the late fifth or early sixth week from the last menstrual period (Batzer et al. 1983), two to three weeks later than hCG. Thus biochemical analysis continues to be the best method for detection of early pregnancy. However, ultrasound has made a major impact on pregnancy monitoring from the seventh week onwards* – i.e. when the fetal heart can be detected (Robinson 1975; Jouppila et al. 1980a, b; Hertz et al. 1980).

Ultrasound Diagnosis of Pregnancy

The human embryo enters the uterus approximately five days after ovulation as a 16-cell morula (Hamilton et al. 1945). After in-vitro fertilisation (IVF) the embryo is replaced two days after retrieval – i.e. at the two to eight cell stage (Riddle et al. 1987). The early morula and blastocyst are 0.1 – 0.2 mm in diameter (Hamilton et al. 1945) – beyond the resolution of ultrasound.

* All gestation lengths will be expressed from the last menstrual period.

The blastocyst starts to implant seven to eight days after ovulation. At this stage endometrial development is detectable by ultrasound and differentiates between pregnant and non-pregnant cycles (Rabinowitz et al. 1986). By the 12th day after ovulation the blastocyst has sunk below the endometrial surface and has started to differentiate, forming an amniotic cavity, primitive yolk sac, extraembryonic mesoderm and primary yolk sac (Hamilton et al. 1945). None of these changes can be visualised by ultrasound, although serum hCG and SP1 are detectable at this time (Lenton et al. 1981; Ahmed and Klopper 1983; Grudzinskas et al. 1986).

The "intradecidual sign" is the earliest indicator of pregnancy and was reported to be present in 91.7% of 36 patients with subsequently confirmed intrauterine pregnancies (Yeh et al. 1986). The intradecidual sign is the finding of the gestational sac or a sonolucent area within a thickened area of endometrium lying to one side of the midline of the uterine cavity. This sign could be visualised 11 days after ovulation; it had a specificity of 100% and an accuracy of 92.7% when used to diagnose ectopic pregnancy. In the same group the "double decidual sac" sign was present in 63.9% of 36 patients, had a specificity of 60% and an accuracy of 63.4% in respect of ectopic pregnancy. Thus high-resolution abdominal sector scans can detect pregnancy at about the same time that a rapid increase in hCG would occur (Lenton et al. 1981; Jones et al. 1983). In our series the earliest time at which a gestation sac was identified was 12 days after oocyte retrieval. However, most gestation sacs were not identified until the fifth week and in some cases not until the sixth week (Table 15.1). In previous studies it was only in the fifth week that the gestation sac could be reliably identified (Batzer et al. 1983), the fetus and an active fetal heart being seen in the seventh week in some cases; 100% accuracy was not achieved until the eighth week. In our series 100% identification of fetus and fetal heart occurred in the seventh week (Table 15.1), probably reflecting the increased quality of current ultrasound equipment. In more than 70% of patients these observations were made one week earlier. Many patients with delayed diagnosis had pelvic scars, which hinder abdominal sonography.

Table 15.1 Detection of 84 normal singleton pregnancies by ultrasound

Weeks of gestation	Gestation sac seen (%)	Fetus seen (%)	Fetal heart seen (%)	*n*
4	48.4	–	–	31
5	92.3	6.4	2.6	78
6	100	81.5	73.6	76
7	100	100	100	77
8	100	100	100	68
9	100	100	100	70
10	100	100	100	68

The presence of the yolk sac indicates fetal development. This structure can be seen from the sixth week onwards, with reliable diagnosis from the seventh week (Crooij et al. 1982). The presence of the fetal heart beat is associated with a favourable outcome, with over 93% of pregnancies reaching viability (Table 15.2) (Jouppila et al. 1980a, b; Hertz et al. 1980).

Table 15.2. Predictive values of ultrasound parameters in 84 normal singleton pregnancies and 42 early pregnancy failures

Weeks of gestation	Presence of fetal heart activity in normal pregnancy (%)	Absence of fetal heart activity in pregnancy failure (%)	Presence of fetus in normal pregnancy (%)
5	100	2.6	83.3
6	93.3	73.6	93.9
7	93.9	100	93.9
8	94.4	100	93.2

Biochemical Diagnosis of Pregnancy

Pregnancy can be diagnosed by means of hCG measurements from the midluteal phase onwards. The most accurate diagnosis is made by taking two samples two to three days apart to allow for the levels to double (Jones et al. 1983). If exogenous hCG has been administered in the luteal phase (e.g., following IVF) (Riddle et al. 1987), eight to 14 days should be allowed before hCG measurement (Grudzinskas et al. 1986; Braunstein et al. 1973). SP1 measurement is an alternative, since this protein can be detected in serum as early as 8 days after ovulation and in all patients after 13 days (Ahmed and Klopper 1983). Other pregnancy proteins, such as pregnancy associated plasma protein-A (PAPP-A), human placental lactogen (hPL) and placental protein 5, appear too late in the maternal circulation to be of value in early detection of pregnancy (Grudzinskas et al. 1986).

Intrauterine Pregnancy Failure

Pregnancies that fail before the expected onset of menses can sometimes be diagnosed with hCG and SP1 assays (Seppälä et al. 1978; Ahmed and Klopper 1983) and have been termed "biochemical" pregnancies (Jones et al. 1983). Transient appearance of hCG has been described in 15% to 60% of potentially fertile cycles (Braunstein et al. 1977; Chartier et al. 1979; Miller et al. 1980; Edmonds et al. 1982; Edwards and Steptoe 1983; Jones et al. 1983; Whittaker et al. 1983; Wilcox et al. 1985; Riddle et al 1987). Hertig (1952), using histology, identified an early pregnancy loss rate of 43%. Intrauterine pregnancy sacs can be seen in some of these preclinical pregnancies by means of ultrasound, but detection is difficult even in the hands of experienced ultrasonographers (Riddle et al. 1986).

The longer a failed pregnancy progresses the more likely it is that complications such as excessive uterine bleeding or infection will occur. In the case of missed abortion or anembryonic pregnancy the emotional aspects of long delays of diagnosis must be considered. Biochemical measurements and ultrasound when used alone cannot make a reliable distinction, and much effort has been devoted to the use of combined ultrasound and hCG measurement (Kadar et al. 1981a,b, 1983; Nyberg et al. 1986a).

Kadar and colleagues have proposed a "discriminatory zone" of hCG at 6000 to 6500 IU/l (Kadar et al. 1981b). The absence of a gestation sac with hCG levels above the zone and the presence of a sac with hCG levels below the zone are suggestive of ectopic pregnancy. Difficulties in confirming these claims (Kadar et al. 1981a, b; Nyberg et al. 1986a; Kadar and Romero 1986; Filly et al. 1986) are probably due to the use of different hCG reference preparations. Using the second international standard (2nd IS) for hCG, it was possible consistently to demonstrate gestation sacs in normal pregnancy when the hCG levels were above 1800 IU/l (Nyberg et al. 1986a). Some 36% of abnormal pregnancies were detected by depressed levels of hCG, and when correlated to gestation sac size 65% of abnormal gestations were detected. Even allowing for the difference in hCG reference preparations, there was a discrepancy in the findings of these two groups that may be explained by the use of higher-resolution ultrasound equipment by Nyberg et al. (1986a). The ultrasound parameters for diagnosis of abnormal intrauterine pregnancy include a small-for-dates gestation sac (Anderson 1980) and absence of fetal heart activity after the seventh week (Robinson 1975; Anderson 1980), but these criteria rely on accurate menstrual dating. Parameters that do not rely on knowledge of gestational length have been divided into major and minor criteria (Nyberg et al. 1986a, b). Major criteria were a gestational sac greater than 24 mm mean diameter with absence of the fetus, a gestational sac greater that 19 mm mean diameter with absence of the yolk sac and a distorted gestational sac. All of these major criteria individually had a specificity and positive predictive value of 100% but a low sensitivity of 10% to 41%. Minor criteria were a decidual reaction of less than 3 mm, weak decidual amplitude, irregular contour, absent double decidual sign and a low implantation site. The individual specificities were 98% or 99%, and 100% specificity was reached when any three minor criteria were demonstrated. With one major and/or three minor criteria, 44 of 83 abnormal gestations (53%) were correctly diagnosed on a single examination, with a positive predictive accuracy of 100% and no knowledge of gestational length.

Biochemical parameters have been applied for many years in the detection of abnormal pregnancy (Niven et al. 1972; Nygren et al. 1973; Garoff and Seppälä 1975; Kunz and Keller 1976; Braunstein et al. 1978; Schultz-Larzen and Hertz 1978; Damber et al. 1978; Jovanovic et al. 1978; Jouppila et al. 1979; Masson et al. 1983a, b; Salem et al. 1984; Westergaard et al. 1985; Yovich et al. 1986). PAPP-A has predictive value for pregnancy failure when compared with hPL, SP1, progesterone, oestradiol, and pregnancy-zone protein (Westergaard et al. 1985). Serial normal PAPP-A levels and ultrasound showing an active fetal heart had specificity of 99% for normal outcome. However, Yovich et al. (1986), in a prospective series of women conceiving after treatment for infertility, found that hCG was superior in differentiating between normal and anembryonic pregnancies as early as the fourth week: PAPP-A did not reach statistical significance until the seventh week. However, only PAPP-A was able to predict pregnancy failure in 11 subjects in whom ultrasound revealed fetal heart activity (Westergaard et al. 1985). Usually the finding of a fetal heart beat is associated with an excellent prognosis (Jouppila 1980b).

Ectopic Pregnancy

In England and Wales the number of ectopic pregnancies per million estimated conceptions increased from 420 in 1970–72 to 520 in 1976–78 but with a decrease in the death rate over the same periods from 12.2 to 9.4 maternal deaths per million estimated conceptions (HMSO 1982). In America the ectopic pregnancy rate doubled between 1965 and 1976 (Sivin and Cooper 1979). Diagnosis of ectopic pregnancy includes the non-invasive techniques of biochemistry and ultrasound (both of which may impose delay) and the invasive techniques of culdocentesis (which has a high false positive rate) (Halpin 1970) and laparoscopy (which itself carries a risk [Chamberlain and Brown 1978]). A positive hCG test indicates the possibility of ectopic pregnancy. If an intrauterine gestation sac is visualised by ultrasound, ectopic pregnancy can usually be excluded (Donald 1965a, b; Nyberg et al. 1983).

The use of discriminatory zones of hCG may help in the diagnosis of ectopic pregnancy (Kadar et al. 1981a, b; Nyberg 1986a), but each unit should define reference levels to allow for variation in ultrasound equipment and for the hCG reference preparations used (Romero et al. 1984; Kadar and Romero 1986; Filly et al. 1986). Depressed levels of hCG have been reported in ectopic IVF pregnancies (Okamoto et al. 1987), being observed as early as two weeks after oocyte retrieval. However, there was a false positive rate of 83%. The role of SP1 in predicting ectopic pregnancy is uncertain, but results similar to those obtained with hCG would be expected (Lenton et al. 1981; Grudzinskas et al. 1986). PAPP-A levels are depressed in ectopic pregnancy and cannot invariably be detected (Sinosich et al. 1985; Grudzinskas et al. 1985). When implantation occurs outside the uterus, PAPP-A synthesis is selectively disrupted. Ultrasound can assist in the diagnosis of ectopic pregnancy. The actual demonstration of an ectopic pregnancy sac is rare but diagnostic (Lawson 1978; Pedersen 1980; Robinson and De Crespigny 1983). Conversely the finding of a normal intrauterine pregnancy excludes ectopic pregnancy (Donald et al. 1965a, b), except in the rare cases of heterotopy, which are estimated to occur in 1 in 30000 pregnancies (Winer et al. 1957; Romero et al 1984). Before the eighth week of pregnancy, intrauterine fluid collections can easily mimic an intrauterine gestation sac (the "pseudogestational sac" [Schoenbaum et al. 1978; Marks et al. 1979; Pedersen et al. 1980; Weiner 1981; Kadar et al. 1981a, b; Bradley et al. 1982; Nyberg et al. 1983]). The double decidual sac sign has a high specificity for normal intrauterine pregnancy (Bradley et al. 1982; Nyberg et al. 1983) but is difficult to establish before the seventh week (Robinson and De Crespigny 1983). The intradecidual sign has a higher specificity and sensitivity (Yeh et al. 1986). The presence of free fluid in the pouch of Douglas is suggestive of ectopic pregnancy (Donald 1965b; Pedersen 1980) but is not diagnostic because free peritoneal fluid has been observed in normal menstrual cycles (Davis and Gosink 1986).

We have evaluated ultrasound findings in 161 clinical pregnancies established by IVF or gamete intrafallopian transfer (GIFT). The pregnancies were monitored weekly throughout the first trimester or until definite failure had occurred. All data was normalised so that day 14 was the day of oocyte collection (Riddle et al. 1987) Within this group, 84 normal singleton, 26 anembryonic and 11 ectopic pregnancies and five pregnancy failures with a fetus present were analysed. Anembryonic pregnancy was defined as no ultrasound sign of a fetus; in all such cases when histology was performed the ultrasound diagnosis was confirmed. Ultrasound examinations were performed with abdominal sector scans using a 3.5 mHz or 5 mHz transducer. Errors of gestational sac volume were minimised by doing the scans when the patient had a full but not overdistended bladder (Baker et al. 1985).

In all normal pregnancies the gestation sac was seen in the sixth week and fetal heart activity in the seventh week. The predictive value of the observation of a gestational sac at weeks four to 10 is shown in Table 15.3. Specificity for ectopic pregnancy was poor because of five cases of heterotopic pregnancy (combined intrauterine and extrauterine pregnancies), all proven histologically. This rare phenomenon is much commoner following ovarian stimulation, IVF or other methods of assisted conception. Of the five heterotopic pregnancies, four were tubal ectopic pregnancies with no fetus and an anembryonic intrauterine pregnancy; and one was a tubal ectopic pregnancy (fetus absent) combined with a normal intrauterine pregnancy. The value in respect of anembryonic pregnancy by sizing of the gestational sac (Table 15.4) was 100% sensitivity from the eighth week but was not 100% diagnostic because of the heterotopic pregnancies. Ultrasound alone cannot be relied on for accurate diagnosis.

Table 15.3. Predictive value of the presence of an intrauterine gestation sac in detecting 84 normal singleton pregnancies and 42 pregnancy failures

Weeks of gestation	Sensitivity (%)	Specificity (%)	False positive (%)	Efficiency (%)
4	48.0	68.4	28.6	56.0
5	94.7	78.6	25.0	73.5
6	100	80.0	22.5	77.3
7	100	75.8	22.3	79.3
8	100	81.8	20.9	80.0
9	100	81.8	14.3	88.1
10	100	76.4	22.5	87.2

Table 15.4. Values of gestation sac volume below 10th centile in detecting anembryonic pregnancy

Weeks of gestation	Sensitivity (%)	Specificity (%)	False positive (%)	Efficiency (%)
4	100	100	0	100
5	40	91.5	5.4	83.0
6	61.9	95.4	13.3	81.2
7	93.3	96.1	22.7	94.5
8	100	97.2	43.0	96.2
9	100	97.2	33.3	97.5
10	100	97.2	33.3	97.5

Conclusion

No single biochemical or ultrasound criterion can reliably differentiate between the various types of pregnancy failure. As new biochemical parameters are identified further knowledge may be gained. The new generation of high-resolution sector scanners with high-frequency vaginal transducers give picture quality that allows most pregnancy parameters to be identified one week earlier than with abdominal scans. In the future, the diagnosis of normal and abnormal implantation will probably be made by a combination of transvaginal ultrasonography and rapid biochemical tests.

References

Ahmed AG, Klopper A (1983) Diagnosis of early pregnancy by assay of placental proteins. Br J Obstet Gynaecol 90: 604–611

Anderson SG (1980) Management of threatened abortion with real-time sonography. Obstet Gynecol 55: 259–264

Baker ME, Mahony BS, Bowie JD (1985) Adverse effect of an overdistended bladder on first trimester sonography. AJR 145: 597–599

Batzer FR, Weiner S, Corson SL, Schlaff S, Otis C (1983) Landmarks during the first forty-two days of gestation demonstrated by the β-subunit of human chorionic gonadotrophin and ultrasound. Am J Obstet Gynecol 146: 973–979

Bradley WG, Fiske CE, Filly RA (1982) The double sac sign of early intrauterine pregnancy: use in exclusion of ectopic pregnancy. Radiology 143: 223–226

Braunstein GD, Grodin JM, Vaitukaitis J, Ross J (1973) Secretory rates of human chorionic gonadotrophin by normal trophoblast. Am J Obstet Gynecol 131: 25–32

Braunstein GD, Karow WG, Gentry WC, Wade ME (1977) Subclinical spontaneous abortion. Obstet Gynecol 50: 415–445

Braunstein GD, Karow WG, Gentry WC, Rasor I, Wade ME (1978) First trimester chorionic gonadotrophin measurements as an aid in the diagnosis of early pregnancy disorders. Am J Obstet Gynecol 143: 25–32

Chamberlain G, Brown JC (1978) Gynaecological laparoscopy: the report of the Working Party of the Confidential Enquiry into Gynaecological Laparoscopy. Royal College of Obstetricians and Gynaecologists, London, pp 105–153

Chartier M, Roger M, Barrat J, Michelon B (1979) Measurement of plasma hCG and beta hCG in the late luteal phase: evidence of the occurrence of spontaneous menstrual abortions in infertile women. Fertil Steril 31: 134–137

Crooij MJ, Westhuis M, Schoemaker J, Exalto N (1982) Ultrasonographic measurement of the yolk sac. Br J Obstet Gynaecol 89: 931–934

Damber MG, von Schultz B, Solheim F, Stigbrand T, Carlstrom K (1978) Prognostic value of the pregnancy zone protein during early pregnancy in spontaneous abortion. Obstet Gynecol 51: 677–681

Davis FA, Gosink BB (1986) Fluid in the female pelvis: cyclic patterns. J Ultrasound Med 5: 75

Donald I (1965a) Diagnostic uses of sonar in obstetrics and gynaecology. Br J Obstet Gynaecol 72: 907–919

Donald I (1965b) Ultrasonic echo sounding in obstetrical and gynecological diagnosis. Am J Obstet Gynecol 93: 935–941

Edmonds DK, Lindsey KS, Miller JR, Williamson E, Wood PJ (1982) Early embryonic mortality in women. Fertil Steril 38: 447–453

Edwards RG, Steptoe PC (1983) Current status of in vitro fertilisation and implantation of human embryos. Lancet ii: 1265–1269

Filly RA, Laing FC, Nyberg DA (1986) Abnormal pregnancy: early diagnosis by US and serum chorionic gonadotrophin levels. Radiology 161: 854–855

Garoff L, Seppälä M, (1975) Prediction of fetal outcome in threatened abortion by maternal serum placental lactogen and alphafetoprotein. Am J Obstet Gynecol 121: 257–261

Grudzinskas JG, Gordon YB, Jeffrey D, Chard T (1977) Specific and sensitive determination of pregnancy specific beta-1 glycoprotein by radioimmunoassay. Lancet i: 333–335

Grudzinskas JG, Westergaard JG, Teisner B (1985) Pregnancy-associated plasma protein A in normal and abnormal pregnancies. In: Bischof P, Klopper A (eds) Proteins of the placenta.

Karger, Basel, pp 184–197
Grudzinskas JG, Westergaard JG, Teisner B (1986) Biochemical assessment of placental function: early pregnancy. Clin Obstet Gynaecol 13(3): 553–569
Halpin D (1970); Ectopic pregnancy: the problem of diagnosis. Am J Obstet Gynecol 106: 227–230
Hamilton WJ, Boyd J, Mossman HW (1945) Human embryology. W Heffer and Sons Ltd, Cambridge, pp 49–76
Hertig AT, Rock J, Adams EC, Menkin EC (1952) Thirty-four fertilized human ova, good, bad and indifferent recovered from 210 women of known fertility. Paediatrics 23: 202–211
Hertz JB, Mantoni M, Svenstrup B (1980) Threatened abortion studied by estradiol 17-β in serum and ultrasound. Obstet Gynecol 55: 324–328
HMSO (1982) Report on confidential enquiries into maternal deaths in England and Wales, 1976–78. HMSO, London, pp 101–105
Jones HW Jr, Acosta AA, Andrews MC, et al. (1983) What is a pregnancy? A question for programs of in vitro fertilization. Fertil Steril 40: 728–733
Jouppila P, Tapanainen J, Huhtaniemi I (1979) Plasma hCG levels in patients with bleeding in the first and second trimesters of pregnancy. Br J Obstet Gynaecol 86: 343–349
Jouppila P, Seppälä M, Chard T (1980a) Pregnancy-specific beta-1 glycoprotein in complications of early pregnancy. Lancet i: 667–668
Jouppila P, Huhtaniemi I, Tapanainen J (1980b) Early pregnancy failure: study by ultrasonic and hormonal methods. Obstet Gynecol 55: 42–47
Jovanovic L, Dawood MY, Landesmann R, Saxena BB (1978) Hormonal profile as a prognostic index of early threatened abortion. Am J Obstet Gynecol 130: 274–278
Kadar N (1983) Ectopic pregnancy. In: Studd J (ed) Progress in obstetrics and gynaecology, Vol 3, Churchill Livingstone, Edinburgh, 286–296
Kadar N, Romero R (1986) Abnormal pregnancy: early diagnosis by US and serum chorionic gonadotrophin levels. Radiology 161: 854
Kadar N, Caldwell BV, Romero R (1981a) A method of screening for ectopic pregnancy and its indications. Obstet Gynecol 58: 162–166
Kadar N, DeVore G, Romero R (1981b) Discriminatory HCG zone: its use in the sonographic evaluation for ectopic pregnancy. Obstet Gynecol 58: 156–161
Kunz J, Keller PJ (1976) hPL, oestradiol, progesterone and AFP in patients with threatened abortion. Br J Obstet Gynaecol 83: 640–644
Lawson TL (1978) Ectopic pregnancy: criteria and accuracy of ultrasonic diagnosis. AJR 131: 153–161
Lenton AE, Grudzinskas JG, Gordon YB, Chard T, Cooke ID (1981) Pregnancy specific beta-1 glycoprotein and human chorionic gonadotrophin in early human pregnancy. Acta Obstet Gynecol Scand 60: 489–492
Marks WM, Filly RA, Callen PW, Laing FC (1979) The decidual cast of ectopic pregnancy: a confusing ultrasonographic appearance. Radiology 133: 451–454
Masson GM, Anthony F, Wilson MS, Lindsay K (1983a) Comparison of serum and urine hCG levels with SP1 and PAPP-A levels in patients with first-trimester vaginal bleeding. Obstet Gynecol 61: 223–226
Masson GM, Anthony F, Wilson MS (1983b) Value of Schwangerschaftsprotein 1 (SP1) and pregnancy associated plasma protein A (PAPP-A) in the clinical management of threatened abortion. Br J Obstet Gynaecol 90: 146–149
Miller JF, Williamson E, Glue J, Gordon YB, Grudzinskas JG, Sykes A (1980) Fetal loss after implantation. Lancet ii: 554–556
Niven PAR, Landon J, Chard T (1972) Placental lactogen levels as a guide to the outcome of threatened abortion. Br Med J iii: 799–801
Nygren KG, Johansson EDB, Wide L (1973) Evaluation of the prognosis of threatened abortion from the peripheral levels of plasma progesterone, estradiol and human chorionic gonadotrophin. Am J Obstet Gynecol 116: 916–922
Nyberg DA, Laing FC, Filly RA, Uri-Simmons M, Jeffrey RB Jr (1983) Ultrasonographic differentiation of the gestational sac of early intrauterine pregnancy from the pseudogestational sac of ectopic pregnancy. Radiology 146: 755–759
Nyberg DA, Filly RA, Duatre Filho BL, Laing FC, Mahony BS (1986a) Abnormal pregnancy: early diagnosis by US and serum chorionic gonadotrophin levels. Radiology 158: 393–396
Nyberg DA, Laing FC, Filly RA (1986b) Threatened abortion: sonographic distinction of normal and abnormal gestation sacs. Radiology 158: 397–400
Okamoto SH, Healy DL, Morrow LM, Rogers PAW, Trounson AO, Wood EC (1987) Predictive value of plasma human chorionic gonadotrophin β subunit in diagnosing ectopic pregnancy after

in vitro fertilisation and embryo transfer. Br Med J ii: 667–670
Pedersen JF (1980) Ultrasonic scanning in suspected ectopic pregnancy. Br J Radiol 53: 1–4
Rabinowitz R, Laufer N, Lewin A, Navot D, Bar I, Margalioth EJ, Schenker JG (1986) The value of ultrasonographic endometrial measurement in the prediction of pregnancy following in vitro fertilization. Fertil Steril 45: 824–828
Riddle AF, Sharma V, Booker M, Campbell S, Ford NT, Mason BA, Parsons J, Whitehead MI (1986) Serial ultrasonography in pregnancies initiated by in vitro fertilization and embryo transfer. Hum Reprod Suppl 2: 39 (abstr)
Riddle AF, Sharma V, Mason BA, Ford NT, Pampiglione JS, Parsons J, Campbell S (1987) Two year's experience of ultrasound directed oocyte retrieval. Fertil Steril 48: 454–458
Robinson HP (1975) The diagnosis of early pregnancy failure by sonar. Br J Obstet Gynaecol 82: 849–857
Robinson HP, De Crespigny L (1983) Ectopic pregnancy. Clin Obstet Gynaecol 10(3): 407–421
Romero R, Jeanty P, Hobbins JC (1984) Diagnostic ultrasound in the first trimester of pregnancy. Clin Obstet Gynecol 27: 287–313
Salem HT, Ghaneimah SA, Shabaan MM, Chard T (1984) Prognostic value of biochemical tests in the assessment of fetal outcome in threatened abortion. Br J Obstet Gynaecol 91: 382–385
Schoenbaum S, Rosendorf L, Kapelman N, Rowan RT (1978) Grey scale ultrasound in tubal pregnancy. Radiology 127: 757–761
Schultz-Larsen P, Hertz JB (1978) The predictive value of pregnancy specific beta-1 glycoprotein in threatened abortion. Eur J Obstet Gynecol Reprod Biol 8: 253–257
Seppälä M, Ronnberg L, Ylostalo P, Jouppila P (1978) Early detection of implantation by pregnancy-specific β_1-glycoprotein in an infertile woman treated by artificial insemination and human chorionic gonadotrophin. Fertil Steril 32: 608–609
Sinosich MJ, Ferrier A, Teisner B, Porter R, Westergaard JG, Saunders DM et al. (1985) Circulating pregnancy-associated plasma protein A and its tissue concentration in tubal ectopic gestation. J Clin Reprod Fertil 3: 311–317
Sivin I, Cooper T (1979) IUD use and ectopic pregnancy rates in the United States. Contraception 16: 151–173
Weiner CP (1981) The pseudogestational sac in ectopic pregnancy. Am J Obstet Gynecol 139: 959–961
Westergaard JG, Teisner B, Sinosich MJ, Madsen LT, Grudzinskas JG (1985) Does ultrasound examination render biochemical tests obsolete in the prediction of early pregnancy failure? Br J Obstet Gynaecol 92: 77–83
Whittaker PG, Taylor A, Lind T (1983) Unsuspected pregnancy loss in healthy women. Lancet i: 1126–1127
Wilcox AJ, Weinberg CR, Wehmann RE, Armstrong EG, Canfield RE, Nisula BC (1985) Measuring early pregnancy loss; laboratory and field methods. Fertil Steril 44: 366–374
Winer AE, Bergman WD, Fields C (1957) Combined intrauterine and extrauterine pregnancy. Am J Obstet Gynecol 74: 170–178
Yeh H, Goodman JD, Carr L, Rabinowitz JG (1986) Intradecidual sign: a US criterion of early intrauterine pregnancy. Radiology 161: 463–467
Yovich JL, Willcox DL, Grudzinskas JG, Bolton AE (1986) The prognostic value of hCG, PAPP-A, oestradiol and progesterone in early human pregnancy. Aust NZ J Obstet Gynaecol 26: 59–64

16. Investigation and Control of Embryo Implantation in an In-Vitro Fertilisation Programme

R.G. Forman, J. Testart and R. Frydman

Introduction

The implantation of "good quality" embryos into a "receptive" uterus is a fundamental requirement for the establishment of pregnancy following both in-vivo and in-vitro fertilisation. However, unlike certain other aspects of reproduction, our knowledge of the peri-implantation period remains limited (Hearn 1986). Although we have been able to promote follicular growth and ovulation in women for nearly 30 years (Gemzell et al. 1958), our capacity to influence implantation in a positive sense is negligible. Scientific experimentation in this area is limited, and the multifactorial nature of embryo-uterine interactions greatly increases the complexity of the subject.

This chapter will discuss how the technique of in-vitro fertilisation (IVF) may affect these interactions and will address the question of what can be done to improve the chances of implantation for IVF embryos.

The Statistical Approach

Clinical data derived from IVF programmes have been analysed to determine, with mathematical models, the probability that the embryo(s) will implant and that the uterus will sustain implantation (Speirs et al. 1983; Walters 1985; Walters et al. 1985). The two-parameter model seems to provide a good fit to the observed data from several large IVF programmes (Walters 1985). This shows that 40%–60% of uteri are receptive and 20%–40% of embryos have the capacity to implant. Our own data for 186 pregnancies have been subject to the same analysis: 46% of embryos and 37% of uteri appear to be favourable (J. Testart, unpublished observations). Using a different approach, O'Neill and Saunders (1984) have estimated that 43% of embryos are viable, as assessed by their ability to produce a platelet activating factor. A similar figure has been proposed by Yovich et al. (1984) from an analysis of implanting embryos.

Table 16.1 shows the outcome of transfer of frozen thawed embryos relative to the result of transfer of fresh embryos derived from the same IVF cycle. A pregnancy rate of 29% was observed after transfer of frozen embryos, whereas the transfer of the fresh "sibling" embryos was unsuccessful. This suggests that the failure of fresh embryo transfer was more likely to have resulted from an unfavourable uterine environment than intrinsic embryonic factors.

Table 16.1. Result of 100 frozen–thawed embryo transfers related to the outcome of the fresh embryo transfer

Outcome of embryo(s) in IVF cycle	Pregnancy rate after transfer of: (% and no. of embryo transfers)	
	Fresh embryo(s)	Frozen–thawed embryo(s)
None transferred	–	24% (41)
Some transferred:		
Pregnant	11%[a] (76)	0 (4)
No pregnancy	0 (76)	29%[a] (55)

[a] $P<0.01$.

These figures for the relative embryonic and uterine contributions to implantation are only approximations, but they provide a baseline upon which the influence of modifications of IVF technique may be judged. It is unlikely that embryonic viability can be altered to any great extent in the near future. Abnormalities arising during oocyte maturation or fertilisation give rise to the majority of non-viable embryos (Edwards 1986), and though it should be possible to select the better-quality embryos using morphological (Mohr et al. 1983) or biochemical criteria (Leese et al. 1986), there is at present no method that unequivocally increases the quality of the embryos.

The uterine factor includes endocrinological and endometrial components. The remainder of this paper will be devoted to an assessment of the uterine contribution to implantation.

The Hormonal Approach

Pharmacological doses of oestrogen have an antinidatory effect on the uterus (Morris and Van Wagenan 1973). Elevated serum estradiol (E_2) induced by follicular stimulation might have a similar effect, but there is little evidence to support this assumption. It has been suggested that the ratio of E_2 to progesterone may be more important than the absolute levels of either hormone (O'Neill et al. 1985; Gidley-Baird et al. 1986). To clarify these issues we have analysed over 800 embryo transfer cycles.

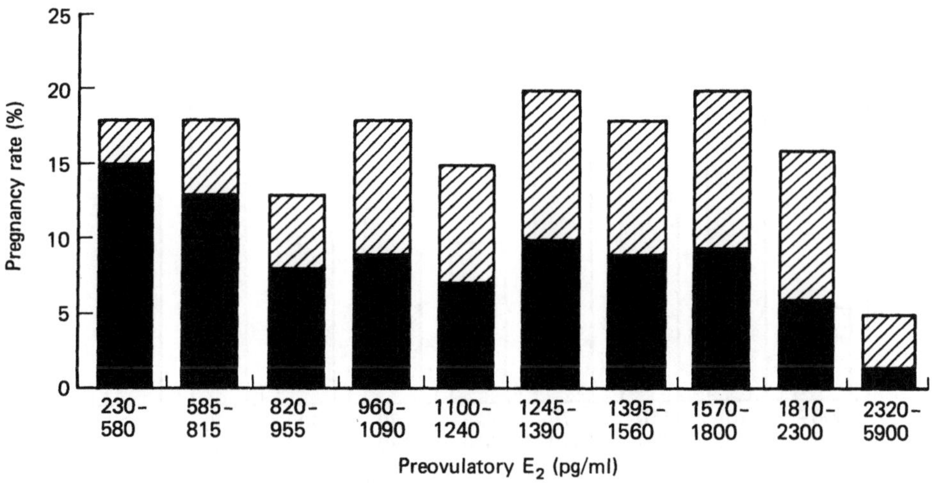

Fig. 16.1. Pregnancy rate following transfer of one (black) and two (*hatched*) embryos in relation to serum E_2 level immediately before hCG administration. This graph is based on 550 embryo transfer cycles. The E_2 concentrations are divided into groups of 10 centiles. Each group therefore contains 55 cycles.

Pregnancy Rate and Preovulatory E_2 Level

In investigating the relationship between pregnancy and the ovarian response to stimulation it is important to control for the number of embryos transferred. The pregnancy rate after the transfer of one and two embryos is shown in Fig. 16.1. There is a significant decrease in implantations when the preovulatory serum E_2 level is greater than the 90th centile for our patient population (2300pg/ml). This is the group with the highest proportion of transfers of two embryos and would normally be expected to have the highest pregnancy rate. The same parameters are shown in Fig. 16.2 after the transfer of three embryos. In this case the pregnancy rate does not vary significantly with preovulatory E_2 level.

Relationship between Follicular and Luteal Phases

Previous reports disagree on the relationship between follicular and luteal phase E_2 and progesterone levels (Cohen et al. 1984; Muasher et al. 1984). These variables were significantly correlated in the present series (Fig. 16.3). This correlation was further strengthened in a subgroup of patients receiving human chorionic gonadotrophin (hCG) for luteal phase support.

Pregnancy Rate and Luteal Phase Hormonal Environment

Fig. 16.4 compares the pregnancy rate following embryo transfer in relation to the mean early luteal phase serum concentrations of E_2 and progesterone and the ratio of E_2/progesterone (calculated from the mean of the hormone concen-

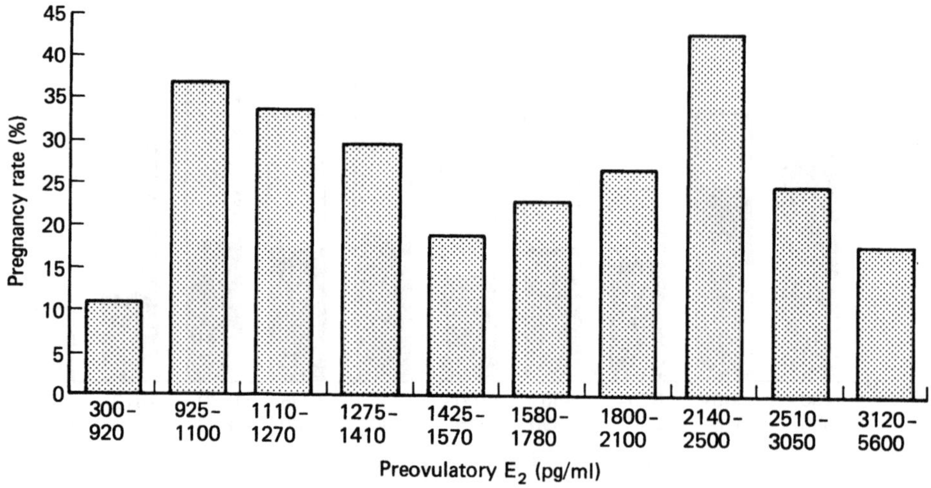

Fig. 16.2. Pregnancy rate following transfer of three embryos in relation to serum E_2 level immediately before hCG administration. The E_2 concentrations are divided into groups of 10 centiles. Each group contains 27 cycles.

trations on days 3 and 6 following oocyte recovery). The pregnancy rate appears to be normally distributed in relation to the luteal E_2. There is a trend towards reduced implantation in association with low and high E_2 concentrations; although this does not reach significance. However, more embryos were transferred in high E_2 cycles than in low E_2 cycles and a higher pregnancy rate would therefore be expected. The pregnancy rate does not vary with the luteal progesterone levels or the ratio of E_2/progesterone.

Implications for Treatment

Thus, the implantation rate following follicular stimulation for IVF is reduced when the preovulatory E_2 concentration is excessive. This applies to the transfer of one and two embryos, but not three embryos. This suggests that the reduced uterine receptivity is relative and may be offset by increasing the number of embryos available for implantation. In this study the adverse effects of high E_2 could not be countered by changes in the E_2/progesterone ratio.

The Endometrial Approach

Histological Aspects

The endometrium in the luteal phase of stimulated IVF cycles is different from that in the natural cycle (Frydman et al. 1982; Cohen et al, 1984). There has been considerable debate as to whether follicular stimulation leads to luteal phase inadequacy (Kerin et al. 1981; Garcia et al. 1981; Feichtinger et al. 1982).

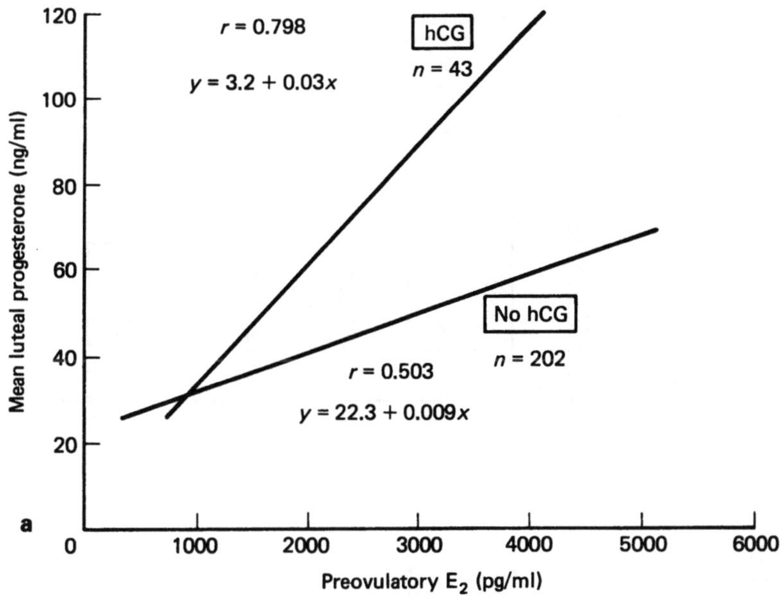

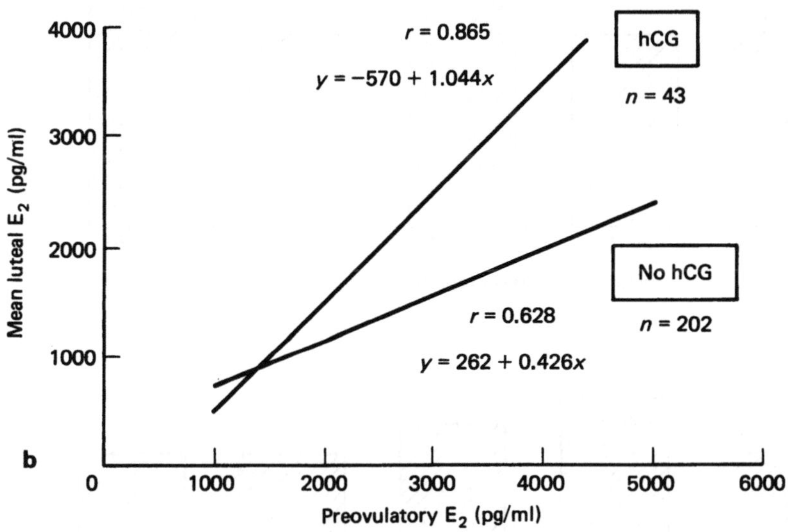

Fig. 16.3a, b. Regression analysis of preovulatory E_2 with mean early luteal progesterone (**a**) and mean early luteal E_2 (**b**) in cycles with and without hCG supplementation of the luteal phase.

However, when the endometrium in the early luteal phase of IVF cycles was dated according to Noyes' criteria, it was shown that the endometrium was at an advanced stage of maturation and that this was related to the progesterone level (Garcia et al. 1984). Our own data (Fig. 16.5) confirm the findings of Garcia et al (1984); it would appear that the follicular response determines the subsequent

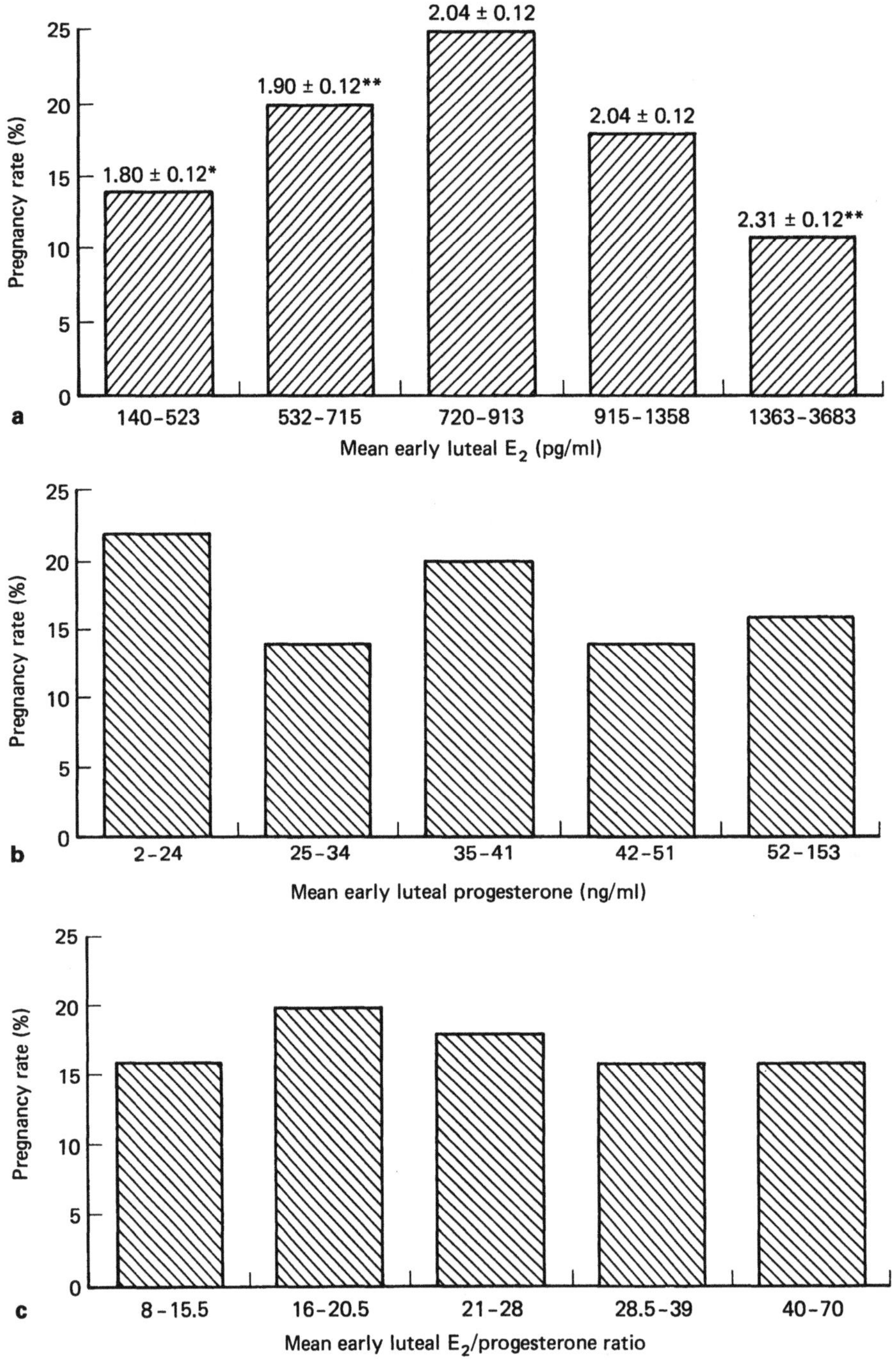

Fig. 16.4a–c. Pregnancy rate in relation to luteal phase concentrations of E_2 (**a**), progesterone (**b**) and E_2/progesterone ratio (**c**). The mean number (±SEM) of embryos transferred per cycle is shown in relation to the luteal E_2. $^*P<0.01$. $^{**}P<0.05$.

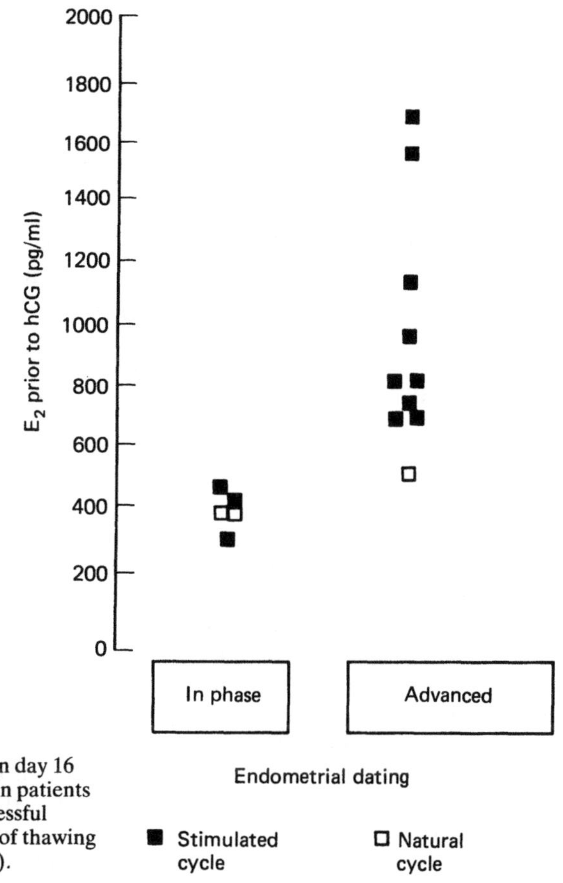

Fig. 16.5. Histology of endometrium on day 16 (day of ovulation normalised to day 14) in patients undergoing oocyte retrieval with unsuccessful fertilisation (stimulated cycle) or failure of thawing of cryopreserved embryos (natural cycle).

degree of endometrial maturation, presumably because of the correlation between preovulatory E_2 and early luteal phase progesterone concentrations.

Ultrastructural Aspects

Luteal phase endometrium has been assessed by means of scanning electron microscopy (Martel et al, 1987). The endometrial cells in stimulated cycles were found to exhibit apical protrusions (pinopodes) two days after ovulation. These pinopodes, which were rarely present at this stage in spontaneous cycles, have a role in the endocytosis of macromolecules and pinocytosis of uterine fluid (Enders and Nelson 1973; Parr 1980). In the rat they are associated with the period of uterine receptivity to the embryos (Psychoyos and Martel 1985).

Biopsies performed six days after ovulation in spontaneous cycles demonstrated abundant pinopode formation, but these structures were greatly decreased in stimulated cycles. This work confirms the more basic histological evaluation in showing that the early luteal phase endometrium, far from being inadequate, may be too advanced for its chronological age.

Endometrial Steroid Hormone Receptors

Circulating steroid hormones act only after binding to their appropriate receptors in the target organ(s) (for review see Agarwal 1983). In the natural cycle E_2 increases the concentration of its own receptor and that of progesterone, whereas progesterone has an inhibitory effect on both types of receptor (Levy et al. 1980a). It is difficult to predict the influence of follicular stimulation on receptor levels in the luteal phase, because the high concentrations of both E_2 and progesterone, plus the administration of anti-oestrogens, may be expected to have antagonistic actions. In one of the few published studies on receptor levels in stimulated IVF cycles a reduction of both oestradiol (ER) and progesterone (PR) receptor concentrations was reported on the day of oocyte retrieval (Ronnberg et al. 1985). This effect was presumed to be due to the continuing anti-oestrogenic action of clomiphene citrate administered early in the follicular phase.

A study is in progress at Clamart to assess the endometrial environment in the early luteal phase of IVF cycles in more detail. Patients in whom oocytes are obtained but fertilisation is unsuccessful because of sperm factors are eligible for inclusion in the trial. Natural cycles are assessed in patients whose cryopreserved embryos failed to withstand the thawing procedure. An endometrial fundal biopsy is done on the day normally scheduled for embryo replacement. The biopsies are stored in liquid nitrogen and subsequently analysed for ER and PR with the technique described by Bayard et al. (1978) and modified by Levy et al. (1980b). Preliminary results suggest that the progestagenic effect predominates (Fig. 16.6). In other words, progesterone down-regulates its own receptor, and this down-regulation cannot be prevented by the high circulating E_2 concentration in the luteal phase. This observation may be of fundamental importance. When preovulatory E_2 levels are excessive luteal serum progesterone levels are also high, and the progestogen receptors will be down-regulated. The administration of endogenous progesterone in these circumstances can only compound the

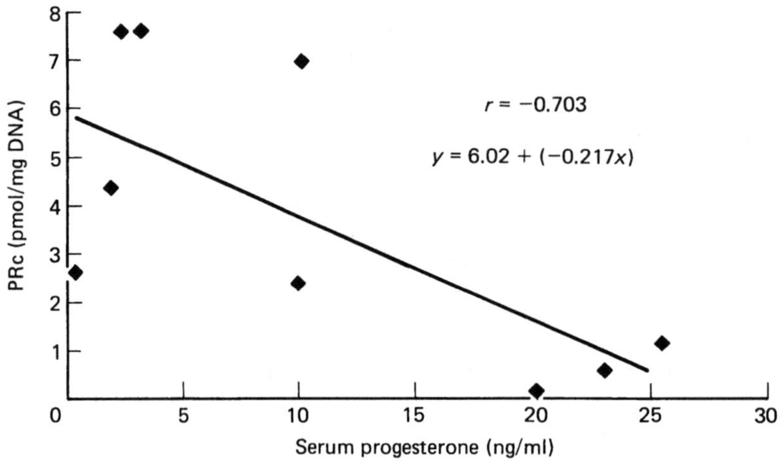

Fig. 16.6. Regression analysis of serum progesterone and cytosolic progesterone receptor (PRc) on cycle day 16.

situation. This would explain our previous findings that a reduced pregnancy rate is related to serum E_2 levels but not to the E_2/progesterone ratio. However, more data are needed before a firm conclusion can be drawn.

Luteal Phase Support in IVF Cycles

Progesterone and progestagenic drugs are widely used in IVF programmes to "boost" the luteal phase. What is the evidence that this treatment is justified? Very few randomised controlled studies have been reported on adequate numbers of patients. Those that do fulfil these criteria fail to show significant differences between treated and control patients (Leeton et al. 1985; Belaisch-Allart et al, 1987, McBain 1987). However, this question can only be resolved by a large multicentre study. Of more pertinent interest is the rationale behind such therapy. The concentration of progesterone in the peri-implantation period is determined by follicular stimulation and is generally abundant. The endometrium has adequately responded to this progestagenic stimulus, as demonstrated by advanced maturity. Furthermore, it is unlikely that it is capable of further response, because the progesterone receptors are probably down-regulated. One of the few arguments in favour of luteal phase support is that the luteal phase may be shortened in stimulated cycles (Edwards 1985). However, this has been disputed (Gronow et al. 1985). There is undoubtedly a fall in E_2 and progesterone levels around the 8th to 10th day of the luteal phase, but whether this is the cause or a consequence of implantation failure is uncertain. Again, there is a need for large controlled trials of progesterone or hCG administration confined to the latter half of the luteal phase.

Embryo Cryopreservation

The pregnancy potential of an IVF cycle is related to the total number of embryos available (Testart et al. 1986). The introduction of long-acting gonadotrophin releasing hormone analogues may enable follicular stimulation to be prolonged and the number of preovulatory follicles increased. This will probably be associated with high circulating E_2 concentrations at the time of induction of ovulation with hCG. Two policies of embryo transfer are possible: either to transfer up to three embryos to offset the adverse effect of high E_2 and freeze the remainder (minimum freezing); or to freeze all embryos for subsequent transfer in natural or only slightly stimulated cycles (maximum freezing) (Frydman et al. 1986). In practice, it is probably desirable on psychological grounds always to transfer one fresh embryo during the IVF cycle. Experience over the past two years has shown that a frozen embryo has, on average, an 11.6% chance of implantation: this figure varies with the developmental stage at freezing but includes all stages. By combining this result with the data obtained with transfer of fresh embryos theoretical tables of pregnancy rates for these two different policies can be constructed (Table 16.2). For example, when five embryos are available the theoretical pregnancy rate is 49.5% with minimum freezing (pregnancy rate for three fresh embryos = 26.3% + [2 × 11.6%]). In the maximum freezing policy the calculated figure is 61.2% (one fresh embryo = 14.8% + [4 × 11.6%]).

Table 16.2. Theoretical pregnancy rates with different freezing policies

No of embryos available	No. of embryos frozen	
	Minimum	All except one[a]
1	14.8%[b]	–
2	19.9%[c]	26.4%
3	26.3%[d]	38.0%
4	37.9%	49.6%
5	49.5%	61.2%
6	61.1%	72.8%

[a] Assuming pregnancy rate for 1 frozen embryo = 11.6%.
[b,c,d] Pregnancy rate following transfer of 1 (n=283), 2 (n=211) or 3 (n=217) unfrozen embryos.

It will be several years before a similar calculation can be made with real rather than theoretical data, because no patient should be included in the analysis until all her stored eggs have been used. Such calculations will, however, have an intrinsic bias because those patients who become pregnant during the period studied will probably not have reclaimed all their embryos.

Embryo cryopreservation is, however, one of the few techniques that enable us to control implantation following IVF, if only by offering the possibility of bypassing a potentially unfavourable uterine environment.

Conclusion

Implantation is the least controllable part of IVF. Embryonic viability cannot yet be altered, and evidence has been presented that progestagenic support of the luteal phase is unlikely to improve uterine receptivity. Regulation of the follicular phase holds the key to controlling the endometrial environment in the peri-implantation period. It is too early to say whether a change in the policy for embryo transfer will alter the pregnancy potential of an IVF cycle, but embryo cryopreservation appears promising in this respect.

Acknowledgements. We thank all the clinical and laboratory staff working in the IVF unit at Hôpital Antoine Béclère, without whom the studies reported in this paper would not have been possible. Special thanks are due to Professor Nessman for her expert histological analysis of the endometrial biopsies and to Dr Robel and Dr Eychenne at INSERM Unit 133 for their invaluable assistance with the steroid hormone receptor investigations.

References

Agarwal MK (1983) General considerations in steroid hormone receptor research. In: Agarwal MK (ed) Principles of recepterology. de Gruyter, Berlin, pp 2–67

Bayard F. Sergio D, Robel P, Baulieu E-E (1978) Cytoplasmic and nuclear estradiol and progesterone receptors in human endometrium. J Clin Endocrinol Metab 46: 635–648

Belaisch-Allart J, Testart J, Fries N, Forman R, Frydman R (1987) The effect of dydrogesterone supplementation in an IVF program. Hum Reprod 2: 183–185

Cohen JJ, Debache C, Pigeau F, Mandlebaum J, Plachot M, de Brux J (1984) Sequential use of clomiphene citrate, human menopausal gonadotropin, and human chorionic gonadotropin in human in vitro fertilization. II. Study of luteal phase adequacy following aspiration of the preovulatory follicles. Fertil Steril 42: 360–365

Edwards RG (1985) Normal and abnormal implantation. In: Edwards RG, Purdy JM, Steptoe PC (eds) Implantation of the human embryo. Academic Press, London, pp 303–333

Edwards RG (1986) Causes of early embryonic loss in human pregnancy. Hum Reprod 1: 185–198

Enders AC, Nelson DM (1973) Pinocytic activity of the uterus of the rat. Am J Anat 138: 277–300

Feichtinger W, Kemeter A, Szalay S, Beck A, Janisch H (1982) Could aspiration of the Graafian follicle cause luteal phase deficiency. Fertil Steril 37: 205–208

Frydman R, Testart J, Giacomini P, Imbert MC, Martin E, Nahoul K (1982) Hormonal and histological study of the luteal phase in women following aspiration of the pre-ovulatory follicle. Fertil Steril 38: 312–317

Garcia J, Jones G, Acosta A, Wright G (1981) Corpus luteum function after follicle aspiration for oocyte retrieval. Fertil Steril 36: 565–572

Garcia JE, Acosta AA, Hsiu J-G, Jones HW Jr (1984) Advanced endometrial maturation after ovulation induction with human menopausal gonadotropin/human chorionic gonadotropin for in vitro fertilization. Fertil Steril 41: 31–35

Gemzell CA, Diczfalusy E, Tillinger G (1958) Clinical effect of human pituitary follicle stimulating hormone. J Clin Endocrinol Metab 18: 1333–1348

Gidley-Baird A, O'Neill C, Sinosich M, Porter R, Pike I, Saunders DM (1986) Failure of implantation in human in vitro fertilization and embryo transfer patients: the effect of altered progesterone estrogen ratios in humans and mice. Fertil Steril 45: 69–74

Gronow MJ, Martin MJ, Hay D, Moro D, Brown JB (1985) The luteal phase after hyperstimulation for in vitro fertilization. Ann NY Acad Sci 442: 391–401

Hearn JP (1986) The embryo maternal dialogue during early pregnancy in primates. J Reprod Fertil 76: 809–819

Kerin JF, Broom TJ, Ralph MM et al. (1981) Human luteal phase function following oocyte aspiration from the immediately preovular Graafian follicle of spontaneous ovular cycles. Br J Obstet Gynaecol 88: 1021–1028

Leese HJ, Hooper MAK, Edwards RG, Ashwood-Smith MJ (1986) Uptake of pyruvate by early human embryos determined by a non-invasive technique. Hum Reprod 1: 181–182

Leeton J, Trounson A, Jessup D (1985) Support of the luteal phase in in vitro fertilization programmes: results of a controlled trial with intramuscular proluton. J In Vitro Fertil Embryo Transfer 2: 166–169

Levy C, Robel P, Gautray JP et al. (1980a) Estradiol and progesterone receptors in human endometrium: normal and abnormal menstrual cycles and early pregnancy. Am J Obstet Gynecol 136: 646–651

Levy C, Eychenne B, Robel P (1980b) Assay of nuclearestradiolreceptor by exchange on a glass fibre filter. Biochim Biophys Acta 630: 301–305

Martel D, Frydman R, Glissant M, Maggioni C, Roche D, Psychoyos A (1987) Scanning electron microscopy of postovulatory human endometrium in spontaneous and stimulated cycles. J Endocrinol 114: 319–324

McBain JC (1987) Do women hyperstimulated with clomiphene/hMG for IVF or GIFT benefit from luteal phase support. In: Naftolin F, DeCherney AH (eds) The control of follicular development, ovulation induction and luteal function: lessons from in vitro fertilization. Raven Press, New York pp 293–300

Mohr LR, Trounson AO, Leeton JF, Wood C (1983) Evaluation of normal and abnormal human embryo development during procedures in vitro. In: Beier HM, Lindner HR (eds) Fertilization of the human egg in vitro. Springer-Verlag, Berlin Heidelberg New York, pp 211–221

Morris JM, Van Wagenan G (1973) Interception: the use of post-ovulatory estrogens to prevent

implantation. Am J Obstet Gynecol 115: 101–106
Muasher S, Acosta AA Garcia JE, Jones GS, Jones HW Jr (1984) Luteal phase serum estradiol and progesterone in in vitro fertilization. Fertil Steril 42: 838–843
O'Neill C, Saunders DM (1984) Assessment of embryo quality. Lancet ii: 1035
O'Neill C, Ferrier AJ, Vaughan J, Sinosich MJ, Saunders DM (1985) Causes of implantation failure after in vitro fertilization and embryo transfer. Lancet ii: 615
Parr MB (1980) Endocytosis in the uterine epithelium during pregnancy. Prog Reprod Biol 7: 81–91
Psychoyos A, Martel D (1985) Embryo–endometrial interactions at implantation. In: Edwards RG, Purdy JM, Steptoe PC (eds) Implantation of the human embryo. Academic Press, London, pp 195–210
Ronnberg L, Isotalo H, Kauppila A, Martikainen H, Vihko R (1985) Clomiphene induced receptor kinetics on the day of ovum collection after ovarian stimulation: a study of cytosol and nuclear estrogen and progestin receptors and 17β hydroxysteroid dehydrogenase. Ann NY Acad Sci 442: 408–415
Speirs AL, Lopata A, Gronow MJ, Kellow GN, Johnston WIH (1983) Analysis of the benefits and risks of multiple embryo transfer. Fertil Steril 39: 468–471
Testart J, Lassalle B, Belaisch-Allart J, Forman R, Frydman R (1986) Cryopreservation does not affect the future of human fertilized eggs. Lancet ii: 569
Walters DE (1985) An assessment of two mathematical models of embryo implantation. In: Edwards RG, Purdy JM, Steptoe PC (eds) Implantation of the human embryo. Academic Press, London, pp 219–226
Walters DE, Edwards RG, Meistrich ML (1985) A statistical analysis of implantation after replacing one or more human embryos. J Reprod Fertil 74: 557–567
Yovich JL, Stanger JD, Yovich JM, Tuvick AI (1984) Quality of embryos from in vitro fertilisation. Lancet i: 457

17. Ectopic Pregnancy: Diagnostic Aspects

I. Stabile, J.G. Westergaard and J.G. Grudzinskas

Introduction

Ectopic pregnancies can be divided into three clinical groups. In the first group, with clear evidence of haemoperitoneum, the only investigation needed is laparoscopy or occasionally laparotomy. In the second group, with a less acute presentation (positive pregnancy test, abdominal pain), the diagnosis is aided by a high index of suspicion and a few additional diagnostic tests. The third group includes women at high risk of ectopic gestation (subfertile patients with a past history of ectopic gestation or tubal surgery and those attending for in-vitro fertilisation (IVF) or gamete intrafallopian transfer (GIFT).

Diagnosis of Ectopic Pregnancy

Clinical Examination

The early stages of ectopic pregnancy can mimic gynaecological conditions such as pelvic inflammatory disease (PID), ruptured corpus luteum, dysfunctional uterine bleeding and threatened or incomplete abortion. The initial clinical diagnosis of ectopic pregnancy proves right in less than 50% of cases. Table 17.1 summarises the physical findings in ectopic pregnancy.

Ultrasound Examination

Ultrasound can exclude ectopic pregnancy by demonstration of an intrauterine pregnancy; a combination of intrauterine and extrauterine pregnancy is exceedingly rare (1 : 30000), except after assisted conception (Riddle et al. 1987). The gestation sac of a normal intrauterine pregnancy becomes visible at six weeks of amenorrhoea.

Table 17.1. Clinical findings in proven ectopic pregnancy

Reference	No. of patients	Abdominal pain (%)	Shock (%)	Mass (%)	Rebound tenderness	Normal sized uterus (%)
Bell & Ingersoll 1985	130	–	15.3	25.4	–	–
Draa &Baum 1951	224	–	8.9	32.6	–	–
Crawford & Hutchinson 1954	300	90.7	23.9	38.3	33.7	75
Armstrong et al. 1951	481	92.7	16.6	67.3	40.1	–
Soisson & Moran 1959	50	98	20	64	74	70
Schiffer 1963	268	37.7	7.8	56.7	–	–
Webster et al. 1965	699	93.5	–	50	–	–
Breen 1970	654	–	48.7	49	52	–
Helvacioglu et al. 1979	284	83.5	1.4	29.9	41.2	74.3
Kitchin et al. 1979	191	–	17.3	42.9	–	85.9
Brenner et al. 1980	300	–	–	53.7	–	70.7

Ultrasound cannot directly identify ectopic pregnancy except when a live fetus is demonstrated outside the uterine cavity. This is possible after seven weeks' gestation but has been reported in only 6% of ectopic pregnancies (Pedersen 1980). Furthermore, interpretation of the ultrasound findings depends on an accurate knowledge of the duration of the pregnancy. Around one-third of patients with ectopic pregnancy do not know the date of their last menstrual period, and in others irregular bleeding may confuse the picture. The experience of the sonographer and the resolution of the equipment are also relevant to the correct interpretation of the sonographic findings.

A pseudogestational sac is seen in 10%–20% of cases, and a definitive diagnosis of a viable intrauterine pregnancy must include demonstration of fetal heart activity (Weiner 1981). A pseudogestational sac is due to collection of blood within the hyperplastic endometrium – i.e., the decidual cast. This is concentrated in the centre of the uterine cavity and can thus be distinguished from an early (before six weeks) gestational sac which is eccentric (Abramovici et al. 1983). In addition, an early intrauterine pregnancy may present a double sac appearance because of the concentric decidua capsularis and parietalis (Bradley et al. 1982). Thus the ultrasonic features of an intrauterine gestational sac include an eccentric, asymmetrical trophoblast ring, which may have a double sac appearance.

Other sonographic signs suggestive of ectopic gestation include visualisation of complex or cystic adnexal masses and free fluid in the peritoneal cavity. Cystic or complex masses have been reported with variable frequency (60%–90%) in ectopic gestation, but these are difficult to differentiate from PID, cyst accidents and endometriosis. In addition, Brown et al. (1978) have reported that the adnexal mass was on the contralateral side in seven of 30 cases. In combination with quantitative hCG estimations this may be used to raise the index of suspicion, as it is rarely seen in intrauterine pregnancies (Kadar et al. 1983).

Prospective use of real-time ultrasound alone gives a correct diagnosis of the presence or absence of ectopic pregnancy in 70%–92% of affected cases (Piironen and Punnonen 1981; Gleicher et al. 1983).

Our recent observations (Stabile et al. 1987) in a group of women suspected of harbouring an ectopic pregnancy have highlighted the importance of careful ultrasonic examination, including assessment of uterine area, endometrial thickness and adnexal mass volume. Uterine area measurements alone (at a cut-off of $<20\ cm^2$) allowed us to diagnose an ectopic pregnancy in 72% of patients (sensitivity) with a specificity of 41% and a positive predictive value of 20%. The negative predictive value of this observation was 79%. Thickening of the endometrium (> 10 mm) was found in 50% of proven cases, in 25% of women with ovarian cysts and 42% of those with PID. The negative predictive value of this observation was 84%. The best single predictor of ectopic pregnancy was an adnexal mass in excess of 10 ml volume separate from the ovary (sensitivity 85%, specificity 37%, negative predictive value 90%). Significant free fluid in the pouch of Douglas was seen in 55% of women with proven ectopic pregnancy, 18% of women with PID and 32% of women with ovarian cysts.

The use of any one of these ultrasound features in women with clinical evidence of ectopic pregnancy will diagnose 62% of cases, with a specificity of 57% and a predictive value positive of 25%. The negative predictive value of any of these observations is 87%; thus the absence of positive findings at ultrasound examination is of value in the exclusion of this condition.

Biochemical Tests

Human Chorionic Gonadotrophin (hCG)

The first detection of hCG in the maternal circulation correlates with the implantation of the blastocyst (Lenton et al. 1982). Ultrasound diagnosis of pregnancy is not possible until two to three weeks later.

hCG examinations remain the mainstay in the differential diagnosis of abdominal pain in women of reproductive age (Norman et al. 1985). The percentage of positive results is related to the sensitivity of the test, which in turn reflects the sample fluid tested and technique used. Most routine urine pregnancy tests (latex agglutination inhibition) have a sensitivity of 700–1000 mU/l and are positive in up to 50% of ectopics (Milwidski et al. 1978). Simple dipstick techniques for the measurement of hCG based on enzyme labelled monoclonal antibodies (Norman et al. 1985; Burford et al. 1986) or immunofluorometric assays (Stenman et al. 1983) have contributed greatly because of their high sensitivity and rapid availability.

The false negative rate of hCG estimations in patients with ectopic pregnancy depends on the cut-off level used. Poulsen et al. (1987) reported eight cases of ectopic pregnancy with hCG <25 U/l, emphasising that a high index of suspicion should be maintained even when serum hCG is undetectable. Olsen et al. (1983) and Romero et al. (1985) report false negative rates for ectopic pregnancy of 0.5% at <10 U/l (International Reference Preparation = 5 U/l (2nd International Standard)) and 0% at <3 U/l.

Ectopic pregnancies generally produce hCG at a lower rate than normal pregnancies (Landesman and Saxena 1976). Normal hCG values are found more often in early and in ruptured ectopic pregnancies (Ackerman et al. 1982). This has led to the theory that viable ectopic pregnancies are more likely to rupture than non-viable ones. In addition, negative (<25 U/l) hCG results are associated with necrosis of chorionic tissue (Poulsen et al. 1987), supporting the concept that the reduction in the functional trophoblast mass (Saxena and Landesman 1975) is due to poor vascularisation and cell necrosis. It has also been suggested that the lower hCG levels in ectopic pregnancy may be related to the high incidence of blighted ova (Jouppila et al. 1980a). Indeed, normal hCG levels are recorded when normally developed embryos are present in the fallopian tube (Jouppila et al. 1980b).

Individual variations in levels of serum hCG are large, such that there is a great overlap between normal and pathological pregnancies. In addition, it is often difficult to determine the exact gestational age. Thus single hCG estimations are of limited value in the diagnosis of ectopic pregnancy (Gruneberger et al. 1977). To improve the diagnostic value of hCG estimations, some workers have analysed the rate of hCG production, correlating it with the sonographic detection of a gestational sac. Kadar et al. (1981) have established a discriminatory hCG zone (6000 to 6500 IU/l) – the levels of hCG at which a gestational sac first becomes visible ultrasonically. By calculating the doubling rates of hCG they have devised an algorithm for the management of suspected ectopic pregnancy based on serial (48h) quantitative hCG estimations.

After a spontaneous abortion, the half-life of hCG is between three and seven days (Schwartz and DiPietro 1980). Thus the information obtained from a single hCG estimation may not reflect the actual state of the trophoblast. The shorter half-life of the alpha subunit of hCG may make it a better diagnostic test. L'Hermite-Baleriaux and her colleagues have shown (1982) that if hCG levels are reduced, beta-hCG measurements are less affected than those of the alpha subunit.

Schwangerschaftsprotein 1 (SP1)

SP1 is a glycoprotein with a molecular weight of 90kd secreted by the syncytiotrophoblast (Horne et al. 1976) and appearing in the maternal circulation shortly after implantation. SP1 measurements have been evaluated in the diagnosis of ectopic pregnancy (Ho and Jones 1980). Despite encouraging initial results (Seppälä et al. 1980), the false positive rate is high (Braunstein and Asch 1983). Nevertheless, SP1 can be used as a pregnancy test in women who have received hCG for ovulation induction (Grudzinskas et al. 1986), and hence it may retain a limited role in high-risk groups.

Human Placental Lactogen (hPL)

hPL is a product of the syncytiotrophoblast, detectable in the maternal circulation by radioimmunoassay from six weeks' gestation. In ectopic pregnancy there is a correlation between hCG and hPL levels (Braunstein and Asch 1983). In ectopic pregnancy hPL values are more frequently normal than the levels of hCG or SP1 (Braunstein and Asch 1983), thus limiting its clinical value.

Prolactin

Prolactin is a pituitary and decidual hormone present in the serum and amniotic fluid of pregnant women. Elevated levels are found in women with threatened abortion, but plasma levels are of little value in the assessment of fetal wellbeing (Biswas and Rodeck 1976). There is no data on prolactin levels in ectopic pregnancy.

Alphafetoprotein (AFP)

AFP is an alpha-1 globulin of fetal origin. The concentration of AFP in maternal serum depends on the total amount produced by the fetal liver and on the area available for diffusion from the fetal to the maternal compartment. There is a single report in the literature of an ectopic pregnancy with elevated AFP levels which the authors attribute to leakage of AFP from the fetal blood into the maternal circulation (Kjessler et al. 1977).

Alpha-amylase

The normal fallopian tube contains a higher alpha-amylase concentration than blood (McGeachin et al. 1958). Isolated reports on elevated serum and peritoneal-fluid amylase levels in ruptured ectopic pregnancy (Flege 1966; Hochberg 1974) have not been confirmed.

Ovarian Steroids

Initial studies showed normal or depressed progesterone levels in patients with ectopic pregnancy (Radwanska et al, 1978; Milwidski et al, 1977, 1984). A report on the use of rapid determination of progesterone levels (within 2–4 h) as an adjunct to clinical diagnosis (Matthews et al. 1986) suggested that in patients with clinical signs a depressed progesterone level (<15 ng/ml) increases the likelihood of ectopic pregnancy. By comparing progesterone levels in normal and ectopically implanted pregnancies with similar hCG levels, Hubinont et al. (1987) have showed that the luteal insufficiency observed in ectopic pregnancy cannot be explained simply by the decrease of hCG. The predictive value positive of a progesterone level > 15 ng/ml at less than eight weeks' gestation in respect of normal outcome is 95%.

Jouppila et al. (1980b) reported depressed levels of oestradiol in ectopic

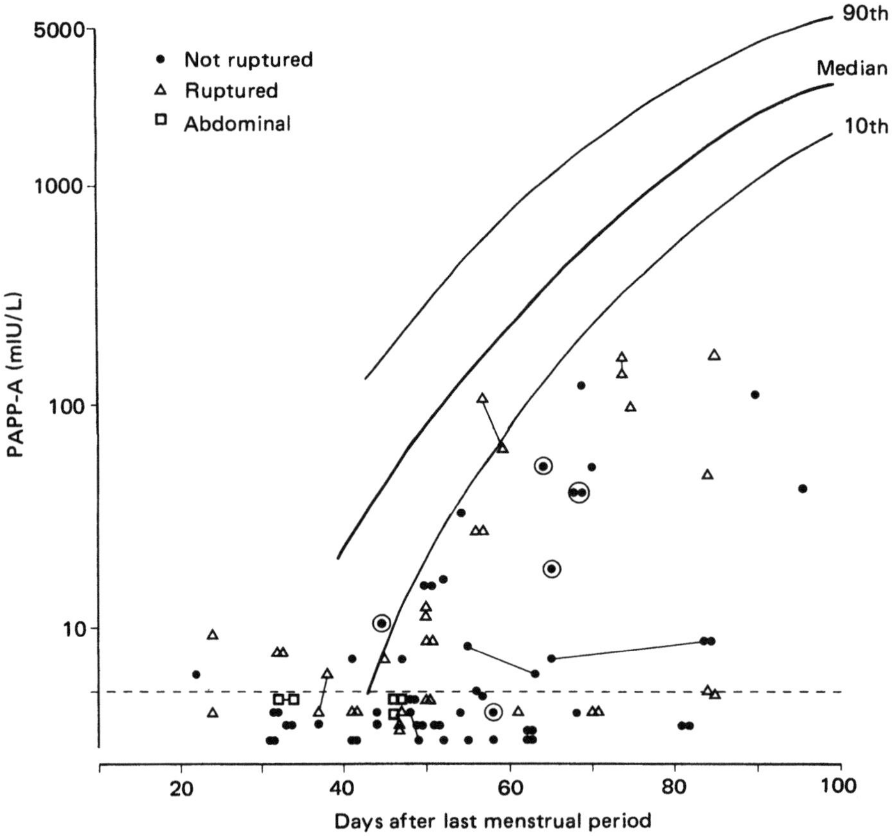

Fig. 17.1. Circulating PAPP-A levels in 60 women with ectopic gestation compared to the 80% confidence of the normal range. Encircled symbols indicate fetal tissue identified.

pregnancy, with the exception of one case where a normal embryo was found in the fallopian tube at eight weeks' gestation. A larger study (Hubinont et al. 1987) showed that 80% of women with ectopic pregnancy had serum oestradiol levels below the 10th centile of the normal range.

Pregnancy Associated Plasma Protein-A (PAPP-A)

Sinosich and colleagues (1983) reported that PAPP-A could not be detected in maternal serum in 9 of 10 women with tubal pregnancy following IVF-ET or natural conception. Although the late appearance (after 6 weeks) of PAPP-A in blood renders its measurement unsuitable for the diagnosis of pregnancy, some studies on PAPP-A suggest that it may be of value in the diagnosis of ectopic pregnancy (Sinosich et al. 1985; Westergaard 1987; Poulsen et al. 1987; Stabile et al. 1987). In 60 women with proven ectopic pregnancy, serum PAPP-A could be detected in only 30, being below the 10th centile of the normal range in 24 women (Westergaard 1987; Fig. 17.1). hCG levels were greater than 25 U/l in 51 of the 60 women, and SP1 was detected in 52 of these patients. It was concluded that hCG and SP1 estimations were of less discriminatory value than PAPP-A. Analysis of qualitative hCG data in conjunction with PAPP-A measurements demonstrated that detectable hCG and depressed or undetectable PAPP-A had the highest relative risk for ectopic pregnancy (Table 17.2).

Thus, PAPP-A synthesis by the trophoblast is severely and perhaps selectively compromised in ectopic gestation. The normal immunohistochemical distribution of hCG, hPL and SP1 in tissues from ectopic gestation contrasts with the absence or deficiency of PAPP-A (Chemnitz et al. 1984; Tornehave et al. 1986).

Table 17.2. Circulating hCG and PAPP-A in prediction of ectopic pregnancy ($n = 54$)

	Sensitivity (%)	Specificity (%)	Predictive value +ve (%)	Predictive value −ve (%)	Relative risk
Depressed (detectable) hCG	57	88	63	85	3.0
Depressed (detectable) PAPP-A)	41	90	59	81	3.1
Depressed (detectable) hCG and depressed or undetectable PAPP-A	57	87	63	85	4.2
Detectable hCG and depressed or undetectable PAPP-A	78	84	63	91	7.0
Normal hCG and depressed or undectable PAPP-A	20	95	61	77	2.65
Detectable hCG and undetectable PAPP-A	37	93	65	81	3.4

Normal = >10th centile; depressed = <10th centile; undetectable hCG = <25 IU/l; undetectable PAPP-A = <5 mIU/l.

Culdocentesis

Culdocentesis is considered positive when non-clotting blood is obtained. A haemoperitoneum is found in 80%–95% of both ruptured and unruptured ectopic pregnancies (Webster et al. 1965; Breen 1970; Helvacioglu et al. 1979). False negative rates of 11%–14% have been reported. In cases of clinical doubt a negative tap should be followed by either a repeat puncture or by culdoscopy (Breen 1970) or colpotomy (Soisson and Moran 1959). In most centres these techniques have been replaced by laparoscopy.

Dilatation and Curettage

The findings of decidua without chorionic villi is strongly suggestive of ectopic pregnancy (Gordon 1980). However, the high false positive and false negative rates reduce the diagnostic value.

Laparoscopy

This is the definitive method to both identify and exclude an ectopic pregnancy. It has been proven to reduce the rate of ruptured pregnancies (Helvacioglu et al. 1979). Missed diagnosis in 3%–4% of cases is limited to very early ectopic pregnancies (Yaffe et al. 1979). Most "false positive" laparoscopies demonstrate conditions for which surgery is often warranted anyway – e.g., haematosalpinx or torsion of a pedicle.

Conclusion

The incidence of ectopic pregnancy appears to be rising, and this includes the increased rate of ectopic pregnancy following successful treatment for subfertility. Although ultrasound technology permits observation of the embryo at an early stage, the earliest indications of ectopic pregnancy are likely to come from the observations of compromised hormone and protein synthesis. At the present time PAPP-A and progesterone measurements hold the greatest promise.

References

Abramovici H, Auslender R, Lewis A, Faktor JH (1983) Gestational-pseudogestational sac: a new ultrasonic criterion for differential diagnosis. Am J Obstet Gynecol 145: 377–379

Ackerman R, Deutsche S, Krumholz B (1982) Levels of hCG in unruptured and ruptured ectopic pregnancy. Obstet Gynecol 60: 13–14

Armstrong JT, Wills, SH, Moore J, Lauden AE (1959) Ectopic pregnancy: a review of 481 cases. Am J Obstet Gynecol 77: 364–370

Bell JW, Ingersoll FM (1950) An evaluation of the symptoms and signs in ectopic pregnancy, based on an analysis of 130 cases. Am J Obstet Gynecol 60: 174–179

Bischof P, Reyes Y, Herrmann WL, Sizonenko PC (1983) Circulating levels of PAPP-A and hCG in intrauterine and extrauterine pregnancies. Br J Obstet Gynaecol 90: 323–325

Biswas S, Rodeck CH (1976) Plasma prolactin levels during pregnancy. Br J Obstet Gynaecol 83: 683–687

Bradley WG, Fiske CE, Filly RA (1982) The double sac sign of early intrauterine pregnancy: use in exclusion of ectopic pregnancy. Radiology 143: 223–226

Braunstein GO, Asch HR (1983) Predictive value analysis of measurements of hCG, SP1, hPL and cystine aminopeptidase for the diagnosis of ectopic pregnancy. Fertil Steril 39: 62–67

Breen JL (1970) A 21 year survey of 654 ectopic pregnancies. Am J Obstet Gynecol 106: 1004–1019

Brenner PF, Roy S, Mishell DR (1980) Ectopic pregnancy: a study of 300 consecutive surgically treated cases. J Am Med Assoc 243: 673–676

Brown TF, Filly RA, Laing FC (1978) Analysis of ultrasonographic criteria in the evaluation for ectopic pregnancy. Am J Roentgenol 131: 967–971

Burford GD, Naftalin AA, Shaw RW (1986) The use of an enzyme linked immunosorbent assay for hCG in gynaecological emergencies. Br J Obstet Gynaecol 93: 87–89

Chemitz J, Tornehorve D, Teisner B, Poulsen AR, Westergaard JG (1984) The localization of pregnancy proteins (hPL, SP1 and PAPP-A) in intra- and extrauterine pregnancies. Placenta 5: 489–494

Crawford E, Hutchinson H (1954) A decade of reports on tubal pregnancies condensed from the literature plus 300 consecutive cases without a death. Am J Obstet Gynecol 67: 567–584

Draa CC, Baum HC (1951) Posterior colpotomy: an aid in the diagnosis and treatment of ectopic pregnancy. Am J Obstet Gynecol 61: 300–311

Flege JB (1966) Ruptured tubal pregnancy with elevated serum alpha amylase levels. Arch Surg (Chicago) 92: 379–398

Gleicher N, Giglia RV, Deppe G, Elrad H, Friberg J (1983) Direct diagnosis of unruptured ectopic pregnancy by real time ultrasonography. Obstet Gynecol 61: 425–428

Gordon PP (1980) Frozen section examination of endometrial curettings: an aid to the rapid diagnosis of ectopic pregnancy. Diag Gynecol Obstet 2: 77–83

Grudzinskas JG, Westergaard JG, Teisner B (1986) Biochemical assessment of placental function: early pregnancy. Clin Obstet Gynecol 13: 553–569

Gruneberger W, Leodolten S, Spona J (1977) The LH/hCG test, a valuable aid in the diagnosis of tubal pregnancy. Gynaecol Obstet Invest 3: 150–153

Helvacioglu A, Long EM, Yang SL (1979) Ectopic pregnancy. An eight year review. J Reprod Med 22: 87–92

Ho PC, Jones WR (1980) Pregnancy specific beta-1 glycoprotein as a prognostic indicator in complications of early pregnancy. Am J Obstet Gynecol 138: 253–256

Hochberg CJ (1974) Tubal alpha amylase. Obstet Gynecol 43: 129–131

Horne CHW, Towler CM, Pugh-Humphries RGP, Thompson AW, Bohn H (1976) Pregnancy specific beta-1 glycoprotein: a product of the syncytiotrophoblast. Experientia 32: 1197–1199

Hubinont CJ, Thomas C, Schwers JF (1987) Luteal function in ectopic pregnancy. Am J Obstet Gynecol 156: 669–674

Jouppila P, Tapanainen J, Huhtaniemi I (1980a) Plasma hCG and ultrasound in suspected ectopic pregnancy. Eur J Obstet Gynecol Reprod Biol 10: 3–12

Jouppilla P, Huhtaniemi I, Tapanainen J (1980b) Early pregnancy failure: study by ultrasonic and hormonal methods. Obstet Gynecol 55: 42–47

Kadar N, DeVore G, Romero R (1981) Discriminatory hCG zone: its use in the sonographic evaluation for ectopic pregnancy. Obstet Gynecol 58: 156–161

Kadar N, Taylor KW, Rosenfield A (1983) The sonographic appearance of the uterus in ectopic pregnancy. AJR 141: 609–615

Kitchin JD, Wein RN, Nunley WC, Thiagarajah S, Thornton WN (1979) Ectopic pregnancy: current clinical trends. Am J Obstet Gynecol 134: 870–876

Kjessler B, Johansson SGO, Lidbjork G, Sherman MS (1977) AFP levels in maternal serum in relation to pregnancy outcome in 7158 pregnant women prospectively investigated during their 14–20 week post last menstrual period. Acta Obstet Gynecol Scand Suppl 69: 25–44

Landesman R, Saxena BB (1976) Results of the first 1000 radioreceptorassays for the determination of hCG: a new, rapid, reliable and sensitive pregnancy test. Fertil Steril 27: 357–368

Lenton EA, Neal LM, Sulaiman R (1982) Plasma concentrations of hCG from the time of implantation until the second week of pregnancy. Fertil Steril 37: 773–778

L'Hermite-Baleriaux AM, Van Exter C, Delville JL (1982) Alteration of free hCG subunit secretion

in ectopic pregnancy. Acta Endocrinol 100: 109–113
McGeachin RL, Hargan LA, Potter BA, (1958) Alpha-amylase in fallopian tubes. Proc Soc Exp Biol Med, 99: 130–134
Matthews CP, Coulsen CB, Wild RA (1986) Serum progesterone levels as an aid in the diagnosis of ectopic pregnancy. Obstet Gynecol 68: 390–394
Milwidski A, Adoni A, Segal S, Palti Z (1977) Chorionic gonadotrophin and progesterone levels in ectopic pregnancy. Obstet Gynecol 50: 145–147
Milwidski A, Adoni A, Miodovnik M, Segal S, Palti Z (1978) hCG (beta subunit) in the early diagnosis of ectopic pregnancy. Obstet Gynecol 51: 725–726
Milwidski A, Segal S, Menashi M, et al. (1984) Corpus luteum function in ectopic pregnancy. Int J Fertil 29(4): 244–246
Norman RJ, Lowings C, Chard T (1985) Dipstick method for hCG suitable for emergency use on whole blood and other fluids. Lancet i: 19–20
Olsen CM, Holt JA, Alenghat E (1983) Limitations of qualitative serum beta-hCG assays in the diagnosis of ectopic pregnancy. J Reprod Med 28: 838–842
Pedersen JF (1980) Ultrasonic scanning in suspected ectopic pregnancy. Br J Radiol 53: 1–4
Piironen O, Punnonen R (1981) The use of ultrasonography in the diagnosis of ectopic pregnancy. Clin Radiol 32: 331–335
Poulsen HK, Westergaard JG, Teisner B, Bolton AE, Grudzinskas JG (1987) Undetectable circulating placental hormones and proteins in some women with ectopic gestation. J Obstet Gynaecol 7: 274–276
Radwanska E, Frankenberg J, Allen EI (1978) Plasma progesterone levels in normal and abnormal early human pregnancy. Fertil Steril 30: 398–402
Riddle AF, Sharma V, Mason B, Ford N, Pampiglione J, Parsons J, Campbell S (1987) Two years experience of ultrasound directed oocyte retrieval. Fertil Steril 48: 454–458
Romero R, Kadar N, Jeanty P, et al (1985) The diagnosis of ectopic pregnancy: the value of the hCG discriminatory zone. Obstet Gynecol 66: 357–360
Saxena BB, Landesman R (1975) The use of radioreceptor assays of hCG for the diagnosis and management of ectopic pregnancy. Fertil Steril 26: 397–404
Schiffer MA (1963) A review of 2668 ectopic pregnancies. Am J Obstet Gynecol 86: 264–270
Schwartz RO, DiPietro DL (1980) Beta-hCG as a diagnostic aid for suspected ectopic pregnancy. Obstet Gynecol 56: 197–203
Seppälä M, Vaesmea P, Rutanen EM (1980) Pregnancy-specific beta-1 glycoprotein in ectopic pregnancy. Am J Obstet Gynecol 136: 189–193
Sinosich MJ, Smith DH, Grudzinskas JG, Saunders DM, Westergaard JG, Teisner B (1983) The prediction of early pregnancy failure by measurement of PAPP-A following IVF and ET. Fertil Steril 40: 539–541
Sinosich MJ, Ferrier A, Teisner B, et al. (1985) Circulating and tissue concentrations of PAPP-A in tubal ectopic gestation. Clin Reprod Fertil 3: 311–317
Soisson FL, Moran JP (1959) Ectopic pregnancy: a review of 50 cases at Tripler US Army Hospital. Am J Obstet Gynecol 77: 352–363
Stabile I, Riddle A, Campbell S, Grudzinskas JG (1987) The diagnosis of early pregnancy failure. In: Teo ES (ed) Advances in fertility and sterility, vol 1. Parthenon Publishing, Lancs, pp 153–160
Stenman UH, Alfthan H, Myllynen L, Seppälä M (1983) Ultrarapid and highly sensitive time resolved fluoroimmunometric assay for chorionic gonadotrophin. Lancet ii: 647–649
Tornehave D, Chemnitz J, Westergaard JG, et al. (1986) Placental proteins in peripheral blood and tissue of ectopic pregnancies. Gynaecol Obstet Invest 23: 97–102
Webster HD, Barclay DL, Fischer CK (1965) Ectopic pregnancy: a seventeen year review. Am J Obstet Gynecol 92: 23–34
Weiner CP (1981) The pseudogestational sac in ectopic pregnancy. Am J Obstet Gynecol 139: 959–961
Westergaard JG (1987) MD Thesis, University of Odense, Denmark
Yaffe H, Navot D, Laufer N (1979) Pitfalls in early detection of ectopic pregnancy. Lancet i: 277

18. Treatments to Enhance Implantation

J.L. Yovich

Introduction

In treating couples with subfertility the clinician will be concerned with: (a) the chance of pregnancy for the treatment mode applied and (b) the chance that the pregnancy will end in the birth of a healthy live infant.

A range of modern techniques enables a much higher chance of achieving conception than was possible a decade ago. These include tubal microsurgery, ovarian stimulation with close monitoring, a variety of techniques involving gamete handling or manipulation, and specific therapies – e.g., for pelvic endometriosis and hyperprolactinaemia (Yovich 1988). Against the improved prognosis for conception, early pregnancy wastage is increased in such patients to up to 27% overall (Yovich and Matson 1988). The type of pregnancy loss relates to the technique applied – e.g., biochemical pregnancies in in-vitro fertilisation (IVF), blighted ovum pregnancies following artificial insemination by the husband and ectopic pregnancies in all groups, particularly prevalent in gamete intrafallopian transfer (GIFT) cases.

In this paper I shall outline a range of studies and treatments aimed at improving the chance of successful implantation.

Implantation Physiology

Fig.18.1 depicts the various influences which bear on human implantation. The likelihood of achieving pregnancy depends on the quality of the embryo that reaches the uterine cavity and the receptivity of the endometrium. Human embryos can be assessed morphologically, and those of poor quality rarely implant (Yovich 1985). There is increasing evidence for embryo/maternal signals passing in both directions – possibly early pregnancy factor, platelet-activating factor (PAF), placental proteins (pregnancy associated plasma protein-A) and endometrial proteins (placental proteins 12 and 14). A localised vascular response occurs at the implantation site, characterised by vasodilation and increased

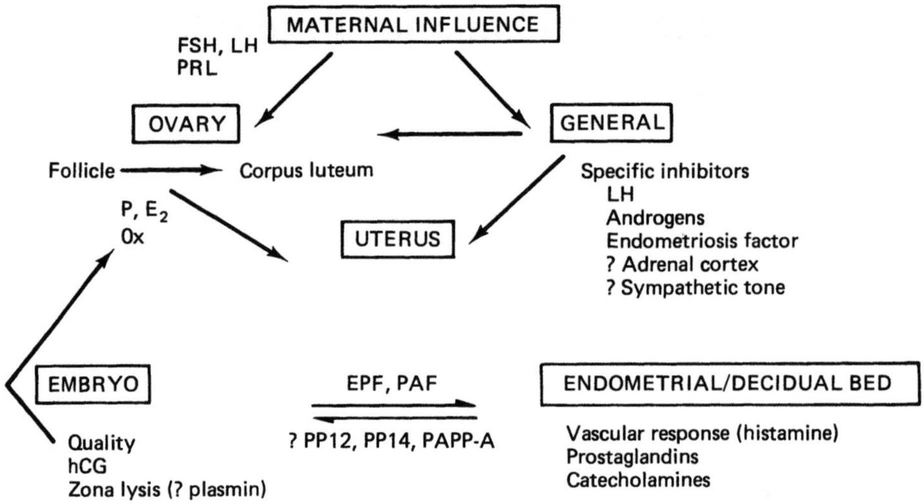

Fig. 18.1. Known and suspected influences on implantation.

capillary permeability, possibly mediated by histamine (Shelesnyak 1959). In addition, prostaglandins and catecholamines are involved in this local reaction and the subsequent decidualisation of the endometrial bed (Psychoyos and Mantel 1985). A number of substances have been postulated to assist zona lysis, including plasmin and a uterine protease. The most important hormone is human chorionic gonadotrophin (hCG), which is secreted by the embryo and has a luteotrophic action and maintains secretion of progesterone (P) and oestradiol-17β (E_2), both necessary for decidualisation of the endometrium. Oxytocin and relaxin are also secreted by follicles and corpora lutea, but their role is unknown.

There are likely to be other substances which stimulate the corpus luteum prior to hCG production. The quality and responsiveness of the corpus luteum in turn is dependent upon preceding events during the follicular phase, which are either under a central influence or are the result of ovarian stimulation regimens. Indirectly, other maternal influences can affect ovarian and uterine physiology – e.g., psychological states, age, total body mass (in particular body mass index < 18 or > 25). Specific inhibitors of implantation may include raised basal luteinising hormone (LH) concentrations, raised androgens, particularly androstenedione, an unknown endometriosis factor, and sympatho-adrenal factors. The latter is implied by studies showing that embryo transfers were more successful if carried out in the evening, when the patient is relaxed. (Edwards et al. 1980). This has not been confirmed during trials at the PIVET Unit.

Luteal Phase Function

Women without ovaries having ovum donation can conceive and maintain their pregnancies on a simple schedule of oestrogen and progesterone replacement during the conception cycle and early weeks of pregnancy. Four such pregnancies have been achieved at PIVET, using the schedule described by Lutjen et al. (1984).

Throughout this paper the luteal phase is measured with respect to the day of oocyte recovery ("ovulation"), denoted as day 0.

Early Luteal Phase (to day 7)

Several workers have reported higher progesterone concentrations in the early luteal phase of pregnant cycles (Lenton et al. 1982; Yovich et al. 1985a; Taylor et al. 1986). The PIVET study (Yovich et al. 1985a) showed that serum progesterone levels on days 1, 2 and 3 were significantly elevated above those of patients who did not conceive. The higher progesterone levels could be reached by the injection of 50 mg of progesterone in oil daily for 5 days, beginning immediately after ovum aspiration. The results suggested an improvement in the pregnancy rate by such treatment but just failed to reach statistical significance. The Norfolk IVF unit (Jones 1985) routinely uses intramuscular progesterone (12.5 mg daily after ovum collection for 16 days and then 17-hydroxyprogesterone caproate each week during pregnancy until 18 weeks. An extra dose of progesterone, totalling 50 mg, is given by intramuscular injection daily for three days after embryo transfer). The Norfolk group have not reported any controlled studies on progesterone support but have indicated that luteal support may be required if patients' response to stimulation is suboptimal. We have analysed luteal phase progesterone levels with respect to the ovarian stimulation regimen (Fig. 18.2). The only significant difference was a higher progesterone level on day 4, following combined stimulation with clomiphene citrate (CC) and human menopausal gonadotrophin (hMG). Those women given the progestagen medroxyprogesterone acetate (MPA) had lower progesterone levels but the difference was not statistically significant. This should be borne in mind in relationship to the late luteal data for MPA. Progesterone levels were further analysed in relationship to the number of follicles (> 1.5 cm) generated (Fig. 18.3). Those women who developed more than four follicles (more likely with the clomiphene/hMG regimen) showed significant elevations of progesterone on days 1 and 2.

A further clue to the early luteal phase is provided in the reports from the Royal North Shore unit in Sydney (O'Neill et al. 1985; Gidley-Baird et al. 1986). They reported a group of women failing to conceive in an IVF programme (despite a good prognosis because the embryos were secreting PAF). This was related to a high E_2/P ratio. No pregnancies were seen in that group, whereas 40% of PAF positive women conceived if the E_2/P ratio was normal (< 18). Furthermore, they showed that 1500 IU of hCG given on alternate days could normalise an abnormal ratio and allow pregnancies to ensue. However, further work by the Sydney group has not confirmed those findings and it appears that the high E_2/P ratios were found in patients in whom hCG was given prematurely. Since modifying their stimulation protocol, they rarely see the abnormal ratios and have not been able to complete a full trial (C. O'Neill and D. Saunders, personal communication). At PIVET, we have not observed pregnancy failure with unusual E_2/P ratios.

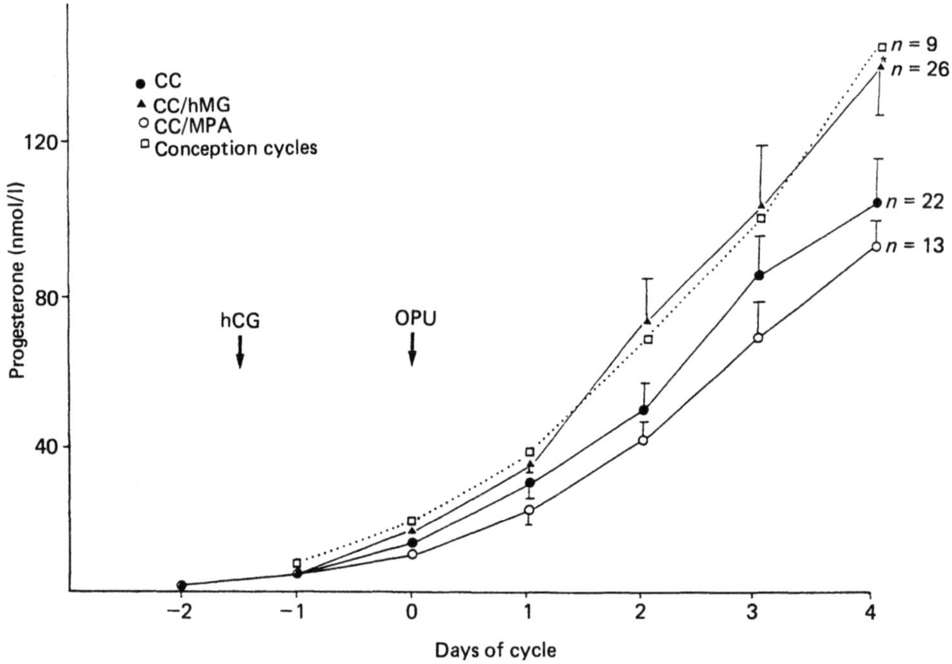

Fig. 18.2. Early luteal phase progesterone levels in relation to ovarian stimulation and the use of MPA (20mg q.i.d. beginning after ovum pick-up (OPU). Results are expressed as mean ± SEM. * CC/hMG vs. CC, $P < 0.05$.

Late Luteal Phase (day 7 onwards)

The diagnosis of luteal phase inadequacy or dysfunction has not found universal acceptance. This is partly because of reluctance to use endometrial biopsy as a method of diagnosis and also because of a prevailing view that normal ovulation and luteal function is the natural consequence of normal follicular development. It would also be argued that short luteal phases are only seen in non-conception cycles and are not a cause of failed conception. At PIVET the view is that of Hull et al. (1982) which indicates that favourable "conception" cycles show mid-luteal progesterone levels as a narrow subset within a wider range accepted for normal ovulatory menstrual cycles. This has been taken further (Yovich et al. 1987), showing that conception cycles have a range of hormonal, ovarian ultrasound and cervical mucus parameters which is quite narrow within the follicular and luteal phases. Isolated luteal phase anomalies do occur and are correctable (84%) with CC or CC/hMG stimulation. The chance of pregnancy was 21%–32%/treatment cycle when ovarian stimulation was combined with hCG to trigger ovulation (initially described by Radwanska and Swyer 1974), with additional hCG support during the luteal phase initially described with gonadotrophin therapy by Townsend et al. [1966] and later with clomiphene citrate by Garcia et al. [1977]).

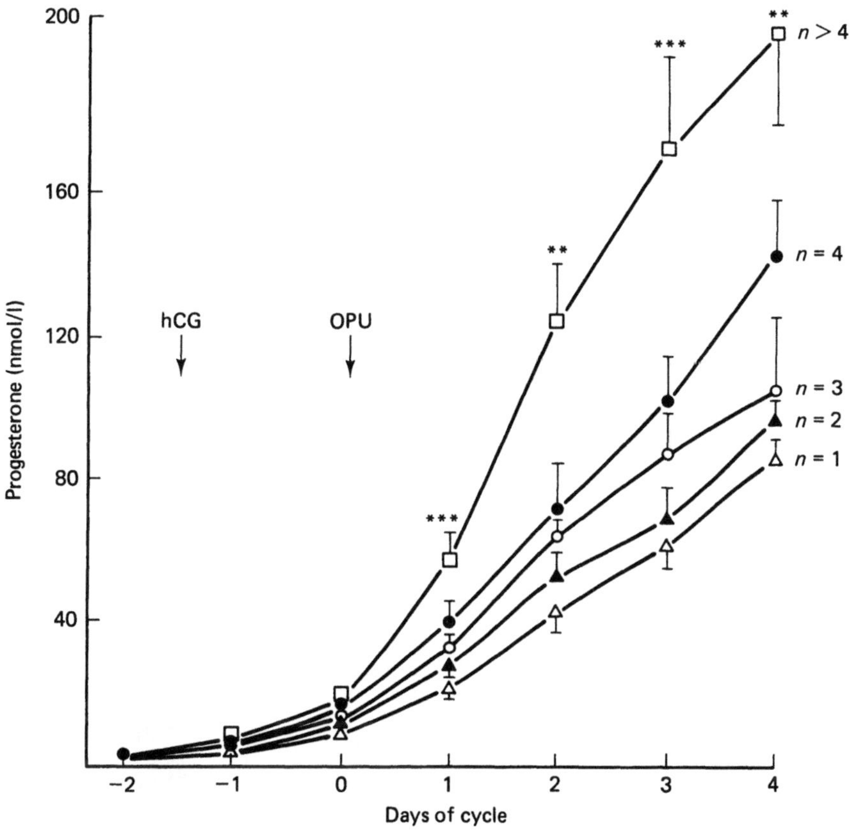

Fig. 18.3. Early luteal phase progesterone levels in relation to the number of follicles aspirated. The higher numbers were more commonly seen in CC/hMG cycles (mean 4.0 ± 0.3) than in CC cycles (2.7 ± 0.2). Results are expressed as mean ± SEM. Stimulation with cc or cc/hMG: ** $P < 0.01$, *** $P < 0.001$.

PIVET studies of luteal phase length are shown in Fig. 18.4. Where clomiphene has been used (with or without hMG), luteal phase lengths are not foreshortened but can be extended further to more than 16 days by hCG. However, where hMG is used alone, in the absence of luteal support the majority of cycles depict a luteal phase of only 9 or 10 days. With hCG given in the protocol described below, luteal phase lengths are corrected to more than 16 days. The shortened luteal phase phenomenon has also been demonstrated by Edwards et al. (1980); this group discarded gonadotrophin stimulation for IVF in 1977 because the luteal phase length was inversely related to the total urinary oestrogen excretion during the ovarian stimulation phase.

Further studies at PIVET on hormonal output in the mid and late luteal phase are shown in Fig. 18.5 for IVF patients having ovarian stimulation with CC or CC/hMG. With the non-conceiving patients the serum concentrations of both E_2 and P are significantly elevated on days 7 and 10, with some prolongation thereafter if hCG has been given. Against this are superimposed the E_2 and P levels of patients who conceived without receiving any hCG support. Their

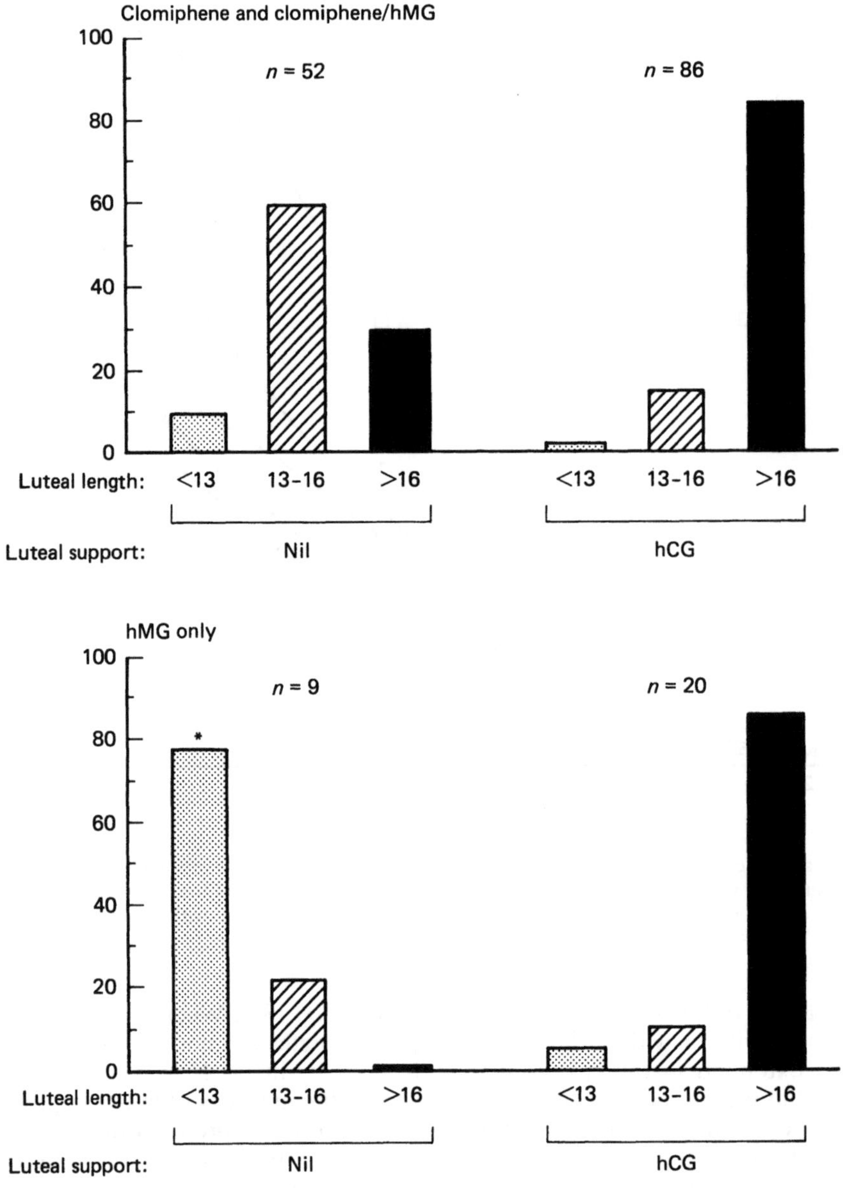

Fig. 18.4. Luteal phase lengths in relation to stimulation protocol and the effects of hCG support. CC and CC/hMG cycles were grouped together, because homogeneity was demonstrated by interaction chi-squared analysis. * The majority of hMG/hCG luteal phases are *c.* 9 days.

levels match the ranges for those given hCG, and this implies that such levels are more compatible with conception.

Further studies on the late luteal phase were undertaken using the progestagen MPA. The luteal phase of such patients is prolonged while the drug is in use, and menses occurred within three days of cessation. Although pregnancies occurred

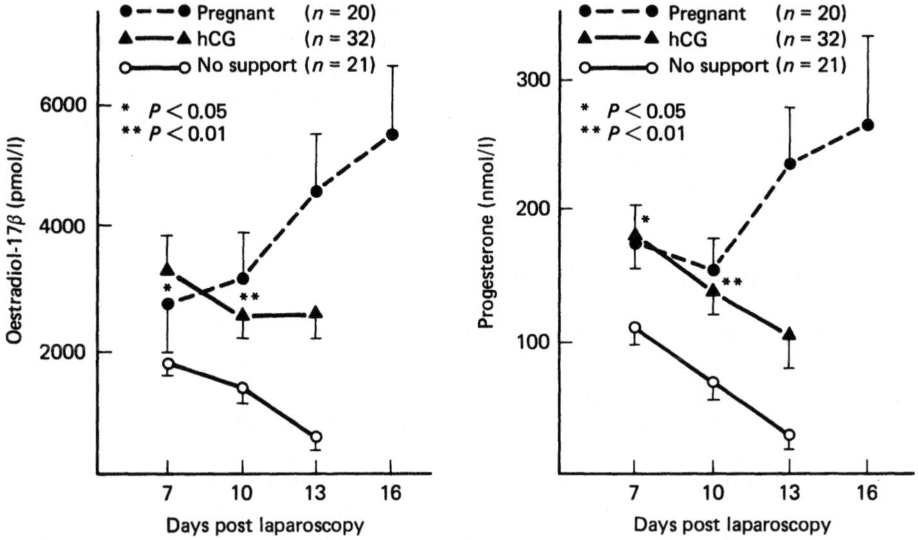

Fig. 18.5. Oestradiol and progesterone levels in the late luteal phase in relation to pregnancy and hCG therapy.

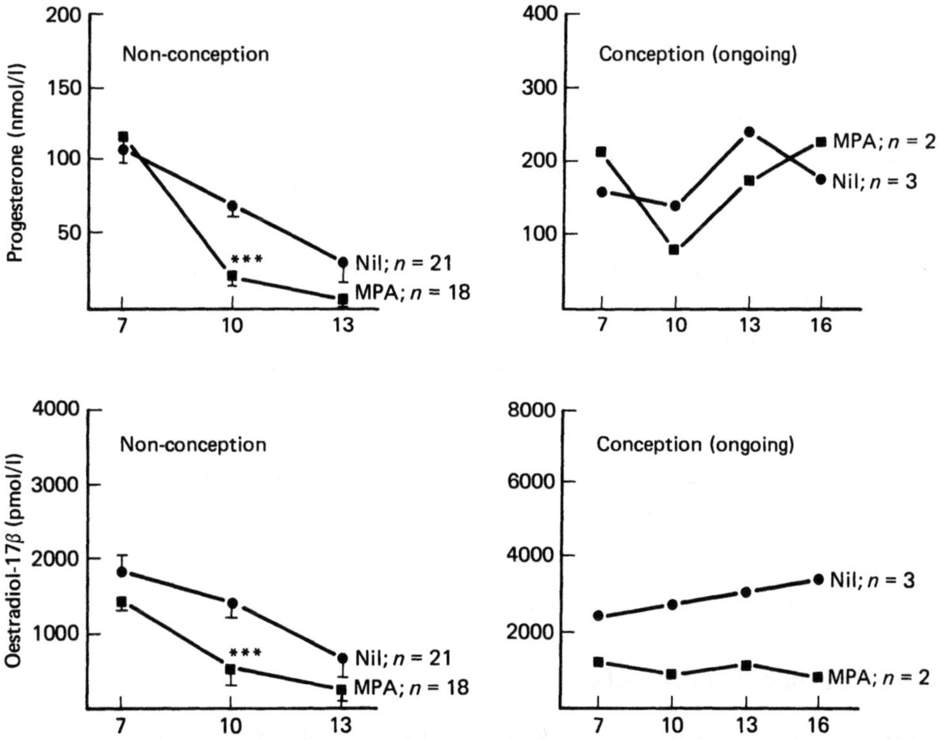

Fig. 18.6. Progesterone levels in the late luteal phase in relation to pregnancy and MPA therapy. $^{***}P < 0.001$.

at the same rate as in untreated patients, use of MPA was associated with luteal suppression, both progesterone and oestradiol being significantly reduced on day 10 (Fig. 18.6). Levels were not significantly reduced on day 7 or earlier (Fig. 18.2), suggesting that the corpus luteum is resistant to lytic or suppressant activity during the early luteal phase. MPA is thought to act by inhibiting the isomerase enzyme system as was previously documented for a range of progestagens by Johansson (1971).

Granulosa Lutein Cell Cultures

Luteinised granulosa cells collected at the time of ovum collection (36 h post ovulation trigger) and cultured in vitro up to 72 h produced progesterone in large quantities (Table 18.1). A significant luteotrophic action was seen with the addition of hCG at concentrations equivalent to therapeutic schedules. However, the addition of MPA at and above therapeutic amounts (15 to 30 nmol/l; Yovich et al. 1985b) had no effect. These findings are in keeping with the clinical data and indicate, first, that the progesterone output from early granulosa-lutein cells is well within their potential reserve and, second, that there is intrinsic resistance to luteal suppressant action at this early stage.

Table 18.1. Progesterone output from luteinised granulosa cells cultured in vitro: effects of MPA and hCG

Agent added	Amount	Progesterone output (ng/μg protein/24 h)
Nil	–	7–19
hCG	⩾ 10 IU/l	Significant increase
MPA	1 – 1000 nmol/l	No effect

Table 18.2. Published reports of randomised controlled trials assessing luteal phase treatments and the effects on pregnancy rates

Treatment	Reference	Dose	Schedule*	Significance
Progesterone	Leeton et al. 1985	50 g im	7–16	NS
	Yovich et al. 1985a	50 g im	0–4	NS
	Trounson et al. 1986	25/50 g im	−1,0,1	NS
hCG	Yovich et al. 1984	1000 IU	4,7,10,13	NS
	Mahadevan et al. 1985	1000 IU	2,5	
		2000 IU	6,9	NS

NS = not significant.
*Days of luteal phase; 0 = day of oocyte recovery.

Luteal Phase Treatments

Published reports using progesterone (Leeton et al. 1985; Yovich et al. 1985a; Trounson et al. 1986) or hCG support (Yovich et al. 1984; Mahadevan et al. 1985) given in a controlled randomised fashion are summarised in Table 18.2.

They do not show any significant benefits. However, all the studies were small series and did not correct for the numerous variables affecting IVF pregnancy rates. At the PIVET Medical Centre cases have been randomly allocated to one of the following regimens:

1. nil
2. MPA (Provera; Upjohn) 80mg/day (20mg q.i.d. orally) from day 0 to 16, beginning immediately after oocyte recovery
3. hCG (Primogonyl; Schering 1000 IU given on days 4, 7, 10 and 13
4. progesterone (Proluton; Schering) 50mg given by intramuscular injection on days 0 to 4 inclusive
5. progesterone/hCG: (3) and (4) combined

MPA was chosen because maternal side-effects are rare, it is readily absorbed orally and appears not to have embryopathic or fetotoxic effects (Yovich et al. 1988). hCG is a purified natural hormone without any known teratogenic properties or negative effects on pregnancy. Progesterone is also a natural hormone thought to be free of teratogenic effects when given therapeutically.

Table 18.3. Pregnancy rates at the PIVET Medical Centre (1982–84) with respect to ovarian stimulation schedules

Regimen	No. of cycles	No. of ETs	Pregnancies	Rate/cycle (%)	Rate/ET (%)
Clomiphene	299	256	45	15.1	17.5
C/hMG	575	494	80	13.9	16.2
hMG	110	91	15	13.6	16.5
Total	984	841	140	14.2	16.7

ET = embryo transfer.
$\chi^2 < 3.84$; no significant differences.

Between 1982 and 1984 the first three regimens were allocated randomly at the onset of the treatment cycle. Once the luteal suppressant action of MPA was revealed, the drug was no longer used in the trial. Subsequently, the Proluton regimen was added as a third mode, together with hCG. There was no difference in the clinical pregnancy rates for the various ovarian stimulation regimens applied during the study (Table 18.3). This is unexpected, because hMG patients are known to provide more oocytes. However, CC was used as the initial agent, and hMG was given on subsequent cycles only to patients who failed to develop at least three follicles on a previous clomiphene cycle. The hMG-alone group tended to be the poorer responders from the first two regimens. The clinical pregnancy rates were significantly higher than in those who had no luteal support (Table 18.4). There was no obvious benefit in adding Proluton to the hCG regimen. The MPA patients conceived at least as readily as the untreated group, despite the known biochemical suppression. It may be argued that the untreated group showed a very low pregnancy rate. The data represent the entire experience during the time of the study, apart from those patients selected for non-randomised treatment. There were marked fluctuations in the overall pregnancy

Table 18.4. Pregnancy rates at the PIVET Medical Centre (1982–84) with respect to luteal phase support treatments

Regimen	No. of cycles	Pregnancies	Rate/cycle (%)
Nil	182	17	9.3
MPA	32	5	13.2
hCG	442	83	18.8[a]
Proluton/hCG	179	35	20.0[a]
Total	841	140	16.7%

[a] Pregnancy rates significantly higher with hCG luteal support regimens.
$\chi^2 > 8.5$; $P < 0.005$.

rates during various phases of the programme and, on occasions, specific laboratory problems were identified. However, no attempt was made to control for these factors in this study, or for other factors which are now known to be relevant.

A further study is under way at PIVET and is applied only within the GIFT programme, where pregnancy rates are more consistent. There is a more careful selection of patients to fulfil the following criteria before inclusion:

age less than 40 years
regular menstrual cycles
good response to stimulation therapy
stimulation by clomiphene or clomiphene/hMG
a normospermic husband with no antispermatozoal antibodies
no previous GIFT attempts
no grade III or grade IV endometriosis
no antispermatozoal antibodies in the woman
no adverse factor such as high LH or androgen levels

Preliminary results of this study are reported in Table 18.5. So far, no significant differences can be shown between the various forms of luteal support.

Table 18.5. Pregnancy rates in a GIFT programme at the PIVET Medical Centre (September 1986–February 1987) with respect to luteal phase treatments applied with strict randomisation at the onset of the treatment cycle

Support	Cycles	Pregnancies	Rate/cycle (%)
Nil	31	6	19.4
hCG	37	8	21.6
PROL	27	6	22.2
PROL/hCG	30	9	30.0
Total	125	29	23.2

No significant differences.

Additional Variables Affecting Implantation

Ovulation Trigger

There is an apparently lower pregnancy rate in both IVF and GIFT programmes if oocytes are recovered following a spontaneous LH surge rather than an hCG injection to trigger ovulation (Table 18.6). This has been a consistent finding at PIVET, although the data presented do not show statistical significance with respect to implantation rates. It is possible that oocytes recovered following a spontaneous LH surge are less likely to be at or near the metaphase II stage, because it is not possible accurately to detect the onset of the surge, and oocytes are not recovered between 9 p.m. and 7 a.m. for logistic reasons. Fertilisation rates are at times significantly reduced following spontaneous LH surges, and it is suspected that embryos derived from such cases may lack the full integrity for implantation.

Table 18.6. Pregnancy rates at PIVET Medical Centre for IVF and GIFT cases categorised with respect to ovum recovery following onset of spontaneous LH surge or following a timed trigger injection of 10000 IU hCG

Treatment	Cycles	Transfers	Pregnancies	Rate/cycle (%)	Rate/transfer (%)
IVF:					
LH surge	71	62[a]	7	9.9	11.3
hCG trigger	435	224	41	17.5	18.3
GIFT:					
LH surge	57	53	10	17.5	18.9
hCG trigger	155	154	40	25.8	26.0

[a] IVF: fewer cases to transfer, $P < 0.025$.

Laboratory Factors

Despite increasing expertise (the PIVET Medical Centre IVF and infertility programme has been functional for five years), laboratory problems are still intermittently identified as a cause of reduced pregnancy rates. The most recent example was the introduction of mixed triple gas ($O_2 : CO_2 : N_2$) delivered to the incubators from outside through copper piping. Excessive concentrations of copper built up in the humidification jars, and subsequently within the culture chambers, affecting fertilisation rates and embryo quality. Fluctuations in water quality and techniques of serum preparation for culture media have also been identified as causing reduced pregnancy rates, despite the reassurance given that mouse embryos were cultured satisfactorily in such media.

Transfer Site

The GIFT technique was introduced at PIVET in December 1985 and immediately appeared to give a better pregnancy rate, with less fetal wastage. It rapidly became the preferred method for all resistant infertility cases where at least one fallopian tube was patent (Yovich et al. 1988). Subsequently, we devised the pronuclear stage transfer (PROST) technique (Matson et al. 1986) as a modification to combine the benefits of IVF (demonstrable fertilisation) with the possible advantage of GIFT (transfer of pronuclear oocytes to the fallopian tubes). Table 18.7 shows the comparative pregnancy rate during a six-month series. It appears that the tubal environment is better for gametes and pronuclear oocytes. It has also been shown to be satisfactory for eight-cell frozen-thawed embryos – the technique described as TEST (tubal embryo stage transfer). At this stage it is uncertain if the significantly improved pregnancy rates relate to the tubal environment or to minimising the intra-laboratory exposure.

Table 18.7. Comparative conception rates for three treatment modes involving gamete manipulation at the PIVET Medical Centre (September 1986–February 1987)

Treatment	Transfer	Cycles	Transfers	Pregnancies	Rate/transfer (%)
IVF	Embryos : uterus	105	102	13	12.8
PROST	Pronuclear eggs : tube	83	53	14	26.4[a]
GIFT	Gametes : tube	123	122	31	25.4[b]

[a] PROST vs. IVF, $P < 0.05$.
[b] GIFT vs. IVF, $P < 0.025$.

Table 18.8. Pregnancy results at the PIVET Medical Centre (September 1986–February 1987) for GIFT treatments with respect to number of oocytes collected

Oocytes	Collections	Pregnancies	% pregnant
1	6	0	0
2	12	0	0
3	16	6	38
4	21	7	33
>4	67	18	27
Total	122	31	25.4

Maximum of four oocytes transferred.

Oocyte and Embryo Numbers

Several groups have shown that the transfer of two or more embryos is better than one in IVF. The PIVET data confirm this and show a similar trend for GIFT (Table 18.8), where the transfer of only one or two oocytes leads to a significantly reduced chance of pregnancy (< 10%), whereas transfer of three or four leads to high rates (> 25%). We are uncertain if this is due to embryos helping each other or whether it relates to the higher follicle numbers and subsequent higher numbers of corpora lutea and the elevated serum progesterone levels in women with more than four follicles (Fig. 18.2). The chance of pregnancy is significantly improved when two or three embryos are transferred into women whose oocytes were derived from more than four follicles (Table 18.9). This is in keeping with the finding of other groups who have noted a higher pregnancy rate in the treatment cycle of those women who have suitable embryos left over for cryopreservation. However, the reason is uncertain, one view supporting the concept of better luteal function, the other that more follicles generate more oocytes and thus a higher proportion of better-quality embryos.

Table 18.9. Pregnancy results at the PIVET Medical Centre (July – December 1984) for IVF treatments with respect to the number of oocyte-bearing follicles. The data are restricted to those women having two or three embryos transferred

Oocytes collected	Two or three embryos transferred	Pregnancies	% pregnant
2	75	10	13.3
3	49	4	8.2
4	64	12	18.8
>4	142	34	23.9[a]
Total	330	60	18.2

[a] $\chi^2 = 6.54$; $P > 0.025$.

Infertility History

IVF data have now been published for a wide range of causes of infertility, including antispermatozoal antibodies, oligospermia, tubal disorders, negative post-coital tests and unexplained infertility (Yovich et al. 1986). Similar pregnancy rates per embryo transfer have been shown for all but pelvic endometriosis, where there appears to be a reduced chance of implantation, regardless of the number of embryos generated and transferred. The finding relates to untreated pelvic endometriosis of grades III and IV severity (Matson and Yovich 1986). It appears not to be the case in patients recently treated by Danazol, and one hypothesis suggests that there is an endometriosis related factor which may inhibit implantation and which can be suppressed by hormonal therapy.

LH and Androgens

It has previously been reported that women with raised basal LH serum levels have reduced fertilisation, poorer quality embryos and virtually no pregnancies following implantation (Stanger and Yovich 1985). This phenomenon has been confirmed in women with raised urinary LH output (Howles et al. 1986). It is thought that the LH acts within the follicle, creating premature aging of the oocyte before recovery. Partial success has been obtained by pituitary suppression for two months before IVF cycling, using pure follicle stimulating hormone (Metrodin; Serono), or continuous gonadotrophin releasing hormone for its down-regulating effect on the pituitary. It appears likely that an analogue may effectively switch off LH output from the pituitary and allow subsequent satisfactory stimulation.

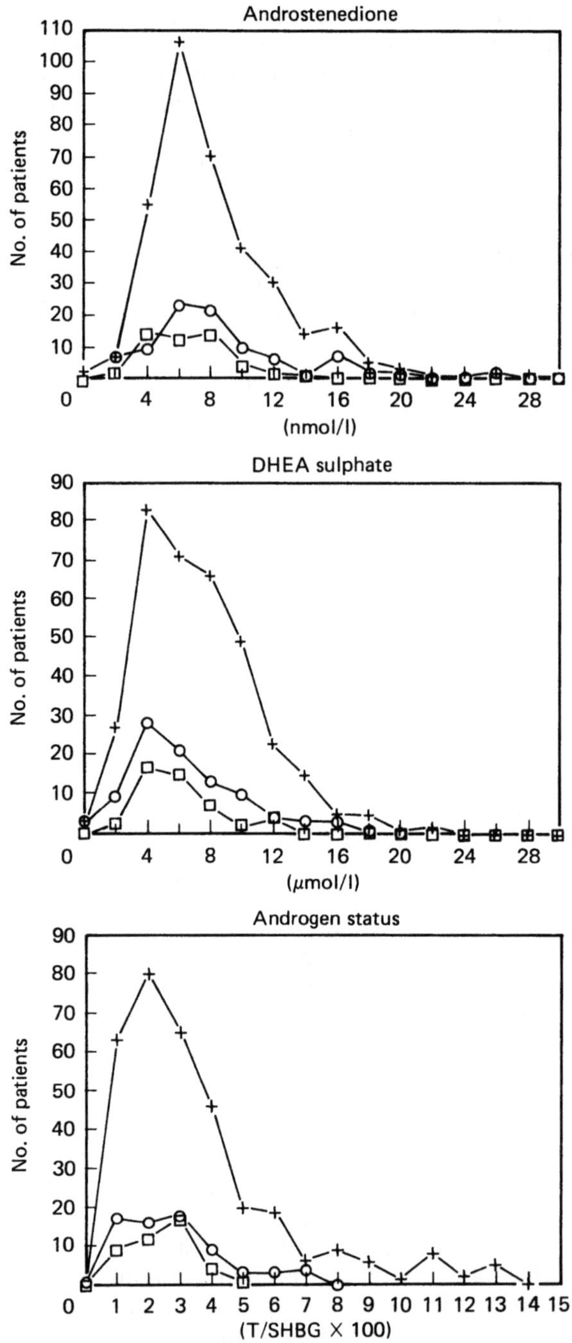

Fig. 18.7. Levels of androstenedione, DHEA sulphate and androgen status measured on day 2 of the follicular phase (before ovarian stimulation) in women undergoing IVF in that cycle. Data are shown for all women who started the treatment cycle and depicts those who were cancelled (○; 93), those who conceived (□; 48) and those who did not conceive but who reached the stage of ovum recovery (+; 351).

We have examined the androgen levels of IVF patients during the early follicular phase. Androstenedione, testosterone, dihydroepiandrostenedione (DHEA) sulphate and sex hormone binding globulin (SHBG) have been compared to cycle outcome and the chance of pregnancy (Fig. 18.7). The most interesting result was that pregnancy did not occur in patients whose androstenedione was > 11 nmol/l or DHEA sulphate was > 13 μmol/l, although more than 10% of patients demonstrate such levels. In addition, the free androgen status (testosterone/SHBG × 100) shows a cut-off level of 5, above which pregnancies were not seen. Approaches to therapy include the use of dexamethasone and spironolactone, with limited success. A trial of cyproterone acetate is now under way.

Conclusions

1. The chance of implantation is a consequence of numerous variables affecting both embryo quality and uterine receptivity.
2. The benefit (with respect to pregnancy rates) from luteal phase support (either early or late) remains unresolved because of the lack of published data from randomised controlled studies with well-matched groups.
3. We believe that there are probable benefits of hCG (± early Proluton), particularly if hMG is used alone for ovarian stimulation.
4. Because of differences between the various clinics dealing with infertility, each programme will adopt its own views regarding the need for luteal phase treatment.

References

Edwards RG, Steptoe PC, Purdy JM (1980) Establishing full-term human pregnancies using cleaving embryos grown in vitro. Br J Obstet Gynaecol 87: 757–768

Garcia J, Jones GS, Wentz AC (1977) The use of clomiphene citrate. Fertil Steril 28: 707–717

Gidley-Baird AA, O'Neill C, Sinosich MJ, Porter RN, Pike IL, Saunders DM (1986) Failure of implantation in human in-vitro fertilization and embryo transfer patients: the effects of altered progesterone/estrogen ratios in humans and mice. Fertil Steril 45: 69–74

Howles CM, MacNamee MC, Edwards RG, Goswamy R, Steptoe PC (1986) Effect of high tonic levels of luteinising hormone on outcome of in-vitro fertilisation. Lancet ii: 521–522

Hull MGR, Savage PE, Bromham DR, Ismail AAA, Morris AF (1982) The value of a single serum progesterone measurement in the mid luteal phase as a criterion of a potentially fertile cycle derived from treated and untreated conception cycles. Fertil Steril 37: 355–360

Johansson EDB (1971) Depression of progesterone levels in women treated with synthetic gestagens after ovulation. Acta Endocrinol 68: 779–792

Jones GS (1985) The role of luteal support in a programme for in vitro fertilization. In: Edwards RG, Purdy JM, Steptoe PC (eds) Implantation of the human embryo. Academic Press, London, pp 285–297

Leeton J, Trounson A, Jessup D (1985) Support of the luteal phase in in vitro fertilization programs: results of a controlled trial with intramuscular Proluton. J Vitro Fertil Embryo Transfer 2: 166–169

Lenton E, Sulaiman R, Sobowale O, Cooke ID (1982) The human menstrual cycle: plasma

concentrations of prolactin, LH, FSH, oestradiol and progesterone in conceiving and non-conceiving women. J Reprod Fertil 65: 131–139
Lutjen P, Trounson A, Leeton J, Findlay J, Wood C, Renou P (1984) The establishment and maintenance of pregnancy using in vitro fertilization and embryo donation in a patient with primary ovarian failure. Nature 307: 174–176
Mahadevan MM, Leader A, Taylor PJ (1985) Effects of low-dose human chorionic gonadotrophin on corpus luteum function after embryo transfer. J Vitro Fertil Embryo Transfer 2: 190–194
Matson PL, Yovich JL (1986) The treatment of infertility associated with endometriosis by in-vitro fertilization. Fertil Steril 46: 432–434
Matson PL, Blackledge DG, Richardson PA, Turner SR, Yovich JM, Yovich JL (1986) Pregnancies after pronuclear stage transfer. Med J Aust 146: 60
O'Neill C, Ferrier AJ, Vaughan J, Sinosich MJ, Saunders DM (1985) Causes of implantation failure after in-vitro fertilization and embryo transfer. Lancet ii: 615
Psychoyos A, Mantel D (1985) Embryo-endometrial reactions at implantation. In: Edwards RG, Purdy JM, Steptoe PC (eds) Implantation of the human embryo. Academic Press, London, pp 195–210
Radwanska E, Swyer GIM (1974) Plasma progesterone estimation in infertile women under treatment with clomiphene and chorionic gonadotrophin. J Obstet Gynaecol Br Cwlth 81: 107–112
Shelesnyak MC (1959) Histamine and the nidation of the ovum. In: Eckstein P (ed) Implantation of ova. Cambridge University Press, London, pp 84–88
Stanger JD, Yovich JL (1985) Reduced fertilization of human oocytes from patients with raised basal LH levels during the follicular phase. Br J Obstet Gynaecol 92: 385–393
Taylor PJ, Trounson AO, Besanko M, Burger HG, Stockdale J (1986) Plasma progesterone and prolactin changes in superovulated women before, during, and immediately after laparoscopy for in vitro fertilization and their relation to pregnancy. Fertil Steril 45: 680–686
Townsend SL, Brown JB, Johnstone JW, Adey FD, Evans JH, Taft HP (1966) Induction of ovulation. J Obstet Gynaecol Br Cwlth 73: 529–543
Trounson A, Howlett D, Rogers P, Hoppen HO (1986) The effect of progesterone supplementation around the time of oocyte recovery in patients superovulated for in vitro fertilization. Fertil Steril 45: 532–535
Yovich JL (1985) Embryo quality and pregnancy rates in in vitro fertilisation. Lancet i: 283–284
Yovich JL (1988) Infertility and fertility: Female. In: Grudzinskas JG, Beedham T (eds) Treatment and prognosis: Obstetrics and gynaecology. Heinemann Medical Books, London, pp 31–48
Yovich JL, Matson PL (1988) Early pregnancy wastage following gamete manipulation. Br J Obstet Gynaecol. 95 (in press)
Yovich JL, Stanger JD, Yovich JM, Tuvik A (1984) Assessment and hormonal treatment of the luteal phase of in vitro fertilization cycles. Aust NZ J Obstet Gynaecol 24: 125–130
Yovich JL, McColm SC, Yovich JM (1985a) Early luteal serum progesterone concentrations are higher in conception cycles. Fertil Steril 44: 185–189
Yovich JL, Willcox DL, Wilkinson SP, Poletti VM, Hahnel R (1985b) Medroxyprogesterone acetate does not perturb the profile of steroid metabolites in urine during pregnancy. J Endocr 104: 453–459
Yovich JL, Yovich JM, Tuvik AI, Junk S, Bootsma B, Matson PL (1986) In-vitro fertilization applied for tubal and non-tubal causes of infertility. Asia-Oceania J Obstet Gynaecol 12: 483–488
Yovich JL, Tuvik AI, Matson PL, Willcox DL (1987) Ovarian stimulation for disordered ovulatory cycles. Asia-Oceania J Obstet Gynaecol 13: 457–463
Yovich JL, Turner Sr, Blackledge DG, Richardson PA, Draper R (1988) Medroxyprogesterone acetate therapy in early pregnancy has no significant fetal effects. Teratology (in press)
Yovich JL, Matson PL, Blackledge DG, Turner SR, Richardson PA, Yovich JM, Edirisinghe WR (1988) The treatment of normospermic infertility by gamete intrafallopian transfer (GIFT). Br J Obstet Gynaecol 95: 361–366

Subject Index

MIX
Papier aus verantwortungsvollen Quellen
Paper from responsible sources
FSC® C105338

If you have any concerns about our products,
you can contact us on
ProductSafety@springernature.com

In case Publisher is established outside the EU,
the EU authorized representative is:
Springer Nature Customer Service Center GmbH
Europaplatz 3, 69115 Heidelberg, Germany

Printed by Libri Plureos GmbH
in Hamburg, Germany